MOVING FORWARD WITH COMPLEXITY

Proceedings of the 1st International Workshop on Complex Systems Thinking and Real World Applications

MOVING FORWARD WITH COMPLEXITY

Proceedings of the 1st International Workshop on Complex Systems Thinking and Real World Applications

Edited by

Andrew Tait (Decision Mechanics, UK/US)
Kurt A. Richardson (Exploratory Solutions, US)

3810 N 188th Ave
Litchfield Park, AZ 85340

Moving Forward with Complexity:
Proceedings of the 1st International Workshop on Complex Systems Thinking and Real World Applications
Edited by: Andrew Tait & Kurt A. Richardson

Library of Congress Control Number: 2011941325

ISBN: 978-0-9842165-9-8

Printed in the United States of America

To Paul...

ABOUT THE EDITORS

Andrew Tait is currently cofounder and Chief Technology Officer of Idea Sciences, a Virginia-based software and consulting firm specializing in the creative use of technology to improve organizational decision-making. During his career he has designed commercial, off-the-shelf, solutions for strategic planning, performance improvement and conflict management. This has led to numerous consulting and training relationships with major commercial and government organizations. Prior to forming Idea Sciences, Andrew held various commercial (technology consulting), government (defense) and academic (business) positions. Andrew's research interests include: decision-making, performance improvement, electronic voting, virtual communities; conflict management; visualization and; improving understanding of complex socio-technical systems.

Kurt A. Richardson, PhD is the CEO of Exploratory Solutions, a small company set-up to develop software to support decision making in complex environments. Kurt also designs and develops application specific integrated circuits for Orbital Network Engineering. He was recently a Senior Systems Engineer for the NASA Gamma-Ray Large Area Telescope (now Fermi). Kurt's current research interests include the philosophical implications of assuming that everything we observe is the result of complex underlying processes, the relationship between structure and function, analytical frameworks for intervention design, and robust methods of reducing complexity, which have resulted in the publication of over thirty journal papers and book chapters, and ten books. He is the Managing/Production Editor for the international journal *Emergence: Complexity & Organization* and is on the review board for the journals *Systemic Practice and Action Research*, *Systems Research and Behavioral Science*, and *Tamara: Journal of Critical Postmodern Organization Science*. Kurt is the author of the recently published *Thinking About Complexity: Grasping the Continuum through Criticism and Pluralism* (Emergent Publications, 2010. Kurt is a qualified spacecraft systems engineer and has consulted for General Dynamics, Lockheed Martin, Raytheon and NASA.

CONTENTS

Editorial: From Theory To Practice....................xix

1. A PRACTICAL METHODOLOGY FOR MANAGING COMPLEXITY

George Rzevski

Introduction 1
Complexity 2
The Seven Criteria Of Complexity 2
Global Market As A Complex System 2
Uncertainty As The Main Issue Resulting From Complexity.................... 4
Managing Complexity 5
Coping with Complexity.................... 5
Tuning Complexity 6
Tools For Managing Complexity—Complex Adaptive Software 7
Collecting And Organizing Domain Knowledge 7
Constructing A Virtual World 8
Connecting Virtual World To The Real World 8
Empowering Virtual World To Manage The Real World In Real Time .. 8
Comparing Multi-Agent Systems With Conventional Software.................... 9
Modeling Complexity 10
Examples Of Commercial Applications 11
Examples Of Engineering Applications 12
A Swarm Of Agents Controlling A Machine Tool 12
Intelligent Geometry Compressor.................... 13
Global Logistics Network 14
A Family Of Intelligent Space Robots 15
A Colony Of Agricultural Machinery.................... 15
Modeling of Exceedingly Complex Systems 15
Conclusions 17
Acknowledgement 17
References 17

2. IDENTIFYING THE MULTI-DIMENSIONAL PROBLEM-SPACE AND CO-CREATING AN ENABLING ENVIRONMENT

Eve Mitleton-Kelly

Introduction 21
Study 1: The Corporate Governance Study.................... 23
Background 24
Methodology.................... 24
Insights on Corporate Governance (CG) Frameworks 27

Insights And Summary Of Study 1 28
Study 2: The Rolls-Royce Marine Study 29
Background 29
Methodology 29
Summary And Insights Of Study 2 30
Study 3: The SSF Study 31
Methodology 32
Insights Of Study 3 34
Study 4: A Longitudinal Study in Two NHS Hospitals 34
The Two Environments In Complexity Terms 34
Complete Change In 2006/7 35
Summary And Insights Of Study 4 37
Study 5: Disaster Risk Reduction In West African States 38
Background 38
The Objective Of The Study 39
Cannot Copy Successful Initiatives 40
Examples Of Local Initiatives 41
Relevant Complexity Theory Principles 42
Summary And Insights Of Study 5 42
Conclusions 43
References 44

3. COMPLEXITY TOOLS FOR SMART GRIDS: PCT AND ABM JOIN FORCES

Liz Varga

Introduction 47
UK Electricity Generation 48
The National Grid 50
Security And Resilience 50
Smart Grid 51
Research Design 52
Personal Construct Theory (PCT) 54
Agent-Based Modeling 55
Conclusions And Further Research 56
Acknowledgements: 57
References 57

4. ON UNDERSTANDING SOFTWARE AGILITY: A SOCIAL COMPLEXITY POINT OF VIEW

Joseph Pelrine

Introduction 63
Getting Comfortable With Complexity—
Sense Making The Agile Way 67
Thinking About Complex Problems 68
Agile As A Technique For Addressing Complexity 69
Success In Software Development Is Only
Retrospectively Coherent 72
Complex Activities Require A Probe-Sense-Respond
Model Of Action 72
Acknowledgements 74
References 74

5. TECS: A BROWSER-BASED TEST ENVIRONMENT FOR COMPLEX SYSTEMS

Kurt A. Richardson

Introduction 77
Test Documentation Creation, Review and Maintenance 79
Operations, Sequences, Tests and Procedures 81
TECS & Authentication 89
TECS & Requirements 89
TECS & Flexibility 89
Test Execution 90
Where TECS Fits In 91
TECS Anomaly Management 91
I&T Program Monitoring 92
Summary 92
Acknowledgement 93
References 93

6. USING PRINCIPLES OF 'HOLISTIC BUSINESS SCIENCE' FACILITATED THROUGH A 'PROCESS AND EMERGENCE TOOL' TO SUPPORT CONVENTIONAL BUSINESSES DEAL WITH COMPLEXITY

Claudius Peter van Wyk

Introduction 95
Holism's Potential Contribution 97
Holistic Business Science—Fantasy Or Reality 99
Complexity And Business Practice 101
Complexity And Ethics 103
Holistic Business Science And The PET 104
Productivizing The Process And Emergence Tool (PET) 105
PHASE 1: Conscious Incompetence 106
PHASE 2: Conscious Competence 106
PHASE 3: Unconscious Competence 107
PHASE 4: Unconscious Incompetence 107
Context 108
The PET—Practical Application 108
Conclusion 109
References 109

7. ASSESSING CAPACITY AND MATURITY FOR CHANGE IN ORGANIZATIONS: A PATTERNS-BASED TOOL DERIVED FROM COMPLEXITY AND ARCHETYPES

Stefanos Michiotis & Bruce Cronin

Introduction 113
The Contribution Of Archetypes 115
Geometry And Meaning 118
Introducing A New Tool And Process 119
Structure Of The Tool 121
Capacity Assessment Process 123
Epilogue 127
References 127

8. THE APPLICATION OF COMPLEXITY THINKING TO LEADERS' BOUNDARY WORK

Alice E. MacGillivray

Context of the Chapter 133
The Nature of Complexity Thinking 133
The Nature of Leadership 134
The Nature of Boundaries 136
Boundary Critique 136
Boundary-Related Research Behind the Edge-Effect™ Tools 138
The Edge Effect™ Kit for leaders 141
Context 141
Edge Effect 142
References 146

9. VALUE CRAFTING: A TOOL TO DEVELOP SUSTAINABLE WORK BASED ON ORGANIZATIONAL VALUES

Sjaña S. Holloway, Frans M. van Eijnatten & Marijn van Loon

Introduction 149
Complexity Concepts 150
Recent Applications Of Work Design 152
Motivation For Empirical Study 155
Value-Crafting Cycle 155
Method used in the Value-Crafting Intervention 157
Results Of The Value-Crafting Intervention 157
Secondary Analysis With Critical Incident Observations 159
Results of the Secondary Analysis with Critical Incidents Observations 161
Discussion 163
Acknowledgements 164
References 164

10. THE MIDDLE GROUND: EMBRACING COMPLEXITY IN THE REAL WORLD

Tim Dalmau & Jill Tideman

Introduction And Context 169
Perspectives From Complexity 170
The Process Enneagram 175
A Whole Of System View 176
A Way Of Engaging With Complex 'Middle Ground' Problems 176
Nine Points Of Inquiry 179
The Green Triangle 180
The Conversation 181
The Process Enneagram In Use 183
Accessing This Approach And Methodology 188
Beyond The Conversation 189
Reprise 191
References 191

11. ENVISAGING FUTURES: AN ANALYSIS OF THE USE OF COMPUTATIONAL MODELS IN COMPLEX PUBLIC DECISION MAKING PROCESSES

Lasse Gerrits & Rebecca Moody

Introduction 195
Modeling Complexity For The Real World 196
Complexity And Public Decision Making 198
Methodology 199
Case Studies 200
Case 1: Morphological Predictions In The Westerschelde (Belgium And The Netherlands) 200
Case 2: Morphological Predictions In The Unterelbe (Germany) 201
Case 3: Flood-Risk Prediction (Germany And The Netherlands) 201
Case 4: Determining The Implementation Of Congestion Charging In London (United Kingdom) 202
Case 5: Predicting And Containing The Outbreak Of Live Stock Diseases (Germany) 202
Case 6: Predicting Particulate Matter Concentrations (The Netherlands) 202
Analysis 203
Complexity In Data 203
Complexity In The Model 204
Complexity In The Decision Making Process 204
Conclusions 208

12. LEADING RADICAL AND RAPID ADAPTABILITY IN A TURBULENT ENVIRONMENT

Ramzi Fayed, Stephen Duns & Gervase Pearce

Introduction 216
Evolution of Adaptive Organizational Leadership Strategies 216
The Adaptive Fractal Leadership Framework 220
Core Activities 221
Support Activities 221
The Fractal Nature of the Leadership Framework 221
Intent Clarification 222
Ongoing Learning 223
Appropriate Divergence 225
Relevant Emergence 227
Timely Convergence 229
Action and Adjustments 231
Conclusion 234
Notes 237
Note 1: Self-Awareness 237
Note 2: Identifying Relevant Stakeholders 237
Note 3: Engaging Stakeholders at Rational and Deeper Emotional Levels 237
Note 4: Mobilizing "Collective Intelligence" 238
Note 5: Bank Queues: A Holistic Systems Thinking Example 239
Note 6: Self Organization 240
Note 7: The Art of Hosting and Harvesting Conversations that Matter (AoH) 241
Note 8: Simplified Approach to Scenario Development (Adapted from Schoemaker 2002 and van der Heijden 1996) 241
Note 9: Argyris and Schon (1978) 242
References 243

13. ETHICAL DECISION-MAKING AND METAPHORS: ENHANCING MORAL CONSCIOUSNESS USING PARABLES AND COMPLEXITY THEORY

Edwin E. Olson

Introduction 249
Virtue Ethics 250
Parables As Antenarratives 251
Parables And Complexity Theory 252
Using Complexity Concepts in Workshops 253
I. Confession—Awareness Of An Injustice: *The Rich Farmer* *253*
II. Worship—Looking For The Good In Others:
The Good Samaritan
(Peterson, 2002) 254
III. Guidance—Including Others In Decisions: *The Feast*
(Peterson, 2002) 255
IV. Celebration—Transforming Systems And Structures:
The Pearl Of Great Price 256
Paraplexity Method 258
Case of the Vineyard Laborers 259
1. Identify Issue 259
2. Choose Relevant Parable 259
Vineyard Laborers
(from Peterson, 2003) 260
3. Encourage Reflection on Parable in Context of Issue 260
4. Present Complexity Perspective 261
5. Amplify Differences in Perspectives 262
6. Apply to Issue and Action Steps 262
Further Research and Application 263
Conclusion 264
References 264

14. INSTITUTIONAL FRAGMENTATION IN METROPOLITAN AREAS AND INFRASTRUCTURE SYSTEMS: GOVERNANCE AS BALANCING COMPLEXITY AND LINEAR TASKS

Jack W. Meek

Introduction 269
The Transportation Challenge In The Los Angeles Region 272
Southern California Association Of Governments (SCAG) 274
The Metropolitan Transportation Authority (MTA) 275
Metrolink (Southern California Regional Rail Authority) 276

Sub-Regional Transportation Initiatives In
The Los Angeles Region 278
The Alameda Corridor-East Project 278
Summary 282
References 283

15. COMMUNITY ENGAGEMENT IN THE SOCIAL ECO-SYSTEM DANCE

Eileen Conn

Introduction 285
The Two Systems Approach 286
The One System Approach 287
Two Systems: Forms And Processes 288
Two Systems In Reality 289
Two Coevolving Systems 290
The Social Eco-System Dance 292
Some Dynamics 292
Emergence In The *Space Of Possibilities*:
A Local Case Example 292
Adjacent Possible In The *Space Of Possibilities* *293*
Nurturing The *Space Of Possibilities* *294*
Hybrid Sub-Systems 295
Vertical And *Horizontal* Systems In The Voluntary And
Community Sectors 296
Community Organizations' Spectrum 297
Typology Of Active Citizen Roles 299
Strengthening The *Horizontal Peer* System 300
Conclusion 301
References 302

16. LANGUAGE, COMPLEXITY AND NARRATIVE EMERGENCE: LESSONS FROM SOLUTION FOCUSED PRACTICE

Mark McKergow

Introduction 309
Acting In A Complex World—Molecules And Meanings 310
The Challenge Of Complexity—Lessons From Life 311
Are Conversations Emergent? Are All Conversations Emergent? .. 313
Psychology, Therapy And Coaching Conversations
Through The Lens Of 'Complicated' 314
An Alternative: The Interactional View 315
Six Simple Principles 317
Solutions—Not Problems 317

Inbetween—Not Individual 319
Make Use Of What's There—Not What Isn't 319
Possibilities From Past, Present And Future 320
Language—Clear, Not Complicated 320
Every Case Is Different—Avoid Ill-Fitting Theory 321
Conversations And Organizations As Emergent 321
Narrative Emergence 322
Conclusions: Lessons For Complexity Practitioners 322
References 323

17. BELLS THAT STILL CAN RING: SYSTEMS THINKING IN PRACTICE

Martin Reynolds

Introduction 328
What Is Systems Thinking In Practice? 330
What Matters In Systems Thinking In Practice 333
Context matters 333
Practitioner Matters 336
Systems Matter 339
Iteration Matters 341
Summary 343
Acknowledgements 345
References 345
Appendix: Assessing a Systems Thinking in Practice Practitioner . 349
Setting The Context: Managing And Leading Change 351

18. A CONSILIENT APPROACH: SUPPORTING LEADERS TO MANAGE AND SUSTAIN SUCCESSFUL CHANGE IN COMPLEX, EMERGENT AND CONTINGENT ENVIRONMENTS

Brian Lawson

The Application Of Complexity Theory To The Management And Leadership Of Change In Organizations 353
Overview 353
The Edge Of Chaos, Emergence And Change 354
Leadership And Complexity Thinking 354
Complexity Thinking And Local Government 355
A Consilient Approach To Change 355
Introduction 355
Resistance To Change And Disrupting Patterns Of Organization And Communication 356

Consilience 358
The Process And Key Elements Of A Consilient Approach To Supporting And Sustaining Change Based On Complexity Thinking 358
Framing And Designing Emergence........ 359
Integral Hosting 360
Mindful Awareness 361
Applying The Consilient Approach To A Real World Situation: The Case Study 362
Background And Overview: A Year Of Change, Emergence And Uncertainty........ 362
Initial Work: November 2009—February 2010........ 363
Creating A Community Of Practice: The March Event, A Taste Of Things To Come 365
Building And Creating Social And Intellectual Capital: March To June 2010 368
From Knowing To Doing: July To September 2010........ 373
From Turbulence To Flow: October 2010 To January 2011 374
Conclusion: Disturbing The Patterns? 378
Final Comment........ 378
References........ 381

INDEX........ 387

Editorial: From Theory To Practice

Andrew Tait[1] *& Kurt A. Richardson*[2]
1 Decision Mechanics, UK/US
2 Exploratory Solutions, US

Alice Munro (n.d.), the Canadian writer, once said, "The complexity of things—the things within things—just seems to be endless. I mean nothing is easy, nothing is simple." The more time we spend studying complexity, the more her sentiments are shared. Of course, the very pervasiveness of this complexity is the reason we gravitate towards it—like basin-dwelling moths to the attractor flame. The increasing number of "complexity"-focused journals stands as a testament to the progress that is being made in this young discipline.

Our passion, however, lies in the possibility of releasing all these ideas into the wider ecosystem. While many of the more beguiling concepts have embedded themselves in everyday language, complexity thinking, as a formal discipline, is clearly much less widespread. There are islands of success, but the intellectual tectonic shifts required to make them continents have not been forthcoming.

Why is this? Maybe it's partially down to the packaging. Complexity thinking is hard. Much of the research draws on sophisticated philosophy. This hinders the broad adoption of the ideas in the professional mainstream. The fact that the amount of research in the area of tools is dwarfed by that in the areas of philosophy and theory serves to compound the problem. A rough analysis of the papers published in the journal *Emergence: Complexity & Organization*, for example, in 2007 shows that less than 10% of them were primarily concerned with the development of tools for practitioners.

There is no doubt that the packaging of complexity into a neat, user-friendly shrink-wrapped "box" is a tall order. It's difficult enough to just describe the damn thing! Maybe this is because we've been gradually increasing the complexity of complexity. As we've experienced the failures of the systems engineering paradigm, and seen the limitations of "new reductionism", our definition of complexity has become increasingly elaborate. Naturally, this has trickled down the pipeline to challenge the tool developers.

But, maybe we can best approach tools from another theoretical direction–and use our understanding of complexity to evaluate and enhance them. Richardson (2008) has discussed the notion of a "modeling culture" where a practitioner uses linear tools in a nonlinear manner. This results in a kind of "cyborg" tool where man is responsible for providing the complex context. However, as complexity researchers, surely we'd like to provide man with more assistance in this area.

In July 2010, a workshop (the *1st International Workshop on Complexity and Real World Applications*) was organized in the UK (Botley, near Southampton) to attempt to further this endeavor. While this edited collection represents the permanent record of that meeting, we hope—no believe—that the bringing together of like-minds to discuss and debate the challenges will pay dividends through subsequent research. We look forward to it with barely contained excitement.

These are early days for this research agenda. We've no doubt that readers will feel some frustration at the lack of "consumer-grade" tools on offer. We'd love to have presented the complexity practitioners' version of Excel to you in these pages. Alas, it was not to be.

However, we, and our complexity colleagues, share your frustrations. This is a vibrant field, and we're only just beginning to see the first shoots. And, let's not forget that the rewards available to those breaking this virgin ground have the potential to be great—always a compelling call to action...

In true complexity fashion, we may have to maintain a broad perspective if we're to track the true impact of complexity thinking and techniques. Just as the field of artificial intelligence has contracted as ideas have been co-opted by other/new disciplines–such as speech processing, collective intelligence and computer-gaming–so the real impact of complexity research may occur in related fields.

One of the most tangible complexity techniques currently in use—and one represented in this volume—is agent-based simulation. Many in that community may not see themselves as complexity scientists, but clearly complexity ideas have had a significant impact on that field.

The first steps on the journey of creating a vibrant community of applied complexity practitioners and tool developers have been taken. Many challenges lay ahead—but, with them a wealth of opportunities.

Let's make the first half of this decade the point where complexity comes of age. We hope that this collection, in some small way, can help to bring that about.

References

Munro, A. (n.d.). Great-Quotes.com, http://www.great-quotes.com/quote/94060.

Richardson, K.A. (2008). "On the limits of bottom-up computer simulation: Towards a nonlinear modeling culture," in L. Dennard, K.A. Richardson and G. Morçöl (eds.), *Complexity and Policy Analysis: Tools and Methods for Designing Robust Policies in a Complex World*, ISBN 9780981703220, pp. 37-53.

1. A Practical Methodology For Managing Complexity

George Rzevski
The Open University, ENG

The paper outlines a methodology for applying complexity science and agent technology to practical large-scale commercial, social and engineering problems. The methodology is derived from a decade of experience in designing and implementing systems for real-time scheduling of taxis, car rentals, seagoing tankers and trucks; dynamic data mining; dynamic knowledge discovery and semantic search. The methodology has also been used for designing adaptive engineering systems and for research into social issues such as eradication of poverty.

Introduction

If future is not given (Prigogine, 2003) it is futile to attempt to predict it, except in terms of probabilities. However predictions in terms of probabilities are of little value for decision makers in charge of commercial or social resources who are responsible for delivering specified results within specified constraints.

How useful is then Complexity Science (Prigogine, 1997) in everyday situations? Could it provide help to a decision maker in charge of delivering crude oil across several continents and oceans at a given cost and within a given deadline? Or, to those whose duty is to protect a bank from cyber attacks?

The answer is: yes, it is extremely useful if it is used for clarifying complexity issues, for planning how to react to unpredictable disruptive events, for designing adaptability into social, business or technological processes at hand and for "tuning" complexity using experimentally derived heuristics.

Let us call these activities "Managing Complexity".

The aim of this paper is to outline a methodology for managing complexity, which has been developed by the author and his team during the last ten years by experimenting with large-scale complex adaptive software applications, which were designed and supplied to businesses in the UK, USA, Germany and Russia.

The methodology has been also successfully applied to the design of adaptive engineering systems, such as robots for space exploration, large compressors, machine tools and tractors, and for maximizing the life of large engineering structures such as airliners.

Finally, the same methodology has been used by the author for researching exceedingly complex issues, as exemplified by the evolution of English language, recent global financial crisis and the eradication of poverty.

Complexity

The Seven Criteria Of Complexity

For a methodology for managing complexity to be practical we must have a concise set of criteria for determining which issue is "complex".

My hypothesis is that complexity is present if all or most of the following *features* are in evidence.

1. INTERDEPENDENCY—A system consists of a large number of diverse components, referred to as Agents, which are interdependent or engaged in rich interaction.
2. AUTONOMY—Agents are not centrally controlled; they are largely autonomous but subject to certain laws, rules or norms.
3. EMERGENCE—Global behavior of the system emerges from the interaction of agents and is therefore unpredictable.
4. FAR FROM EQUILIBRIUM—Global behavior of the system is "far from equilibrium" because frequent occurrences of disruptive events do not allow the system to return to the equilibrium between two disruptive events.
5. NONLINEARITY—Relations between agents are nonlinear, which occasionally causes an insignificant input to be amplified into an extreme event (butterfly effect).
6. SELF-ORGANIZATION—A system is capable of self-organization in response to disruptive events. Self-organization may be initiated by the system autonomously in response to a perceived need, a feature that may be termed creativity.
7. COEVOLUTION—A system irreversibly coevolves with its environment.

Global Market As A Complex System

The Internet-based global market is perhaps the best example of complexity. Let us discuss the seven criteria of complexity as applied to the global market.

Global market consists of an exceedingly large number of suppliers, consumers, investors, lenders, savers, traders, etc. who are engaged with each other in trading.

Market participants are autonomous but subject to national and international laws, regulations and established norms of behavior; there is no central planning system.

Global distribution of supply to demand emerges (Holland, 1998) from local transactions. It is unpredictable although certain patterns of behaviors can be discerned.

Markets clearly operate far from equilibrium (Beinhocker, 2007); new transactions get done and agreed transactions get changed with such a frequency that the market has no time to reach equilibrium.

Extreme events are occasionally occurring as exemplified by "black Wednesday" and the recent financial crisis; as complexity of the Internet-based global market increases, extreme events may become more frequent and/or dangerous (Taleb, 2008).

Perpetual self-organization is in evidence as participants react to any disruptive event by changing or cancelling transactions. Creativity is exercised when participants decide to initiate changes to improve their positions.

Coevolution of society, economy and technology is illustrated in Figure 1 below. Tools aimed at improving quality of life change economic activities, which in turn change society; invented tools become available only if society decides to invest in them and use them.

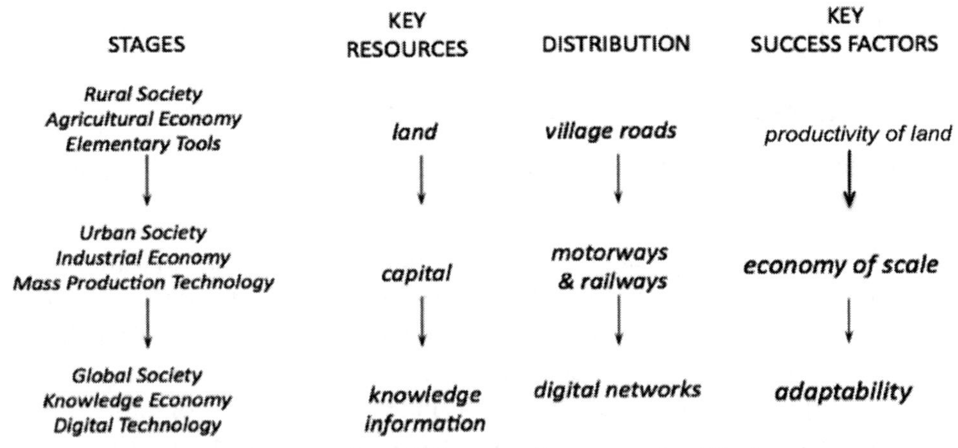

Figure 1 *Coevolution Of Society, Economy And Technology*

It is important to notice that as the economic system evolves so do key economic success factors. Economy of scale, the undisputable key success factor during Industrial Economy, is less and less important as complexity (and dynamics) of Knowledge Economy increases. The new key success factor is *adaptability*, the ability to rapidly produce a positive response to unpredictable changes in the market.

Coevolution advances in steps as shown in Figure 2 below. The process is similar to paradigm shifts in science (Kuhn, 1970).

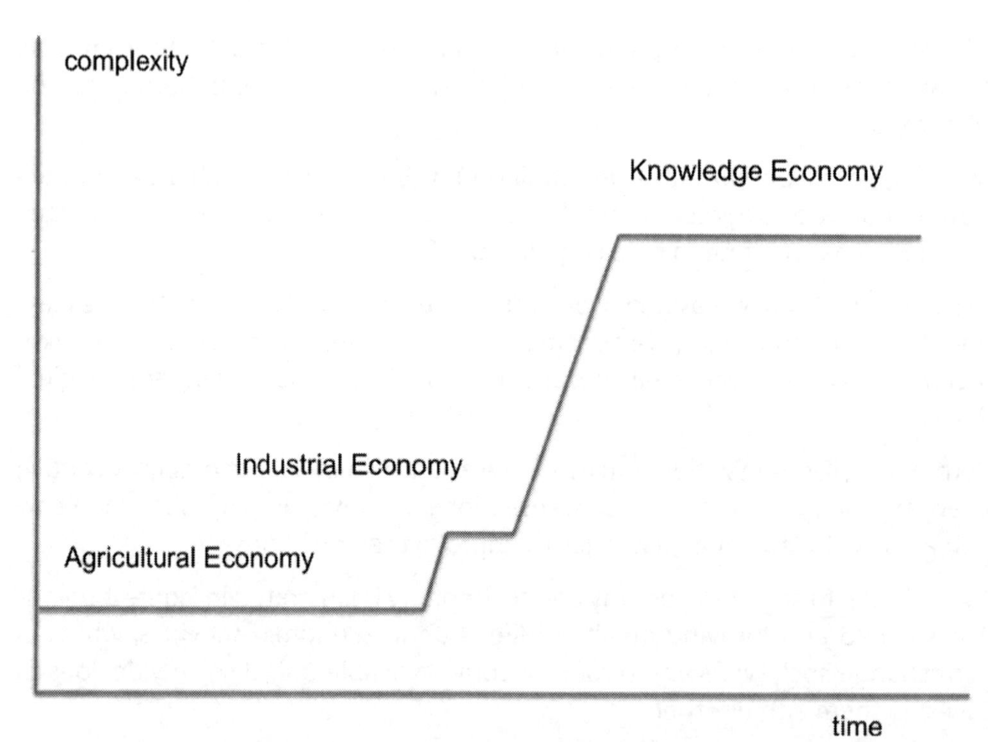

Figure 2 *Step-Wise Increase In Complexity Of Markets Over Time.*

Uncertainty As The Main Issue Resulting From Complexity

The unpredictable behavior of complex environments in which we live and work creates *uncertainty*.

Until recently serious changes in our environment were occurring infrequently and we are brought up by the educational system to expect a stable world operating in a steady state, at least during the life span of a generation. We are steeped in Newtonian Science postulating determinism and predictability. We are expecting that reductionism will triumph and that, sooner or later, a single universal law of physics will explain everything. It is comforting to be able to plan our lives and live to see our expectations fulfilled; uncertainty on a large scale feels menacing. The reaction is often: Let us stop this nonsense. Who is in control? Who is behind this unexpected event that affects me so badly? What is government doing about it? Is there a conspiracy? And, of course, there are no ready available answers to these questions and there is no possibility to stop the world and get off. Some individuals can "give up the rat race" and escape to remote areas, which are not yet affected, but is this the best we can do? What if we all attempt to do it?

Managing Complexity

If we accept that "to control" means to specify a desirable behavior of a system and to steer the system towards achieving it, then complex systems cannot be controlled. The very concept of "emergent behavior" precludes controllability.

Let us establish what else we cannot do.

We cannot eliminate uncertainty by simplifying complex situations and if we attempt to impose rigid structures on complex situations, these structures will, sooner or later, break down (remember centrally planned economies?).

We cannot rely on sophisticated mathematical prediction methods to tell us the future. If future is not given, we cannot predict it.

We cannot expect rigid planning to work under conditions of frequent occurrence of unpredictable disruptive events. Under such conditions plans soon loose any connection with reality.

What can we do then?

The best we can do is to "manage complexity", which means to "cope" with external complexity and "tune" internal complexity.

Coping with Complexity

Coping with complexity is defined here as a means of achieving desirable results under conditions of complexity that is not under our control (in other words, external complexity).

The ability to cope with external complexity is extremely important, for example, for businesses that sell to global markets.

The best strategy for coping with complexity is to develop a capacity for self-organization that will neutralize or reduce consequences of disruptive events when they occur.

Processes with capacity for self-organization are *Adaptive*, that is, capable of achieving their goals when operating in complex environments.

For self-organization to be possible we must have the following elements in place:

1. A range of optional actions that may be necessary to undertake when a disruptive event occurs.
2. Decision-making technology capable of autonomously and rapidly choosing which action to undertake when a disruptive event occurs, and implementing the selected action before the next event.
3. Strategic redundancy of resources to support planned options.

Dynamic forecasting of the occurrence of disruptive events, based on learning and instantaneous updating of forecasts with what actually happened, may help.

Building the capacity for self-organization into systems in which we live and work amounts to designing complexity into our life, which is counterintuitive. Common sense suggests we should attempt to simplify the complexity of the environment, which is not practical because by definition our environment is not under our control.

Tuning Complexity

In cases where we are in charge of a complex system we can impose regulations, which will keep its behaviors most of the time, but not necessarily always, within specified limits.

Methods for tuning complexity are particularly important for those in charge of complex systems, for example, authorities that regulate financial services, healthcare, education, law and order, security and fraud detection as well as managers in charge of business processes such as logistics.

Whilst we cannot do much about complex physical and chemical systems, which are guided by natural laws, we can certainly affect the behavior of social, sociotechnical, administrative and business systems that are guided by law, social norms, constitutions, statutes, policies, rules and regulations.

Emergent behavior of such systems can be kept within certain limits by ensuring that regulations are sufficiently unambiguous to prevent random behavior and yet sufficiently flexible to allow system certain freedom to experiment when facing new challenges (Rzevski & Skobelev, 2007). There exist evidence that the best strategy is to introduce variable regulations—tighter when the system operates in a normal mode and much looser when the system is recovering from effects of an extreme event, which is opposite to what financial authorities did after the recent financial crisis (Rzevski, 2010a).

It is important to note that regulations cannot prevent system nonlinearities to create occasional extreme events. To reduce severity and frequency of extreme events we must use additional heuristics.

There is evidence that reducing the frequency of occurrence and intensity of extreme events is possible by reducing propagation of oscillations through system connections, which can be achieved by increasing the "resistance" to propagations in system links and by partitioning the system into regions that are weakly interconnected with each other in order to prevent extreme events created within a region to spread to other regions, as shown in Figure 3.

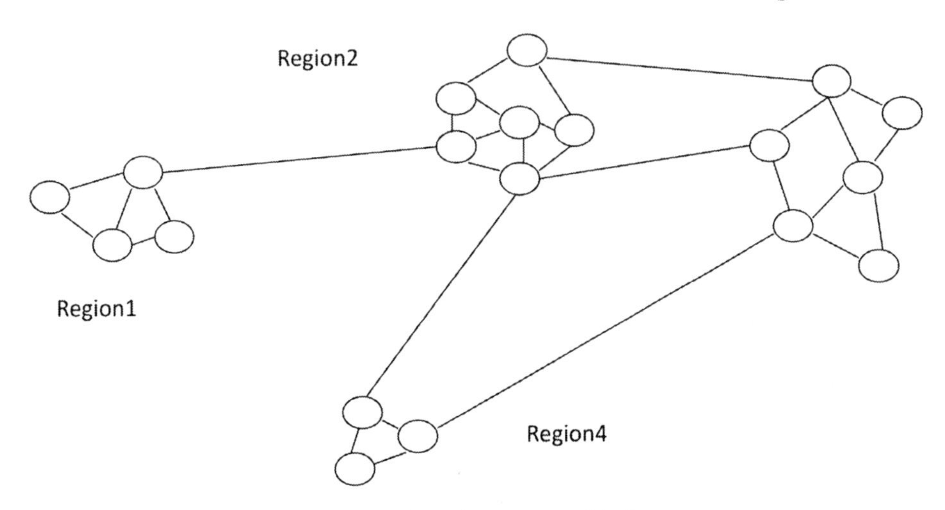

Figure 3 *Partitioning A Complex System To Contain The Occurrence Of Extreme Events.*

Tools For Managing Complexity—Complex Adaptive Software

Self-organization in business processes is feasible only if decisions how to respond to disruptive events are made and implemented rapidly—the decision what to do as well as the action, which implements the decision must be completed in between two consecutive disruptive events.

Here are some examples. In production of cars the frequency of disruptive events is one in two hours whilst in a large-scale metropolitan taxi operation (say, 2,000 vehicles)—one every few seconds.

It is obvious that for scheduling decisions to be done with such a speed we cannot rely on humans. We need *complex adaptive software*. Conventional software is not of much help because it requires a re-start from scratch whenever a disruptive event occurs and requires 8 to 10 hours to re-schedule a typical factory.

To exhibit adaptability software must have an extensive Knowledge Base and built-in artificial intelligence. Let us outline a methodology for developing adaptive software.

Collecting And Organizing Domain Knowledge

The first step is to collect and organize Knowledge on the domain of the Real World that is being investigated.

The most effective method of representing knowledge on a complex domain is to construct a network in which nodes are Classes of Objects of the domain and links are Relations between them. For example, for an airline relevant Object

Classes include: Flight, Passenger, Aircraft, Pilot, Maintenance, Seat Price, Route and Network. Each Object Class is characterized by attributes and scripts describing its behavior. Such domain knowledge representation is called Ontology.

Constructing A Virtual World

The next step is to build a Virtual World capable of representing the Real World that is under consideration.

The Virtual World consists of concrete Objects (instances of Object Classes from domain Ontology) and their Relations. For an airline, a Virtual World will be a network in which nodes are concrete Object such as: Passenger P1, Passenger P2, ... Flight F1, Flight F2, ... Seat S1, Seat S2, ... Aircraft A1, Aircraft A2, ... etc, and links are "S1 is allocated to P3", "A1 is allocated to F2", etc. Complex systems, such as supply chains of large international organizations, and Virtual Worlds that represent them, may contain millions of objects, attributes, rules and relations. To construct Virtual Worlds for such complex problems one requires powerful multi-agent software tools.

Connecting Virtual World To The Real World

Then, there is a need to connect the Real World to the Virtual World.

The Real World (i.e., a complex situation that is being modeled) is perpetually changing. The changes are represented by the occurrence of events exemplified, in an airline, by: Seat Booking, Flight Departure, Flight Delay, Flight Cancellation, Airspace Closure, Aircraft Failure, etc. The occurrence of every Real Event must be communicated instantly to the Virtual World where an equivalent Virtual Event is created, causing the affected part of the Virtual World to adapt to changes originated in the Real World. Every change (adaptation) of the Virtual World must be communicated back to the Real World and implemented before the occurrence of the next Event.

Empowering Virtual World To Manage The Real World In Real Time

Finally, the Virtual World must be empowered to manage the Real World.

A Software Agent (a small computer program) is assigned to every node of the Virtual World with responsibility to maintain its integrity. For example, if a Virtual Aircraft breaks down, the Aircraft Agent sends messages to Agents of all affected nodes letting them know that this Virtual Aircraft does not exists for a time being. The message provokes a flurry of activities among affected Agents who try to accommodate the failure by searching for a replacement. As soon as a solution is found, it is conveyed to the Real World for implementation, ensuring that the two worlds coevolve (change in unison). See Figure 4 below.

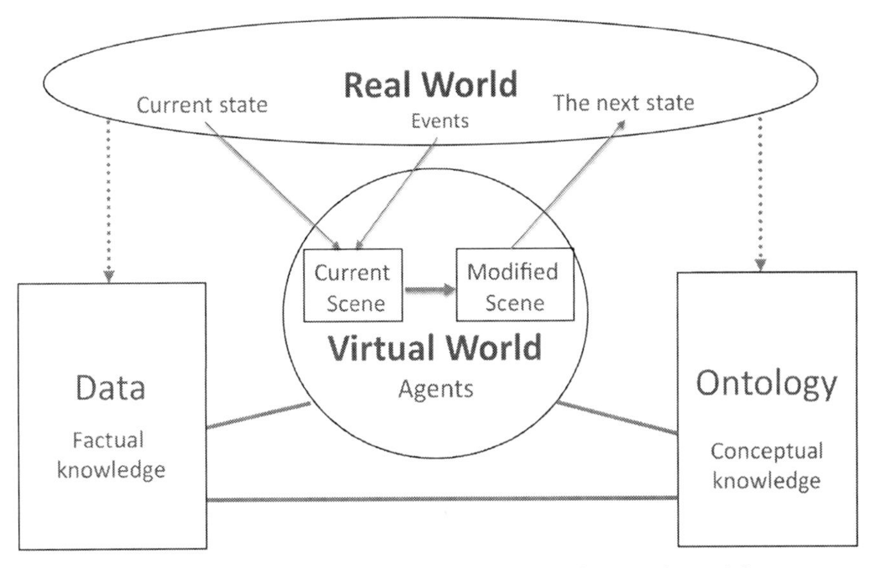

Figure 4 *A Virtual World Manages The Real World*

At present the only technology suitable for constructing complex adaptive software *is multi-agent technology* (Rzevski *et al.*, 2007a).

Comparing Multi-Agent Systems With Conventional Software

Conventional programs allocate resources to demands following pre-programmed algorithms in a sequential manner and therefore, when dealing with a large number of resources and demands, they require a long time to find the optimal allocation. Whenever resources or demands change, these programs start the allocation process from the beginning and if changes are frequent, they "oscillate" and cannot reach the optimal solution. Centralized intelligent systems are more flexible since they are normally driven by heuristics (rules derived from experience). Nevertheless they still solve the allocation problem sequentially and therefore cannot handle frequent changes effectively.

In contrast, high granularity multi-agent systems execute the allocation of resources in parallel (quasi-parallel, in sequential machines). Typically, hundred of thousands of agents located on a single server or workstation work concurrently, and if the problem is distributed over many servers and workstations, the number of concurrent allocation processes can rise considerably. This explains how multi-agent systems can rapidly arrive at a near-optimal allocation of resources in real time. In cases where changes are infrequent and therefore the optimal allocation is possible, agents will systematically reconsider each concluded resource-demand matching with a view to reducing the overall cost function. This is a time consuming process, which agents will carry out in addition to any re-allocations due to changes in market conditions. Agents work solving client's problems 24 hours a day and will continue re-negotiating partially matched deals until the best possible match is achieved or time runs out.

Agents do not have to wait for instructions. They plan and execute tasks autonomously and are capable of deciding when to compete and when to cooperate with each other. They react to any change in demand or supply without being prompted. Agents representing resources will pro-actively try to place them by searching for potential customers, offering discounts, cross-selling, making special offers and/or cooperating with other agents. Agents representing customers will actively search for resources that match their requirements and will ring their clients or send them emails when they obtain satisfactory allocations.

An architecture of complex adaptive software realized using multi-agent technology is shown in Figure 5.

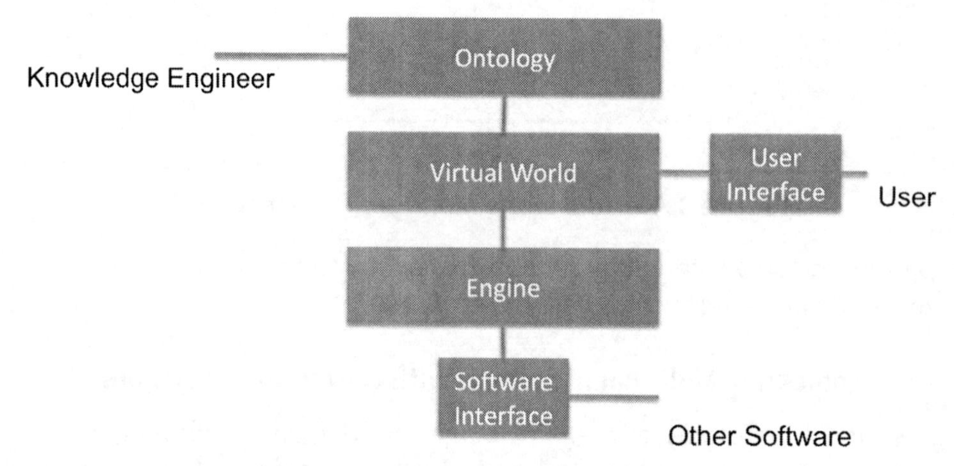

Figure 5 *Architecture Of A Multi-Agent System.*

Modeling Complexity

Only complex models can be used to represent complex systems.

Complex systems change as we attempt to construct their models and these changes must be incorporated in the model as they occur. Models of complex systems must be capable of adapting to these changes and the adaptation must be autonomous (without waiting for instructions from the modeler), which is only possible if models have capabilities of self-organization. In other words, *models must be able to coevolve with situations that they model.*

The key implication of the thesis is that conventional top-down planning methods are not suitable for complex situations. Current-generation computer programs cannot be used as models for complex situations because they are not adaptable. They cannot change by themselves—they must be instructed to do so by programmers. At present, only multi-agent software technology can support adaptation.

Complexity of certain unresolved critical issues (such as global warming, poverty and population growth) is so high that we can expect, at best, to build much-simplified models of these issues with a view to gaining some insight into their resilience to our attempts to resolve them. Let us refer to such issues as *Exceedingly Complex*. Exceedingly complex systems nevertheless can be successfully modeled. An example of a brilliant work on modeling of the population growth is described in (Kapitza, 2006).

The author of this paper is investigating ways of eradicating poverty by modeling "poverty issue" using complex adaptive software models. He has also investigated systemic failures that have contributed to the current global financial crisis using the same methodology (Rzevski, 2010a).

In contrast, there are many complex issues where constituent components and their interactions can be identified and modeled by constructing a Virtual World (a model) almost as complex as the Real World that is being modeled. Nearly all complex business situations and many complex social, economic, security and technological problems, as well as the evolution of urban systems, fall into this category. I shall refer to such issues simply as *Complex or Large-Scale Complex*.

Once a suitable Virtual World is constructed in software, it could be used to simulate behaviors of a complex issue in the Real World under different states of its environment, e.g., studying behavior of a supply chain under varying market conditions (Rzevski *et al*., 2003).

Examples Of Commercial Applications

During the ten-year period, 1999-2009, the author and his close collaborator, Petr Skobelev have developed a very large number of complex adaptive software products using multi-agent technology, which are in commercial use. All these systems have one feature in common—they have succeeded in solving problems, which were considered too complex for generally available conventional methods and tools.

Examples of successfully developed and implemented complexity management systems, include:

1. Managing in real time a fleet of 2,000 taxis, for a transportation company in London (Glaschenko *et al*., 2009).
2. Managing in real time a large fleet of car rentals, for one of the largest car rental operators in Europe (Andreev *et al*., 2009).
3. Managing in real time 10% of the world capacity of crude oil sea-going tankers, for a tanker management company in London (Rzevski *et al*., 2007).
4. Real-time scheduling of a large fleet of trucks transporting parcels across the UK (Rzevski *et al*., 2006; Andreev *et al*., 2007).
5. Selecting relevant abstracts for a research team using agent-based semantic search, for a genome-mapping laboratory in the USA (Rzevski & Skobelev, 2009).

6. Creating contract templates for car insurance using multi-agent based text understanding and dynamic clustering, for a logistics company in the UK (Minakov *et al.*, 2007).
7. Dynamic pattern discovery for a logistics company in the UK (Rzevski *et al.*, 2007b).
8. Managing social benefits for citizens supplied with electronic id cards, for a large region in Russia
9. Agent-based simulator for modeling the airport and in-flight, RFID-based, catering supply chain, luggage handling processes, and passenger processing, for a research consortium in Germany.
10. Resolving clashes in aircraft wing design for the largest commercial airliner in Europe

Examples Of Engineering Applications

Adaptability is the key property of engineering systems operating in complex environments and the most effective way of making artifacts adaptive is by designing them to be complex (Rzevski, 2010b). Several examples of projects in which complexity were designed into engineering systems that the author initiated or in which he was involved are described below (Rzevski, 2003).

A Swarm Of Agents Controlling A Machine Tool

The prototype was developed as an early experiment in testing usefulness of complexity in manufacturing. The first design step was to select agents that will participate in controlling the machine tool, in this case, a simple metal cutting machine. The guiding principle was to employ one agent per task, where tasks are selected such that each requires a specialized but narrow range of know-how. In the case of our machine tool, suitable tasks were considered to be: controlling processing speed, scheduling, condition monitoring, ensuring safety and security and record keeping and reporting.

Thus a team of 5 agents was formed as follows. The *Performance Agent* was given the task of selecting and maintaining the optimal metal cutting speed and the *Maintenance Agent* was given the task of monitoring the condition of the tool. In the case of a crack appearing, the Maintenance Agent was programmed to initiate a negotiation with the Performance Agent whether to terminate the process, slow down, or continue and replace the tool at the next opportune time, depending on the seriousness of the tool damage. The *Scheduling Agent* negotiated the loading of the machine tool with other Scheduling Agents by a kind of auction in which capacities were matched with orders. The *Safety Agent* monitored the immediate environment of the machine tool making sure that operators, or mobile robots, would not enter the danger zone. The *Bookkeeping Agent* kept records and sent reports on the machine operation.

The system Knowledge Base contained up to 10 rules per agent and a reasonable fund of facts. For example, knowledge relevant to the Performance Agent contained data on characteristics of a selection of metals and rules for choosing optimal processing speeds for these metals. Knowledge related to the Maintenance Agent contained data on typical damaged tools and probabilities of each particular type of damage causing a tool breakdown.

The multi-agent system proved to be simple to develop and highly flexible. What was however even more important, the group of five agents showed clearly an *emergent* behavior, which was far more sophisticated than the behavior of each individual agent. Needless to say the system functioned without central control.

Intelligent Geometry Compressor

Axial turbo compressors are used in many areas of industry where large quantities of air or gas have to be moved or compressed. Typical examples are jet engines, the larger gas turbines and gas pumping. All turbo compressors are limited in their performance by the aerodynamic phenomena of stall and surge, where the flow of the gas becomes unstable and can reverse in direction. Stall and surge, if allowed to develop, can cause significant mechanical damage to the compressor. Present designs of axial compressors use fixed geometry rotor blades and fixed geometry stator vanes, with a limited capability to vary vane angles against pre-set limits, using simple control algorithms. Conventionally, the operating point for the compressor is designed to give an adequate safety margin from the surge line, therefore avoiding the possibility of stall or surge in operation.

To reduce safety margins without endangering the compressor, the compressor model was designed containing considerable complexity—variable geometry, where intelligent agents assigned to each movable component negotiate among themselves the ever-changing optimal geometrical configuration under continuously changing aerodynamic conditions. The overall behavior of the compressor emerges from the interaction of agents.

Simulation experiments confirmed that the added complexity enabled the machine to self-organize whenever aerodynamic conditions changed always giving optimal or near-optimal compressor efficiency (Morgan, et al., 2004) Implications for reliability are staggering. Utilizing embedded processing power it becomes feasible to design into the compressor self-diagnosing (monitoring compressor conditions and identifying faults when they occur), self-repair by reconfiguration (isolating faulty parts and thus making them harmless) as well as graceful degradation of performance (repositioning remaining healthy parts to achieve a reduced but acceptable level of performance; also, in case of a serious failure of critical elements, such as actuators, agents can revert to the fixed geometry mode of operation). A compressor with intelligent geometry represents a fundamentally new approach to the design of mechanical structures, a genuine example of the new design paradigm.

Global Logistics Network

Global economy is characterized by a frequently changing demand, which puts a particular pressure on manufacturers who deliver their products to a variety of geographical locations. Purchasing air cargo capacity, hiring road transport, renting warehousing space, monitoring flows of parts and equipment and delivering just in time are business transactions that are now more difficult to execute than ever before. Even internally, within a manufacturing plant, logistics is a nightmare with customers frequently changing their mind about optional features of their purchases.

A feasible solution to this problem is to consider transportation capacity as a commodity. Suppliers of parts and transportation service providers could trade in options and futures. After acquiring transportation capacity options suppliers would send a stream of Intelligent Parcels into Exchange Nodes of the Global Logistics Network (GLN). Each Parcel would have an Intelligent Tag containing a simple Intelligent Agent built into a chip implanted into the packaging material. Parcel Agent would have knowledge about Parcel destination, expected time of arrival, transportation path, storage and handling conditions, Parcel weight and dimensions. Each Element of the GLN (Stores, Transporters and Exchange Nodes) would have resident Intelligent Agents capable of communicating with Intelligent Parcels in their charge. As they travelled through the GLN, Parcels would be regularly updated regarding changes in their destinations and times of their arrival. Once the new destination is known, Parcel Agents and GLN Agents would negotiate new paths through the Network for each individual Parcel. If necessary, Parcel Agents would purchase further transportation options that are required for them to complete their journey. Agents will be capable of making these purchases without any reference to Parcel owners.

This scenario envisages the development of a genuinely distributed, self-organizing logistics network offering a reduction in transportation times and costs. In addition, it would offer an ecologically sound solution to the current environmental damage caused by the excessive movement of goods across the globe. A modest prototype of such a system could be developed within a year. Further developments would be incremental. Suitable Intelligent Stores, Transporters and Exchange Nodes could be added to the Network as and when required. The whole idea could be easily extended to cover intelligent tagging and charging for purchased goods in retail.

A Family Of Intelligent Space Robots

Although the UK space effort has been reduced almost to zero, there are some small funds available for innovative work in space exploration. Several years ago a project concerned with the development of technologies for the autonomy and robustness in space, involving three universities and two spacecraft manufacturers from the UK was supported by research funds. The task was to research a prototype robot for the exploration of Mars. Following the complexity management principles described in this paper, a family of five much smaller intelligent robots has replaced the originally envisaged single robot. Each member of the family had limited intelligence and could undertake simple tasks such as placing scientific instruments onto correct location and, in addition, provide a variety of services to other members of the family, e.g., cleaning their solar cells if they get covered by the space dust and helping them to get out of small crevasses. The cost and weight per unit performance for this family was below the cost and weight of an equivalent single robot. Their size offered an important advantage in packaging for launch and delivery.

A Colony Of Agricultural Machinery

Current agricultural tractors are far too heavy and expensive. Because of their weight they tend to damage the soil structure, which requires expensive remedies, and because of their costs only large agricultural complexes can use them. According to some estimates by agricultural experts almost a half of the tractor weight and cost is due to the requirement to have a cabin and associated equipment, whose sole role is to protect operators and provide them with reasonably comfortable working conditions. Operators are not satisfied with their working conditions primarily because they are bored. It is perfectly feasible with current technology to replace heavy tractors with colonies of small and agile agricultural machinery equipped with sensors and actuators and controlled by networks of agents. A colony of such machinery would behave not unlike a colony of ants. Because of their scalability such colonies would be suitable for work on small and large fields. They could help us to go back to small farming and as a result enjoy healthier food.

Modeling of Exceedingly Complex Systems

Among this class of projects perhaps the most interesting was an investigation into designing stable global financial systems (Rzevski, 2010a). According to analysis presented in this paper the current global financial network has the following *systemic* faults.

Regulations imposed upon financial institutions are not flexible—the degree of freedom given to bankers to innovate was excessive under conditions of stability and insufficient under conditions of the crisis.

The excessive speed of financial transactions and dense interconnectivity of financial institutions increased the risk of instability and may have contributed to the occurrence of extreme events. The solution may include:

- Partitioning of the sector in a number of logical or geographical regions to limit the risk of the propagation of extreme events throughout the whole global system.
- Developing Financial Services Ontology for each region, incorporating the best available expertise on banking, investments, financial trading and regional peculiarities.
- Constructing a set of multi-agent simulators, perpetually evaluating the performance of regional and total financial systems.
- Instituting intelligent regulations, possibly different for different regions, based on a continuous evaluation of system performance.

These conclusions were derived by analogy rather than by direct investigation of the financial system and are therefore tentative. However, a study of the global financial system from the standpoint of Complexity Science would most probably yield similar results.

Complexity science has been also used to study the evolution of English language. The language is continuously changing but these changes are very small in comparison with two stepwise increases in complexity, which occurred when Chaucer and later Shakespeare introduced significant innovations, as depicted in the Figure 6 below.

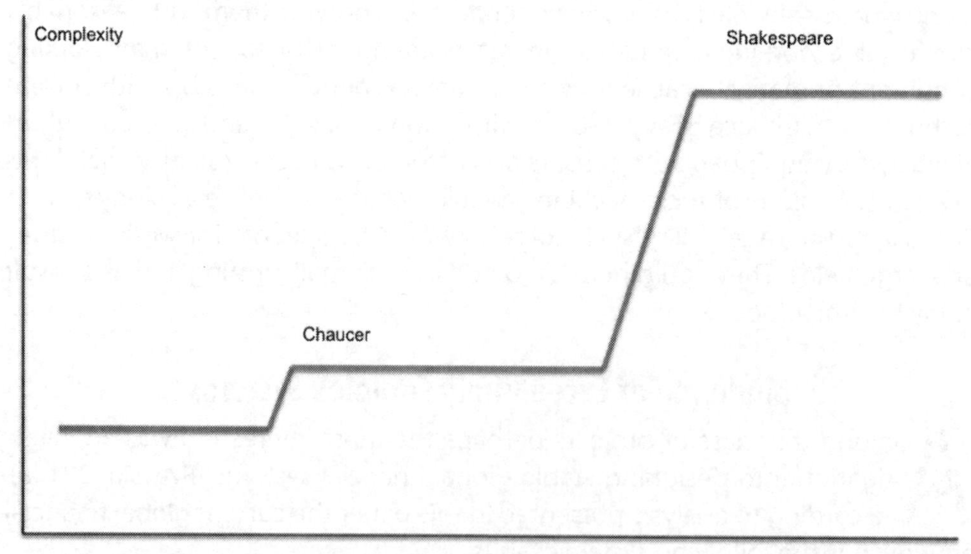

Figure 6 *The Impact Of Chaucer And Shakespeare On English Language.*

Conclusions

We live and work in a complex world and complexity of our environment is perpetually increasing. Current mathematical methods and conventional software are inadequate for modeling complex situations that are characterized by a high diversity of constituent components, very high frequency of unpredictable, disruptive events and occasional occurrence of unpredictable extreme events.

During the last ten years the author with his coworkers has developed a simple and practical methodology for managing complex situations. The methodology is supported by powerful tools consisting of advanced multi-agent technology that exhibit emergent intelligence and is capable of making rapid autonomous decisions in real time. A large number of commercial and engineering applications, as well as studies of exceedingly complex social issues, show the power of the methodology.

Acknowledgement

In 1990, when I was full-time Professor, Design and Innovation Department, the Open University, UK, and Director of the Centre for the Design of Intelligent Systems, I was invited by Professor Vladimir Vittikh of the Soviet Academy of Sciences to give a series of lectures on multi-agent technology in Samara, Russia. This visit marked the beginning of a long-term collaboration with Vladimir Vittikh and Petr Skobelev, a talented researcher and software developer who attended my lectures. With Petr Skobelev we have jointly founded and provided technological leadership for several commercial organizations in London and Samara that have developed software for many projects described in this paper. My special thanks go to Vladimir Vittikh for numerous research ideas that he shared with me during our long and fruitful collaboration and lasting friendship.

References

Andreev, S., Rzevski, G., Shveykin, P., Skobelev, P. and Yankov, I. (2009). "Multi-agent scheduler for rent-a-car companies," *Holonic and Multi-Agent Systems for Manufacturing: Forth International Conference on Industrial Applications of Holonic and Multi-Agent Systems*, ISBN 9783642036668, pp. 305-314.

Andreev, M., Rzevski, G., Skobelev, P., Shveykin, P., Tsarev, A. and Tugashev, A. (2007). "Adaptive planning for supply chain networks," *Holonic and Multi-Agent Systems for Manufacturing: Third International Conference on Industrial Applications of Holonic and Multi-Agent Systems*, ISBN 9783540744788, pp. 215-225.

Beinhocker, E. (2007). *The Origin of Wealth: Evolution, Complexity and the Radical Remaking of Economics*, ISBN 9780712676618.

Glaschenko, A., Ivashenko, A., Rzevski, G. and Skobelev, P. (2009). "Multi-agent real-time scheduling system for taxi companies," in Decker, Sichman, Sierra, and Castelfranchi (eds.), *Proceedings of 8th International Conference on Autonomous Agents and Multiagent Systems.*

Holland, J., (1998). *Emergence: from Chaos to Order*, ISBN 9780198504092.

Kapitza, S., (2006). *Global Population Blow-Up and After: The Demographic Revolution and Information Society*, ISBN 9783980972352.

Kuhn, T., (1970). *The Structure of Scientific Revolutions*, ISBN 9780226458045.

Minakov, I., Rzevski, G., Skobelev, P. and Volman, S. (2007). "Creating contract templates for car insurance using multi-agent based text understanding and clustering," *Holonic and Multi-Agent Systems for Manufacturing, Forth International Conference on Industrial Applications of Holonic and Multi-Agent Systems*, ISBN 9783540744788, pp. 361-370.

Morgan, G., Rzevski, G. and Wiese, P. (2004). "Multi-agent control of variable geometry axial turbo compressors," *Journal of Systems and Control Engineering*, ISSN 0959-6518, 13(218): 157-171.

Prigogine, I. (2003). *Is Future Given?* ISBN 9789812385086.

Prigogine, I., (1997). *The End of Certainty: Time, Chaos and the new Laws of Nature*, ISBN 9780684837055.

Rzevski, G., Skobelev, P., Batishchev, S. and Orlov, A. (2003). "A Framework for multi-agent modelling of virtual organizations," in L.M. Camarinha-Matos and H. Afsarmanesh (eds.), *Processes and Foundations for Virtual Organizations*, ISBN 9781402076381, pp. 253-260.

Rzevski, G., (2003) "On conceptual design of intelligent mechatronic systems," *Mechatronics*, ISSN 0957-4158, 13: 1029-1044.

Rzevski, G., Himoff, J. and Skobelev, P. (2006). "Magenta technology: A family of multi-agent intelligent schedulers," Workshop on Software Agents in Information Systems and Industrial Applications (SAISIA), Fraunhofer IITB.

Rzevski, G. and Skobelev, P. (2007). "Emergent intelligence in large scale multi-agent systems," *International Journal of Education and Information Technology*, ISSN 2074-1316, 2(1): 64-71.

Rzevski, G., Skobelev, P. and Andreev, V. (2007a). "Magenta toolkit: A set of multi-agent tools for developing adaptive real-time applications," *Holonic and Multi-Agent Systems for Manufacturing. Third International Conference on Industrial Applications of Holonic and Multi-Agent Systems*, ISBN 9783540744788, pp. 303-314.

Rzevski G., Skobelev, P., Minakov, I. and Volman, S. (2007b). "Dynamic pattern discovery using multi-agent technology," *Proceedings of the 6th WSEAS International Conference on Telecommunications and Informatics*, ISBN 9789608457607, pp. 75-81.

Rzevski, G. and Skobelev, P. (2009). "Agent based semantic web," Semantic Universe Website.

Rzevski, G. (2010a). "Using tools of complexity science to diagnose the current financial crisis," *Optoelectronics, Instrumentation and Data processing*, ISSN 8756-6990, 46(2): 210.

Rzevski, G. (2010b) "Using complexity science framework and multi-agent technology in design," in K. Alexiou, J. Johnson and T. Zamenopoulos (eds.), *Embracing Complexity in Design*, ISBN 9780415497008, pp. 61-72.

Taleb, N.N. (2008). *The Black Swan: The Impact of the Highly Improbable*, ISBN 9780141034591.

George Rzevski is Emeritus Professor of the Open University, Milton Keynes, UK, engaged in research into Complexity Science Applications and Multi-Agent Technology. He is also a Founder of a network of advanced technology ventures, which design, develop and sell large-scale intelligent multi-agent systems for applications such as logistics, e-commerce, semantic search and knowledge discovery. Before founding commercial ventures and undertaking the leading role in the design of large-scale multi-agent software for commercial applications, George was a full-time academic with The Open University, Milton Keynes, where he was Professor and Director, Centre for the Design of Intelligent Systems. His Centre was well funded by grants from government and industry and his Department was rated 5 out of 5 in the two UK Research Assessment Exercises. George has edited nine books and published a very large number of papers in the areas of Applied Complexity Science, Design, Artificial Intelligence, Multi-Agent Technology and Information Society.

2. Identifying The Multi-Dimensional Problem-Space And Co-Creating An Enabling Environment

Eve Mitleton-Kelly
London School of Economics, ENG

The chapter was written for the *1st International Workshop on Complexity and Real World Applications focusing on Using the Tools and Concepts from the Complexity Sciences to Support Real World Decision-Making Activities,* held at Southampton, UK, in July 2010. What the chapter wished to demonstrate was that tools and methods by themselves are necessary but not sufficient either to support decision making or to make fundamental change in an organization. By contrast understanding organizations as complex social systems and co-creating enabling environments using the principles of complexity can bring about fundamental change. There is no 'black box' or magic in this process; it is simply a matter of addressing complex challenges in an appropriate way. However to do this effectively implies a shift in perspective and a different way of thinking. This is what can bring about changes in relationships, in behavior, and in organizational culture. The chapter uses five different studies to argue that such understanding is essential; while at the same time illustrating some tools and methods used to identify the problem-space; and that complex problems can only be addressed effectively by creating multi-dimensional enabling environments, which address all the inter-related dimensions at the same time.

Introduction

Complex social problems *appear* intractable because they are often approached in a linear and simplistic way. Although such problems are multi-dimensional, the favoured approach and solution, in most organizations, is frequently mono-dimensional focussing on one aspect of the problem such as finance or information systems or restructuring. It is therefore not surprising that the 'solution' does not work.

A complexity approach acknowledges and attempts to identify the multiple interacting dimensions of a complex problem-space, such as the social, cultural, physical, technical, economic, political and other dimensions. The chapter will present five studies with organizations in the private and public sectors as well as with whole countries. Each illustrates a different focus but all use the same logic and a methodology which first identifies the multi-dimensional problem-space and then helps the problem owners to co-create an enabling environment to address the challenge. The key is to enable policy and decision makers to see their organizations as complex systems, to understand their characteristics and to work with those characteristics, instead of constraining them.

Study 1 was an ESRC-funded (Economic and Social Research Council) project involving five Universities including the London School of Economics (LSE), which looked at a new framework for Corporate Governance using a complexity theory perspective. This and the other four studies employed a methodology developed by the Complexity Group at LSE; this study is used to describe four types of analysis. A cross-section of employees from a global pharmaceutical company were interviewed, as well as a set of stakeholders outside the organization. Two broad sets of perspectives were therefore obtained and two sets of reflect-back workshops were given to the respective set of interviewees and others. The methods used in the study were semi-structured interviews, individual and group analyses and two reflect-back workshops. A set of recommendations were made and published.

Study 2 was an EPSRC-funded (Engineering and Physical Science Research Council) project, called ICOSS (Integration of COmplex Social Systems), with several partners from business and government. One of the business research partners was Rolls-Royce Marine (RRM). The distinctive feature of this study is that it involved 14 volunteers from the RRM Accelerated Leadership Development programme as part-time researchers, who did most of the interviews in the USA and the Nordic countries. They also took an active role in the analysis and in formulating the recommendations, all of which were implemented by Rolls-Royce Marine. The RRM case (Mitleton-Kelly E. & Puszczynski L.R. 2006) shows conclusively that the use of complexity principles and the methodology can be used quite effectively by non-academics to identify the problem-space, with only some basic training and introduction to the theory, as most of the methods are familiar. It is their combination and particular perspective which is different, as well as the use of the theory as an explanatory framework.

The other distinctive feature of the RRM study was the use of a quantitative tool based on psychology, which identifies preferences. It showed that the apparent cultural differences between the UK and Nordic managers were not significant; and that other factors were far more important in creating a lack of integration following the acquisition of the Nordic companies by Rolls-Royce, to create Rolls-Royce Marine.

Study 3 was with another business partner in the ICOSS project. This study illustrates the use of art as a discourse enabling tool, to address very sensitive issues which were very difficult to discuss.

Study 4 was a longitudinal study with two London NHS (National Health Service) hospitals funded by the NHS Institute for Innovation and Improvement. It is used to illustrate the changes needed in thinking, in ways of working and in relationships, to create a sustainable endogenous enabling environment. Sustainability in this context is seen as part of the enabling coevolutionary dynamic.

Study 5 was an analysis of Disaster Risk Reduction in West African States. The analysis was undertaken by a group of researchers at both King's College London and LSE. The recommendations were submitted to the President of ECOWAS (Economic Community of West African States), who had commissioned the study. This study is used to show that the methodology can be employed not only with organizations, but also with whole countries.

What the chapter will try to show is that it is an understanding of organizations or states or societies as complex social systems, which is of primary importance, not the use of specific tools. Many tools may be used, as long as they conform to the logic of complexity, and take the multi-dimensional nature of organizational problems into account. The other point is that complex problems cannot be explained using mono-causal explanations, that is why it is essential to identify the multiple interacting dimensions, which together create and re-create the problem-space. These multiple causalities coevolve and change the problem-space. Any 'solution' must therefore also coevolve, hence the importance of co-creating an endogenous enabling environment that will coevolve with its exogenous broader social ecosystem. A final point is that an organization need not be paralysed by the 'complexity' or multiplicity of aspects which need to be taken into account. In-depth interviewing of key responders and thorough analysis can identify clusters of multiple dimensions, which provide an adequate starting point to build an enabling environment. The critical factor is to engage the problem owners, to help them see their organizations or countries as complex social systems, and to provide them with a robust base on which to build a coevolving enabling environment.

Study 1: The Corporate Governance Study

Despite the widespread adoption of corporate governance codes, financial scandals and collapses still occur and the aim of the ESRC-funded project was to rethink corporate governance frameworks from a complexity theory perspective; and to offer some fresh theoretical and practical insights.

Five Universities (Birmingham, Cardiff, London School of Economics, University College London and Jerash University, Jordan) were involved in the study; a report was submitted to the ESRC and a book has been published based on the findings (Goergen et al, 2010). The field work involved a global UK pharmaceutical company, and its stakeholders including institutional shareholders, the Trades Union Congress (TUC), the London Stock Exchange, the Financial Reporting Council, the Financial Standards Authority, the Treasury, and the Department of Trade and Industry. The data were generated though semi-structured interviews and workshops with the company employees and the stakeholders.

The corporate governance study is used in this chapter to illustrate four types of qualitative analysis based on interview transcripts.

Background

The project looked at two types of actors: internal actors, which include the directors and employees, and external actors which include the shareholders and wider stakeholder groups, plus the legal, social, political influences and structures. Complexity theory emphasizes the relationships, connectivity and interdependence between the internal and external actors, and the various structural influences on the environment in which they operate. By identifying, and analyzing, the multiple elements that interact and influence each other, we hoped to gain a deeper understanding of the evolution of governance frameworks.

A working hypothesis was that all human systems and institutions are complex in the sense that they are multidimensional with social, cultural, political, physical, technical, economic and other dimensions which interact and influence each other. Complex behavior arises through interaction, with specific characteristics which include self-organization, emergence, coevolution, exploration of the space of possibilities, and many others. Not only can complex systems *adapt* to, and *coevolve* with changing conditions; they are also able to *create new order*. In a corporate governance context, new order may take the form of new rules, regulations, corporate governance frameworks, or a new culture; it may also include the creation of *enabling environments* that facilitate good corporate governance within individual organizations as well as industries and economies.

Complexity theory therefore studies social systems holistically; it does not focus exclusively on one aspect such as the legal framework or compliance, but at the *multiple, inter-acting* dimensions that together, create the *social ecosystem*.

By taking a holistic approach, complexity theory looks at all related entities or in this case all corporate governance (CG) actors, which interact and influence each other, within the entire corporate governance environment of *corporate governance social ecosystem*.

Methodology

The methods used in the study were semi-structured interviews, individual and group analyses and two reflect-back workshops. Two sets of interviews were undertaken between December 2006 and February 2007, with eight employees in the PLC from a cross section ranging from company directors to relatively junior employees; and eleven external 'stakeholders' that included regulators, institutional investors, and gatekeepers[1].

---

1. Gatekeepers refer to external third parties who may provide a service of reputational veri-

The interview data, in the form of full transcripts, were analysed using four sets of analyses described in the chapter. The integrated findings on corporate governance were presented at two Reflect Back workshops to the interviewees for validation and further discussion

Semi-structured interviews were based on eight topics that were chosen to stimulate reflection on corporate governance in the past, present and future and on the strengths and weaknesses of different approaches, their implications and consequences, and on the internal/external and interactional governance processes. The questions prompted interviewees to reflect upon what changes to governance took place, how they were introduced and why. The interviewees however responded with answers that also gave insights into changes in company operations, and the data also became useful in understanding executive decision-making.

The interviews took 1.5 hours each, they were recorded with the express permission of each interviewee and most were conducted by two interviewers. The interviews were led by either Prof. Mitleton-Kelly or Prof. Mallin; and all team members took an active role in interviewing. The lead interviewer asked the questions (not seen by the interviewee) and invited comments and reflections, while the second interviewer explored any broader issues. Each interviewee was sent an interviewee briefing in advance to describe the project, explain the process and offer a short introduction to complexity theory. The briefing also contained short biographies of all five team members. The interviewees were also asked at the beginning of each interview if they had any questions about the project and were offered some background. Strict ethical guidelines on the protection of confidentiality and the identity of individual interviewees were observed.[2]

Four Types of Analyses were made based on the interview transcripts. Analysis [1] used direct and full transcripts, as the language used by the interviewees was an important element in the process, and identified the following:

a. Common themes or patterns across all interviews;
b. Dilemmas, i.e., equally desirable objectives that appeared not to be achievable at the same time;

---

fication for the company, and in this position may be able to detect corporate abuse or misdemeanour, and may also ultimately be called upon to bear corporate liability if found to be negligent. See Coffee (2006); Kraakman (1984, 1986).

2. For example, all researchers and the transcriber were required to sign a non-disclosure agreement, agreeing that no information regarding the organization and the interviewees could be discussed with anyone outside the research team. Further, all names and identities were removed and all quotations were non-attributable, used only to illustrate key points made by the interviewees.

c. Critical questions expressed by the interviewees on corporate governance frameworks and on future research in the field, as well as questions raised by the researchers;

d. Underlying assumptions (these surfaced the usually unspoken beliefs of the interviewees) that influenced the way they saw corporate governance.

These were captured within different dimensions including sociocultural, economic, legal, political, technical, and other. The individual entries were then cross linked to capture the connectivities and interdependencies between them. These clusters formed the basic building blocks for the other analyses and for the presentations. Analysis [2] identified the enablers and inhibitors to effective corporate governance. Analysis [3] complemented the other two analyses using a range of other data and expertise contributed by the interdisciplinary team. This included insights from legal and regulatory frameworks as well as quantitative data and financial insights. Analysis [4] used all the other findings and interpreted them using the principles of complexity theory to gain fresh insights into the main corporate governance challenges. The insights from the four analyses and their integrated findings on corporate governance were presented first to the company employees and subsequently to the stakeholder interviewees at two Reflect Back workshops. The discussion from the workshops was then incorporated into the final report submitted to the company and the ESRC.

In terms of the analysis process for Analysis [1], the interviews were first transcribed by a professional transcriber. All the transcripts were read by all members of the team, but they were divided between team members for analysis. Each team member analysed at least 8 interviews and all interviews were analysed by at least three researchers to reduce interpretation bias. Each researcher worked on a first level analysis individually, identifying the common themes, dilemmas, questions and underlying assumptions. All the researchers then met for two days, on 17 and 18 April 2007, for a joint group analysis. The group analysis covered Analyses [1], [2] and [3] in depth, and provided some examples to illustrate the relevant complexity principles for Analysis [4].This process allowed the team to deepen understanding of the issues; and the different members of the team could contribute insights from their specific expertise on corporate governance. A joint analysis was made of all the interviews and helped identify the cross links between the different themes, dilemmas and underlying assumptions and produced joint findings, which were more than the sum of the parts. Interpretation bias was reduced by questioning each other's individual interpretation.

There were two **Reflect-Back Workshops**: one for the PLC on 30 April 2007 (Workshop 1) and another for the stakeholders on 1 May 2007 (Workshop 2). The initial, non-attributable findings (individual respondents were not identified) were presented to the interviewees of the company for validation at Workshop 1. The findings offered a "mirror" to the organization, while the workshop process allowed the researchers to discuss the findings with the interviewees and other attendees at the workshop from the PLC to test their interpretation. In this case, as participants included those that had not been interviewed, but could offer valuable insights, the process enhanced understanding and enriched the findings.

The three-hour Stakeholder Reflect-Back Workshop 2 took place the following afternoon on 1 May. It included most of the stakeholder interviewees and the whole research team. The process was interactive and was used not only to validate interpretation but also as a forum for the stakeholders to discuss some of the key issues. The whole workshop was recorded and the discussion was transcribed as this was part of the overall process and the transcription was thus part of the research protocols.

The integrated findings from the analyses and workshops were submitted to ESRC and the report is at www.lse.ac.uk/complexity.

Insights on Corporate Governance (CG) Frameworks

Regulators; institutional investors; listed companies, their employees, customers and suppliers; the financial press and analysts; and other such CG actors, are intimately inter-connected through positive and negative feedback processes. They influence each other through formal and informal relationships in such a way that they change their behavior; in complexity language, they coevolve. It is therefore essential to look at all the related actors within the whole corporate governance social ecosystem. It is not enough to focus on the relationship between the board and the shareholders, as this provides only a partial view of the relevant dynamics that influence good corporate governance.

Another insight was that changes are most effective when they are not purely top-down, through government regulation. The logic of complexity argues that the most effective approach is both bottom-up and top-down, or at both the micro and macro levels of interaction. A related insight arises when a company wishes to achieve two distinct, but closely related, objectives. One objective might be to improve the ethical standards both within the industry and within the organization to avoid stricter additional external regulation. The second objective might be to look at risk as part of good governance and to make awareness of potential risk something that permeates the whole organization. Both objectives are related to good corporate governance in the broader sense of good corporate responsibility. What this insight illustrates is that there is no single universally applicable solution that remains successful over time. It is therefore essential to allow responsible exploration of alternative approaches or solutions to maintain and sustain good governance.

Following from the above point, another insight was that good corporate governance is more likely to emerge if there is a supportive enabling environment in a company. Such an environment may be created both at industry level through an industry association, which would include all the major companies, and at individual company level. An enabling environment thus needs to be created at all scales from individual companies, via each industry, to national government and international levels. An enabling environment can only be robust if it is supported and subscribed to at multiple levels, and if it is sufficiently flexible to adapt and to coevolve within its own ecosystem and to adjust appropriately and sustainably to changes in the broader external environment.

The key is the creation of a coevolving enabling environment, which supports and encourages good governance. The study also identified a set of *enablers and inhibitors* of a good corporate governance model that together contributed to an *enabling environment.*

Insights And Summary Of Study 1

Study 1 is used to illustrate the qualitative methods used, of four types of analyses and Reflect Back Workshops to report back, to validate the findings and to deepen insights both of the stakeholders and the researchers. This sets the scene for the other studies in terms of the qualitative methods used. However, the emphasis of the chapter as a whole is to show that tools and methods by themselves are not enough to bring about fundamental change. What Study 1 achieved was to use a complexity theory perspective to question some fundamental tenets of corporate governance. Traditionally, corporate governance frameworks focus on the relationship between the board and the shareholders; the study questioned that emphasis and demonstrated that other stakeholders need to be included as the dynamics of influence within the corporate governance ecosystem are not confined to the board and the shareholders. Such an emphasis brings about a narrow focus on financial return, which is only one dimension in a multi-dimensional complex environment, and can lead to a skewed and partial view of good corporate governance.

Another strongly held view was that the only way to influence and change corporate governance frameworks, was top down through government regulation. This view was again challenged and the study showed that influence can be exerted effectively bottom up at organizational and industry levels. Another related view was that there can only be one universally applicable solution. Complexity theory challenges the single solution view and supports a distributed micro-strategy approach, which not only allows but encourages the exploration of different possible solutions locally.

The study finally proposed an alternative approach based on the idea of an enabling environment at multiple scales to encompass individual organizations and whole industries. This idea of enabling environments will be a common thread argued within different contexts throughout the chapter.

The key is to use the insights provided by in-depth analysis and complexity theory to look at the problem-space from a different perspective and to build on that to co-create an enabling environment. This next stage is illustrated by Study 2.

Study 2: The Rolls-Royce Marine Study

The method of analysis used in the Corporate Governance study had also been used with Rolls-Royce Marine (RRM). Two major differences, being illustrated by this study were (a) that 14 volunteers from their Accelerated Leadership Development programme took part in the project as part-time researchers and did most of the interviews in the USA and the Nordic countries. They also took an active role in the analysis and in formulating the recommendations, all of which were implemented by Rolls-Royce Marine; (b) The RRM study also used a tool based on psychology, called Landscape of the Mind (LoM, developed by Kate Hopkinson) which identifies individual and group preferences. It was used as a quantitative tool, based on questionnaires and it showed that the apparent cultural differences between the UK and Nordic managers were not significant; and that other factors were far more important in creating a lack of integration following the acquisition of the Nordic companies, by RR to create Rolls-Royce Marine (RRM).

Background

RRM had acquired an organization made up of small firms that had already gone through a series of mergers and acquisitions (M&As), in a different but related market to its main operations. Rolls-Royce (RR), the parent company was primarily a UK organization while the acquired company was distributed throughout the Nordic countries (Norway, Sweden and Finland). There were therefore *apparently* significant cultural differences and in the first two years after the merger these tended to overshadow other differences in business processes and procedures. At this point RRM joined an EPSRC-funded project led by the LSE Complexity Group called ICOSS.

Methodology

When RRM joined the project, the HR (Human Resources) Director organized a workshop with the Accelerated Leadership Development (ALD) group, which were relatively young potential future leaders of the company. As a result, 14 ALD members joined the LSE research team. Together they formed four teams and interviewed the top three management layers in the Nordic countries, the UK and the USA. The RRM teams were guided by the LSE research team in conducting semi-structured interviews and in analyzing the findings from the transcripts using a similar method to that described in Study 1.

At a 2-day workshop the four teams identified a set of common themes, dilemmas, key questions and underlying assumptions, captured according to different dimensions, and cross linked to form clusters. A set of recommendations, developed by the ALD and the LSE team, based on the findings, was presented to the Executive Board and all recommendations were adopted. The interview findings were supported by the 'Landscape of the Mind' (LoM) tool, which identifies preferences in decision-making, knowledge acquisition and sharing, etc. The LoM analysis is based on an email questionnaire completed by the participants; in this case 70 senior managers from the UK and all the Nordic countries had answered the LoM email questionnaire and the results were quite surprising. They showed that there was no significant difference in the preference profiles of the Nordic and the UK teams, hence many of the problems associated with the national cultural differences were more apparent than real.

The findings showed that the national cultures issue had become a smoke screen that was used to hide all the other difficult issues that were impacting the relationship between the parent company and the acquire company. Eight overarching themes were identified at the 2-day workshop; different national cultures were only one of them:

1. Complicated organizational structure;
2. Human behaviors;
3. Relationships;
4. Cultures;
5. Communication;
6. Matrix interfaces;
7. Leadership/role of central team/management, and;
8. Identity.

On the basis of these findings, the LSE-ALD teams made a set of recommendations, all of which were accepted and implemented by the Board.

Apart from the interviews, LoM findings, workshop and Board meeting, the LSE Group also attended meetings, conferences and joint presentations. In addition a second set of interviews was conducted with the high flyer ALD team that had acted as interviewers; in addition an agent-based model was built based on an email questionnaire.

Summary And Insights Of Study 2

The RRM study (Mitleton-Kelly E., 2003, 2006; Mitleton-Kelly E. & Puszczynski L.R., 2006) showed conclusively that the use of complexity principles and the LSE methodology can be used quite effectively by non-academics to identify the multi-dimensional complex problem-space, with only some basic training and introduction to complexity theory, as most of the methods were familiar. It was the combination of methods, and particular perspective, which was different, as well as the use of the theory as an explanatory framework. The use of the psychology LoM tool which provided hard quantitative data, based on questionnaires undertaken by 70 managers in the UK and Nordic companies, helped to provide the evidence which an engineering company found acceptable and accessible. Once this evidence was presented they were prepared to pay attention to the 'softer' qualitative findings based on the interview data. The combination of the quantitative and qualitative tools and methods together with the involvement of 14 young and active leaders meant that there was a fully engaged group which could act as ambassadors throughout the organization. In addition the support of the HR Director and a senior Project Manager, who helped to manage the project with RRM, provided additional credibility when presenting the findings and recommendations to the Board. The key however was a fundamental change in perspective. The Board, the senior managers involved and the young leaders found the evidence compelling and shifted from 'a single cause' view to a more complex one acknowledging that the lack of integration, had multiple dimensions and therefore had to be addressed along all those dimensions simultaneously. By implementing all the recommendations, the organization co-created an enabling environment based on complexity theory principles, which addressed multiple inter-related issues at the same time.

Eighteen months into the project, the relationship between the parent and the acquired company had improved and the new company had increased its market share. The integration process in RRM was successful, partly due to the recognition of a problem early on and an openness to experiment with a different approach. They did so, by identifying the problem-space and by creating an appropriate enabling environment to address it, informed by complexity theory.

Study 3: The SSF Study

Mitleton-Kelly (2006) describes two cases following a merger and acquisition. One was RRM and the other was SSF (to retain its anonymity this company is referred to as a Service Sector Firm), which had gone through a very thorough due diligence process and the market considered the merger process a great success. Two years after the merger, however, SSF was suffering from severe dysfunctional relationships. Part of the problem was that individuals were uncertain of the boundaries of their authority and autonomy, which had been clear in the pre-merger firms. SSF knew they were good at the pre-merger process and expected that a successful post-merger process would follow automatically. The LSE Complexity Group started work with SSF two years after the merger.

Methodology

The analysis in Mitleton-Kelly (2006) was based on 22 semi-structured interviews conducted with a wide cross-section of interviewees in SSF, on many meetings with individuals and groups and on a Reflect-Back Workshop, which validated the findings.

The reason for including this study is to illustrate the use of an unusual tool; a large poster-sized painting which illustrated the dysfunctional behavior of the organization, shown in Figure 1 below, and was used as a discourse enabling tool.

One of the main themes which emerged from the interviews and analysis was that too many projects were claiming equal high-level priority and all were competing for limited resources, time, funding and energy. Many interviewees appeared exhausted with the constant battle to meet so many high priority deadlines. *"We try to do too much all the time, constantly failing." "We have so many projects clambering for resources at the moment. All the projects are classed as priority one and so it's difficult to know what to work on first."* Overall there had been just too much change and they were suffering from change-fatigue *"they are punch drunk from all the change"*; with the corollary that it was difficult to believe that the transformation would take place.

This was indicative of the lack of direction and focus. They needed clarity to help them concentrate their efforts on the important issues; they also felt the lack of leadership, which was not readily available as the Executive team was not *"knitted together"* and did not have a *"consensus view".*

Weak leadership coupled with a controlling attitude, lack of a clear vision and direction, all contributed to becoming a dysfunctional organization. This was not something that could be discussed with the Executive team. We therefore put up the painting, a larger version of Figure 1, on the wall while having coffee before the workshop. When the SSF participants first looked at it they were bemused; there was a difficult atmosphere until someone recognized what the picture depicted, found it amusing, laughed and started talking about it, trying to interpret it. Once that happened then the research team were able to actually talk about extremely sensitive issues, which otherwise would have been extremely difficult to talk about. Even with less sensitive issues, a picture can be used as a catalyst for discussion or as a discourse enabling tool.

Another reason for using art is to capture a complex problem with many elements in one picture. It helps us visualize several elements at once, which we could not keep in mind at the same time. Finally, it can act as a transitional object, if it helps us to move from a familiar artefact to a complex abstract idea.

Figure 1 Illustrated Of The Dysfunctional Behavior Of An Organization
(Image created for LSE Complexity Group by image 2010@delta7.com, julian@delta7.com)

Insights Of Study 3

Study 3 is not offered as a full study, only as an example of using an unusual tool; art was unexpected and it facilitated a necessary and very sensitive shift in perspective at the workshop.

Study 4: A Longitudinal Study in Two NHS Hospitals

Mitleton-Kelly (2011) discusses a longitudinal study in two London NHS hospitals, both of which faced a significant deficit. To address that challenge they had to change fundamentally their way of thinking, some of their ways of working and many of their relationships, to become organizations capable of learning and changing.

What the case shows are two different styles of leadership, creating totally different 'enabling environments'.

The methods used were again in-depth analysis, Reflect-Back Workshops and the development of a relationship with the lead teams over a 2 year period, introducing them to complexity thinking.

Two new Chief Executives, appointed to help each hospital address its financial deficit, had helped create with their management teams, two different environments. In Hospital X addressing the financial deficit was seen as offering possibilities for improvement in the service; the approach was collaborative and everyone participated and tried their best to contribute to making a difference. They were anxious about the future nature of the hospital, but this was a manageable anxiety, which manifested itself almost as a curiosity about the future. The approach was participative with a dual top-down and bottom-up process to change, involving everyone in the hospital at all levels and skills.

In Hospital Y anxiety about an uncertain future, was uppermost in their mind and obscured everything else. There was a strong impression that they were not being given the space to contribute; the management team talked about participation and contribution, but in practice this was seriously constrained, as the approach was primarily top-down.

The Two Environments In Complexity Terms

Hospital X had facilitated *self-organization, exploration-of-the-space-of-possibilities, active feedback, coevolution* and *emergence* in *the following ways.* Staff felt that they had 'permission' to *self organise* to try out ideas locally and to *explore the space of possibilities* by experimenting with alternative procedures and processes to improve the patient journey. They could discuss the outcomes openly and honestly within their group and share it more widely with others. They also initiated cross-directorate projects, which helped to bridge the tight boundaries between specialities. The cross-over was not always successful, but the possibility of bridging boundaries was present. As each team made a change the others were influenced to a varying degree, and the active feedback did make a difference encouraging other directorates in the cross-over projects to also try out some new ideas. The reciprocal influence resulting in changes in the interacting entities is part of the *coevolutionary process* which was facilitated. Furthermore, no one knew in advance precisely what would happen and the outcome was primarily *emergent*, in the sense that it was unpredictable; it arose through interaction and was systemic with the whole being more than the sum of the parts. This latter is important; it was collaboration and working in teams and joint projects which brought about emergent change. The process was systemic and could not be reduced to individual parts or components or specific individuals on their own. That is part of the reason why 'best practice' cannot be copied. The process is systemic, emergent and context dependent. It cannot be reduced to 'building blocks' which can simply be re-assembled in a different context and give rise to an identical outcome.

Hospital Y talked about facilitating *self-organization*, but constrained it; for example new ideas had to be given the 'go ahead' from the senior team and no one had 'permission' to just try out a new idea locally. The whole atmosphere was one of constraint rather than encouragement with very limited opportunities to *explore the space of possibilities*. Hospital Y was not beset by problems (Hospital X had a much larger financial deficit) and they did have significant successes in some specialities such as the maternity service, which was growing beyond all expectations. There was however, no active learning from these successes and the focus was very much on attaining financial balance. There was also little active *feedback*, and few opportunities for staff to get together to review performance and reflect in an open, relaxed and informal atmosphere. Reviewing was done formally in terms of performance management. By restraining self-organization and exploration and by not actively reflecting on the outcomes the learning environment was constrained.

In terms of sustainability, it is of fundamental importance to understand the dynamics of complex social systems, whose outcome is unpredictable and relatively fragile.

Complete Change In 2006/7

Following the Reflect Back Workshop after the first phase of the study, there was a fundamental change in Hospital Y. The atmosphere had changed dramatically and the following comments were recorded at a workshop in October 2007:

> *...this organization just feels as if it's much stronger, it's a better place ... not necessarily because the people are different, it's the fact that they've been given the opportunity ... the responsibility and the authority to get on with it.*

> *... we have never been so strong financially, yet the external environment has never been so weak, so why are we so good? ... I think it's because divisions have really taken it on board, ... it feels very different than the way it used to be, it was always a bit hit and miss ... Whereas now it's very firm, you will deliver on this, and people are given the authority to get on and do it. I mean we're not quite there yet, but that does feel different.*

They started to explore the space of possibilities and recognized that change is not about spelling out what everyone has to do, but in creating the right enabling environment. *"It's not really telling people—I think it's creating the organization that takes the responsibility itself.* It is about distributed intelligence and distributed responsibility. They introduced a new management structure which enabled Divisions to take responsibility for their own plans, within a clear overall strategy for the hospital. Not all Divisions and Departments, however, were able to successfully adapt to the new regime and inefficient departments were helped to improve.

All these changes were made against significant external changes in the 'health ecosystem'. Payment by Results had been introduced which meant higher salaries for Hospital Consultants and GPs (General Practitioners or family doctors), which increased costs. The NHS (National Hearth Service) Agenda for Change therefore had the effect of increasing costs, while lowering productivity *"we haven't got better productivity through paying people more, we've actually got less..."* The European Working Time Directive and external societal changes, had changed the shift pattern and had reduced the amount of time which junior doctors worked. District General Hospitals were also losing some of their traditional work, as the NHS restructured itself and some work was to go to Specialist Centres and other work to GPs. All these external and imposed changes made it necessary for hospitals to rethink their role. To survive the changes in the health ecosystem, Hospital Y had to be aware of what was happening and address it fully; they had to learn how to scan the landscape and *identify the emergent patterns.* They could not just *adapt* to the changes, but had to find radically different ways of working, by *exploring the space of possibilities.* They also had to develop new relationships with the independent sector, with GPs and with PCTs (Primary Care Trusts), and to develop new *connectivities,* and establish new feedback processes. They had to use resources differently; and make active use of their *distributed* leadership, distributed intelligence and distributed expertise, by facilitating *local autonomy* and *self-organization.*

They also had to adopt a different way of thinking such as working in partnership with the independent sector; marketing their services further afield (outside their immediate catchment area); redress the balance of having to deal with the difficult long stay patients, by bringing in more lucrative work; do the core emergency work really well; think more in business terms: *"if something is making a profit, we should expand it. If it's making a loss we need to sort it out and make it more efficient"* e.g. maternity and cardiology may be profitable, while rheumatology, because of the large drugs bill, can make a loss *"but you can't not do these things"*; and finally they had to address the conflict between the political and the cultural, for example whether to continue to provide a service that is needed for the local community, accepting that such a service makes a loss.

They also had to develop a different way of working. The Government had imposed targets and performance management such as a 4 hour target in A&E (Accident & Emergency) and reducing the length of stay. Hospital Y had met the A&E target by streamlining the process and by opening a Medical Assessment Centre. As a result the patient pathway had changed. They described their new attitude as *"can do with a smile"*. In addition they worked with a greater cultural mix with multiple faiths, beliefs, and backgrounds.

In complexity theory terms, changes in the ecosystem had pushed the hospital *far-from-equilibrium* in the sense that they could no longer operate under their existing regime using established norms and procedures. They reached a critical point and had to either do things differently or go downhill. They therefore *explored their space of possibilities* and developed new ways of working such as greater autonomy and self-organization; working better as a team, supporting each other and acknowledging their inter-dependence. They also developed a different way of thinking as well as learning a new business language, which enabled them to work with the independent sector. Finally, they developed new relationships or new patterns of connectivity internally and externally.

Summary And Insights Of Study 4

The distinctive characteristic of complex systems is their ability to create new order; that is, a different way of working, thinking and relating. Hospital Y created and *continued to create new order.* In other words they understood their role in the coevolutionary process, which isn't merely a question of adaptation but of influencing the environment through their changed behavior, while continuing to actively respond and change. Sustainability is this continuous process of coevolution. It is neither a one-off change which remains static nor a reversion or adherence to the status quo. This means understanding and working with (not constraining) the characteristics of organizations as complex social systems. In the case of Hospital Y they became more tolerant and comfortable with *emergence*, unpredictability and uncertainty by co-creating a new *coevolving enabling environment* which by definition had become sustainable.

The case illustrates some essential elements of sustainability. Leadership and the creation of an enabling environment are necessary but not sufficient; the changes have to become embedded within the organizational culture, through a different way of working, relating and thinking. Both successes and failures, while exploring the space of possibilities, need to be shared within a learning environment. Successes cannot be copied, but if the underlying principles are understood, then they can be adapted to a new context. In both hospitals what worked was full engagement of the staff at all levels. Major sustainable improvement can only (a) be achieved and (b) persist if there is active involvement of employees and innovation flourishes if the application of their distributed intelligence is encouraged. This is not a theoretical notion. It has been observed repeatedly in organizations in both the private and public sectors.

What Study 4 illustrates is the importance of a fundamental and sustained shift in thinking throughout the organization facilitated by the different perspective offered by complexity theory and by developing distributed leadership throughout the organization and encouraging full engagement of all stakeholders.

Study 5: Disaster Risk Reduction In West African States

Background

The President of the ECOWAS (Economic Community of West African States) Commission had requested the study to look at potential humanitarian crises. Interviews were conducted between October and December 2008 in Nigeria, Sierra Leone and Mali. The data gathering phase proved difficult as interviewees did not wish to be recorded and only hand written notes were permitted. It also proved very difficult to spend adequate time with each interviewee. The usual period of 1.5 hours with each interviewee, for an in-depth interview, was not possible and in most cases the researcher was only given approximately 20 minutes. However, since she was able to conduct almost 30 interviews, we were able to undertake a partial analysis. The final report was based on that analysis and the expertise and deep background knowledge of the research team.

The research team included researchers from the Humanitarian Futures Programme at King's College London (KCL) with in-depth knowledge of Africa and from the Complexity Group at LSE, bringing expertise in complexity theory and the methodology being used, based on in-depth group analysis. All researchers came together over 3 days to look at the findings as a group. The group analysis reduces interpretation bias and provides a deeper understanding of the issues involved. Both KCL and LSE researchers participated in the group analysis. Although this is a rigorous process, much depends on the quantity and quality of the original data, and for reasons explained above, the set of interviews did not provide the depth normally required, this weakness however was partially redressed by the specific knowledge brought by the KCL researchers.

The reasons for using Study 5 is to show that the methodology used for organizations can also be used to identify the problem-space in whole States; and that rigorous research methods can overcome some difficulties in field research, but the difficulties have to be acknowledged. A third reason is the questioning of the assumption that effective action can only be taken top down. As will be seen below the research discovered some local initiatives, which were effectively addressing specific issues locally by creating the right enabling environment and to propose a method of learning from such success.

The Objective Of The Study

The Study's objective was to identify some of the key issues in the political, economic and environmental context or problem-space as perceived by the interviewees; and to offer some practical recommendations to address them.

The main focus of the report was potential *humanitarian* crises and how the ECOWAS Commission might address them; none were unknown or new and other studies had focussed on specific crises. What this study attempted to do was to put them in their broader context, to identify their inter-relationship and to suggest that political, economic and environmental crises do not exist in isolation from each other. They are inter-dependent and any solution will need to address all three categories simultaneously.

As might be expected, climate change was at the heart of the environmental concerns, and no single issue stood in isolation, as the following comments from interviewees illustrate: *"Dry seasons are getting longer"* and *"In 20 years it will be the most serious problem in West Africa—other consequences ... (lack of) water will be a major challenge. Changes in weather and climate, which is in part responsible for geothermal and hydrothermal ... Then you have the problem of UV rays, cancer gets worse."*

Rainfall patterns have changed causing both drought and floods. This in turn affects *"food security, agriculture. All agricultural productivity is based on rainfall. If this trend continues (reducing rainfall), it will be minimum to plant but not enough to sustain crops to maturity"* and *"the normal expectation of rain is changing. People plant at the wrong times."* Farmers not only plant at the wrong time, they may also be planting the wrong seeds. They may need to plant crops, which are more drought resistant and need less water. Information on changing rainfall patterns and on appropriate drought-tolerant crops is available, but not widely known; and one of our recommendations to the ECOWAS Commission was that local communities be made aware of this vital information on planting times and drought resistant crops, coupled with practical advice.

This could be done in at least two ways: (a) making sure that the correct information and advice are available and well communicated; while (b) engaging with local communities to both impart that information and advice, and at the same time, to learn if and how anything is being done locally that addresses the changing rainfall pattern and the use of drought-tolerant crops. Engagement with local communities is paramount and the only way to ensure fast and effective implementation. It was also recommended that the Commission would need to identify similar initiatives both within the ECOWAS States and in other countries and to learn from these initiatives.

Cannot Copy Successful Initiatives

It was pointed out that simple identification of successful initiatives was not enough; that was simply where the process started. Extensive research by the LSE Complexity Group with major organizations in the private and public sectors has shown quite clearly that 'good practice' cannot simply be copied. Complexity theory explains why not; when something is copied it is the 'what' and the 'how' that is reproduced in a totally different context. Because of the different context (or different initial conditions) the outcome is likely to be different; it can never be the same. At the same time what and how things are transferred is often partial and this again militates against successful reproduction of success. We have seen time and again successful initiatives in one part of a large organization or 'best practice' in another organization, being transferred and not working.

The key is to identify and to understand the underlying principles of success in a particular context. Why did it work in that context? What would have stopped it working well in that context? These are the key questions that need to be raised and explored. Once those underlying principles have been identified (and explained using complexity theory), then they can be adapted (not slavishly copied) in a new context.

Some of these local initiatives had been identified by the KCL & LSE project and were given as examples. They were local, either to a Member State (e.g. in Mali and Sierra Leone) or to a community; and they involved local people and communities. The reason for highlighting these initiatives is twofold. One is to provide a few examples of positive action being taken locally and to redress the balance in a positive way; the second is to explain how these can be used and to suggest that a concerted effort to identify other local initiatives would be worth while.

The value of these initiatives is not in copying them. Their value lies in the inspiration they provide and in learning why they were successful or why they failed. The relevant questions therefore become 'why did it work in that context?' and 'what would have stopped it working in that context?' From this understanding can be derived underlying principles, which are transferable. It is this deeper understanding, which can then lead to successful adaptation of local initiatives into a new and different context.

The recommendation was for the Commission to take positive action through public education, the provision of relevant information and a network of local contacts to help share the learning. Not only could the Commission encourage local communities to make these initiatives known, it could also use its local representatives and contacts to help identify successful initiatives. The idea was to create an enabling learning environment based on local initiatives, using a dual bottom-up and top-down approach.

Examples Of Local Initiatives

Illiteracy

The following quotation from Sierra Leone captures what in complexity theory is known as 'self-organization', which is enabling people to take action and responsibility themselves without external direction. What the respondent is saying is that public education is one way of opening up that possibility, by actively using the community's 'distributed intelligence'. This acknowledges that informed and appropriate action is not confined to external experts or senior civil servants or even formally appointed leaders. That knowledge, expertise and leadership can be found distributed within local communities.

> *We have massive illiteracy. In Sierra Leone 70%; Liberia and Guinea, it's the same problem. The issue of risk reduction through public education is critical. There's poverty, but communities need to take responsibility. We advocate this in Sierra Leone. Community people should take it forward. That self help will go a long way ... local people still have capacity. We try to identify their capacity. Use your knowledge to address issues. They need to develop strategies. Some disasters may be too far. By the time experts reach, it's too late. That's why we advocate communities first.*

Counter-Deforestation Measures

Sierra Leone has put in place:

> *[C]ounter-deforestation measures. ... [W]e have national police and army to patrol the western area to control firewood and timber. We tie that with massive sensitisation. We go to community schools and meet traditional rulers. We target community radios. We take professionals from environment, water and discuss issues of impact and how to deal with it. These are short term measures. The long term measure is to train a good number of forest rangers—we can't have the army out all the time.*

Peace Building—Based On Distributed Intelligence And Local Contributions

> *There are indigenous mechanisms going on for traditional peace building, mediation and dialogue.*
>
> *ECOWAS should develop a solid partnership with civil society organizations, the ones at its base who work with populations at the grass roots.*

All these initiatives need to be studied to understand what helped them succeed and what to avoid and this information needs to be made widely available to other communities. Success has a re-enforcing effect through positive feedback, the better it works the more others are inclined to try out similar initiatives; and the more is known about why they worked, the next iteration will be improved, and so on. The whole thing creates a positive environment of 'can do', which counters the consistent negative messages generated by the possibility of imminent crises.

Relevant Complexity Theory Principles

There are two other principles from complexity theory which are relevant here and which will help policy makers understand why supporting and learning from these initiatives is so important. One is *'exploration of the space of possibilities'*; when a complex system (in this case a group, an organization or a country) is facing a crisis, and when past solutions are no longer effective, it searches for new options, new solutions and creative alternatives to address the problem. Not all the options will work, but some will be effective and successful. By making these successful attempts more widely known, the Commission could help to reduce the number of unsuccessful attempts.

Furthermore, when a new idea is being tried, both the idea and the people involved, evolve and change in the process. When the change is reciprocal and all those involved influence and change each other, then the process becomes *coevolutionary*. When change is only in one direction, e.g. when individuals or a group change *in response to* changes in their environment, then this process is called *adaptation*. However, when adaptation in time affects the environment and changes it, then both the adapting entity and its environment have influenced each other and changed in the process. This reciprocal coevolutionary process is very powerful; that is why it is not always necessary to make major interventions in order to bring about significant change.

Social, cultural, political, economic and environmental processes tend to be coevolutionary. As interacting humans we constantly influence each other in varying degrees. The process also applies between the physical environment and people. We influence and change that environment through our actions; in time the physical environment also changes and may affect us in an adverse way. This is part of the argument for human-induced climate change. Understanding these dynamics is essential for policy makers and complexity theory offers an explanatory framework, which many policy makers in the private and public sectors have found invaluable.

Summary And Insights Of Study 5

Study 5 has been included in this chapter for the following reasons:

a. It demonstrated that the method developed to identify the problem-space in organizations, can be used effectively to identify societal and national issues and to develop recommendations on addressing potential economic, political and environmental crises in order to stop them becoming humanitarian disasters.

b. Field research is rarely trouble-free. This particular case was extreme in its difficulties, but a sound research approach can overcome many such difficulties.
c. The study was very broad ranging and this chapter could not give details on the entire project; it therefore focused on highlighting how enabling learning environments can be facilitated using relatively low levels of funding and technology. Both of these issues are highly relevant in African countries.
d. It questioned the dominant view of top-down intervention being the only approach and recommended a dual bottom-up and top-down approach, working with local communities and civil society groups.
e. Finally it highlighted a few local initiatives which illustrate some important complexity principles.

Conclusions

The argument of this chapter, based on the logic of complexity theory, is that tools and methods are necessary but not sufficient in bringing about organizational and societal change. They can, however, be used to develop an in-depth understanding of the multi-dimensional problem-space. No complex issue, challenge or problem can be addressed effectively by focusing on a single issue or dimension, as they do not exist in isolation; the different inter-related dimensions interact and influence each other. Once the complex problem-space has been identified that understanding can be used to co-create with the relevant stakeholders an enabling environment, which addresses all the relevant inter-related issues at the same time. Complex problems often do not have a single definitive solution, but an enabling environment which is responsive and coevolving with its changing broader social ecosystem, is much more likely to address the issues effectively. This is a key message both for policy makers in the public sector and decision makers in the private sector. The approach applies equally to organizations and to societal and national issues.

Another key point is that a variety of traditional and new tools may be used in combination. It is the approach which makes the difference. The approach developed and used by the LSE Complexity Group uses several qualitative and quantitative tools and methods including agent based models, simulations, psychology tools, qualitative analysis and art. The criterion for their use is that they are underpinned by the logic of complexity theory, and take the multi-dimensional nature of social entities (organizations, countries or societies) into account. The chapter has demonstrated several of these tools.

However, it is the understanding of organizations as complex social systems and the co-creation of enabling environments using the principles of complexity, which is more likely to bring about fundamental change, if it brings about a shift in perspective and a different way of thinking.

The chapter used five different studies to argue that such understanding is essential; while at the same time illustrating some tools and methods used to identify the problem-space; and that complex problems can only be addressed effectively by creating multi-dimensional enabling environments, which address all the inter-related dimensions at the same time. One critical factor is engagement of the problem owners, to help them see their organizations or countries as complex social systems, and to provide them with a robust base on which to build a coevolving enabling environment.

References

Coffee, J.C. (2006). *Gatekeepers: The Role of the Professions and Corporate Governance*, ISBN 9780199288090.

Goergen M., Mallin C., Mitleton-Kelly E., Al-Hawamdeh A., Hse-Yu Chiu I. (eds.) (2010). *Corporate Governance and Complexity Theory*, ISBN 9781849801041.

Kraakman, R.H. (1984). "Corporate liability and the costs of legal controls," *Yale Law Journal*, ISSN 0044-0094, 93: 857-898.

Kraakman, R.H. (1986). "Gatekeepers: The anatomy of a third-party enforcement strategy," *Journal of Law, Economics, & Organization*, ISSN 8756-6222, 2(1): 53-104.

Mitleton-Kelly, E. (2003). "Complexity research: Approaches and methods - The LSE complexity group integrated methodology," in A. Keskinen, M. Aaltonen, and E. Mitleton-Kelly (eds.), *Organizational Complexity*, ISBN 9789515641083, pp. 56-77.

Mitleton-Kelly, E. and Puszczynski, L.R. (2006). "An integrated methodology to facilitate the emergence of new ways of organizing," in A.A. Minai, D. Braha and Y. Bar-Yam (eds.), *Unifying Themes in Complex Systems, Proceedings of the Fifth International Conference on Complex Systems*, ISBN 9783642176340.

Mitleton-Kelly, E. (2006). "Coevolutionary integration: The co-creation of a new organizational form following a merger and acquisition," in *Emergence: Complexity & Organization*, ISSN 1521-3250, 8(2): 36-47.

Mitleton-Kelly, E. (2011). "A complexity theory approach to sustainability: A Longitudinal study in two London NHS hospitals," *The Learning Organization*, ISSN 0969-6474, 18(1): 45-53.

Eve Mitleton-Kelly is Founder & Director of the Complexity Research Programme, LSE; visiting Professor at the Open University; SAB member to the 'Next Generation Infrastructures Foundation', TU Delft; on Editorial Board of *Emergence: Complexity & Organization*; was Coordinator of Links with Business, Industry and Government of the European Complex Systems Network of Excellence, Exystence (2003-2006); Executive Coordinator of SOL-UK (London) (Society for Organizational Learning) 1977-2008; and Policy Advisor to European and USA organizations, the European Commission, several UK Government Departments; Scientific Advisor to the 2012 World Forum on Public Governance and to the Governments of Australia, Brazil, Canada,

Netherlands, Singapore and UK. EMK's research has concentrated on addressing apparently intractable problems at organizational, national and global levels and the creation of enabling environments based on complexity science. She has led, and participated in, projects funded by the EPSRC, ESRC, AHRC, the European Commission, business and government to address problems associated with: IT-business alignment; organizational integration post M&A; corporate governance; leadership, sustainable development, organizational learning, innovation, etc. Publications etc. at www.lse.ac.uk/complexity. She has developed a theory of complex social systems and an integrated methodology using both qualitative and quantitative tools and methods. The theory is being used for teaching at universities around the world, including three EPSRC-funded short courses at LSE, to train researchers and two courses at Beijing to train senior government officials. Her first career between 1967-83, was with the British Civil Service in the Department of Trade and Industry, where she was involved in the formulation of policy and the negotiation of EU Directives.

3. Complexity Tools For Smart Grids: PCT And ABM Join Forces

Liz Varga
Cranfield University, ENG

This paper presents an innovative way of combining existing complexity tools to offer a better description of the diffusion of distributed electricity generation and Smart Grid. These complexity tools are Personal Construct Theory (PCT) and Agent Based Modeling (ABM). The primary value of ABM is that it permits multiple iterations of a model using slightly different input conditions (or variables), demonstrating how small changes can be amplified as the dynamics of the system evolve. The challenge for ABM is to accurately capture the behavior (or rules) of individuals (or agents). Agent rules and associated variables in existing models are determined using a wide variety of research methods which are not usually stated explicitly. The culminating rules are often mechanical, treating agents as automaton, and so do not reflect the complex emergence of new variables in the system being modelled. PCT, and its associated structured interview methodology called Repertory Grid Technique, elicits from respondents a set of constructs which reflects how individuals understand their behavior (towards electricity generation and consumption), consciously or otherwise. The constructs of multiple respondents provide a legitimate set of increasingly mature constructs which describe evolving rules for differentiated behavior in Smart Grid adoption. These agent rules, together with environmental information, such as the technological trajectories of products, political/regulatory incentives/taxes, economic wealth, population structure, climate information and housing stock, allow the ABM to be developed for a particular region or country. Running the ABM will show barriers to adoption and allow interventions to be tested in order to speed an emerging landscape of increased distributed energy generation and more use of renewables. Importantly, interventions will be linked to both sustained, behavioral change and contextual settings making for a more robust description of the diffusion of Smart Grid.

Introduction

Understanding the changes needed to consumer behavior is critical to understanding how electricity production can meet demand over the next 40 years or more. UK electricity demand has reduced over recent years largely reflecting poor recent economic conditions (DECC, 2010) but overall demand is set to grow and importantly, demand is uncontrolled. At the same time, central non-renewable generating capacity is under pressure due to ageing power plants, onerous emissions legislation, stretch carbon targets and increasingly more economically difficult to find non-renewable sources. In order for the UK to avoid a dependency on imported electricity and related economic, political, security and other risks, electricity demand needs to match the capacity it can supply. Capacity can grow by the use of smaller scale, locally generated electricity and through greater awareness of consumption (the Smart Grid) which alleviates the pressure on central generation, but is not without its own challenges.

This paper examines the nature of electricity generation in the UK and the drivers for change. It describes how Smart Grid is dependent on changes in behavior of some proportion of the UK population in order to balance supply and demand. Personal Construct Theory (PCT) is then discussed as a means for eliciting constructs and Agent Based Modeling (ABM) as a means for investigating changes in the system which could bring about significant transitions. An argument is put forward for using the methods as part of a larger research design, which also captures aspects of the environment (political, technological, regulatory, etc). Next steps are to implement the proposed methodology in a community study within the UK.

UK Electricity Generation

Total supply of electricity in the 2009 in the UK amounted to 380 TerraWatt hours (TWh) (DECC, 2010). A further 565TWh was generated and lost due to conversion, transmission and distribution losses (*ibid.*). Of the 380TWh, the domestic sector is the largest electricity consumer accounting for around 32%. Industry consumes 29%, the commercial sector 25%, and the remainder by the energy industry and transport. Electricity generation is responsive to demand; power stations are fired up centrally to meet anticipated demand peaks as if electricity was a freely abundant resource. This is because the costs of energy consumption are often invisible (Berkhout *et al.*, 2000) and global de-carbonization of fuels is motivated not only by real and perceived environmental concerns but also to improve energy consumption efficiencies (Economides & Wood, 2009). UK legislation in the Climate Change Act, 2008, requires the Secretary of State to reduce net UK carbon account for the year 2050 by 80% of the 1990 baseline (UK parliament, 2008). Estimated carbon dioxide emissions from UK electricity generation are shown in Table 1 but these exclude emissions created during the production and distribution of electrical devices intended for domestic use and the emissions from their use.

Fuel	Emissions (tonnes of carbon dioxide per GWh electricity supplied)		
	2007	2008	2009
Coal	913	903	915
Oil	623	730	633
Gas	400	404	405
All fossil fuels	626	608	598
All fuels (including nuclear and renewables)	500	496	452

Table 1 *Estimated CO_2 Emissions From Electricity Generation 2007 To 2009 (DECC, 2010: 123).*

Technology, such as carbon capture storage or carbon sequestration (Holt, 2009) may remove carbon dioxide and so help to deliver carbon targets, but the present pressure is to reduce the use of fuels which contribute to the cumulative level of emissions.

But for now all ten of the largest power stations are fuelled by non-renewables, largely coal, and importantly all were built in the late 1960s and early 1970s (The Daily Telegraph, 2009). Some are already due to close and the investment needed to replace or upgrade power stations to conform to Industrial Emissions Directive (to reduce sulphur and nitrogen oxide emissions) and compensate for potential disadvantages of the carbon trading regime of the EU's Emissions Trading Scheme, requires assurances via government guarantees that National Grid will buy minimum production capacity.

Such plants are dependent upon non-renewable sources of energy, largely, coal, oil and gas. Globally, proven reserves are diminishing: oil should last 45 years given our current consumption rates (BP, 2010), gas some 20 years longer, and coal around 150 years (The Independent, 2007). In terms of production, the UK's electricity is almost entirely home produced from a variety of resources across England, Scotland and Wales. Net imports of electricity in 2009 amounted to only one per cent of electricity supply (DECC, 2010) reducing from three percent in 2008. By contrast net imports of gas accounted for a third of gas input into the transmission system (DECC, 2010), with crude oil at 45% of indigenous energy production and coal at 70% so the UK is dependent on imported non-renewable fuels

Conventional oil production may soon go into decline as supply challenges are exacerbated by rising demand and strengthening environmental policy (Owen *et al.*, 2010). Supply and demand are likely to diverge between 2010 and 2015 and will have an increasing negative effect on the macro-economy. Natural gas is the only hydrocarbon source of energy with an economically viable opportunity to reduce carbon intensity through reduced carbon emissions and discovered unexploited discoveries remain plentiful (Economides & Wood, 2009).

The National Grid

Electricity production and supply are balanced through a single public limited company in the UK: the National Grid. The transmission system of England and Wales operated by the National Grid was owned by the 12 privatized regional electricity companies and floated on the Stock Exchange in 1995. The transmission system is linked to continental Europe via an inter-connector to France under the English Channel. Prior to March 2005 the UK electricity industries of each country operated independently although inter-connectors joined all three grid systems together. In addition to economic and security benefits, environmental benefits from inter-connection are possible, see for example Zhu *et al.*(2005).

In April 2005 under the British Electricity Trading and Transmission Arrangements (BETTA) which introduced the Energy Act 2004, the National Grid became the sole operator of all three transmission systems (DECC, 2009), although Scottish Power, and Scottish and Southern Energy continue to own the two high voltage transmission systems in Scotland. The National Grid provides "continuous real-time matching of demand and generation output, ensuring the stability and security of the power system and the maintenance of satisfactory voltage and frequency" (The National Grid, 2011). The BETTA arrangements created a single wholesale electricity market for Great Britain, a single transmission system and independence from generators and distributors (DECC, 2011). Economic competitiveness was a driver for BETTA, and by 2009 over 53% of domestic users had switched to a non-home supplier (DECC, 2009). The use of forwards and futures markets provide a mechanism enabling the National Grid to balance the system and provide a settlement process (DECC, 2009).

Generator capacity between 2005 and 2009 grew by 5.9TWh, whilst new capacity of 28.4TWh will be created during 2010-14 and another 11.5TWh by 2016 (The National Grid, 2010Ch 3, tables 3.5 & 3.6). Of this new capacity, 12.6TWh is attributable to wind power and none to household-scale distributed energy. The increased growth in generator capacity is a reflection of anticipated growth in demand, and has implications for renewal of parts of the transmission system in order not to reach fault levels (The National Grid, 2010).

Security And Resilience

Capacity growth, demand growth and sources of fuels are not the only challenges of the electricity system: energy security and resilience are ongoing concerns. Curbing the use of fossil fuels is the main lever to a more secure energy economy, since large generators are targets for attack. Scale factors of supply, conversion and transport infrastructure are inversely related to energy services security (Jansen & Seebregts, 2010) so many smaller generators would aid energy security but may be less economically efficient. Increased diversification of energy sources and of the location of imports (and related political stability) contribute to increased resilience (Jansen & Seebregts, 2010). Five key issues have been identified by the energy regulator which arise from current energy arrangements in the UK (Ofgem, 2010):

1. Need for sustained unprecedented levels of investment in difficult financial conditions and a climate of increased risk and uncertainty;
2. Uncertainty in future carbon prices impacting upon low carbon technology investment;
3. Short-term price signals do not reflect the need for more supply security and peak energy supplies;
4. Exposure to additional risks from interdependence with international markets, and;
5. Increasing costs may stretch consumer affordability and so impact on industry competitiveness.

Ofgem (2010) has suggested a range of increasingly transformational options to address doubt over secure and sustainable energy supplies.

Given the issues of efficiency losses, demand growth, environmental impact and energy security, which are global in nature, a new electricity system is emerging.

Smart Grid

The unifying theme of the smart grid (European Communities, 2006) encompassing distributed energy generation is posited as the future model to meet the electricity demand of the UK. It has the potential to reduce issues with the existing system of electricity supply and demand which is largely embedded in non-renewables, centralized control of market structures, and supply management. Smart grid is defined by The European Technology Platform SmartGrids (2010) as "electricity networks that can intelligently integrate the behavior and actions of all users connected to it—generators, consumers and those that do both—in order to efficiently deliver sustainable, economic and secure electricity supplies." This definition embraces a systems' perspective of the integrated behavior of agents who supply and/or demand electricity. The new smart grid system is envisaged as a dynamic system in which new, largely small-scale producers of electricity can emerge to produce electricity for their own consumption or to supply to the grid for re-distribution in return for appropriate reward.

The first step in the transition to Smart Grid intends to combat energy demand through the provision of information about consumption and the use of smart meters and intelligent devices. This opens up opportunities for IT and network organizations (IHS Global Insight, 2009; Jones, 2009; Nair & Zhang, 2009). Plans to roll out around 50 million smart meters will be directly managed by Department of Energy and Climate Change (DECC) rather than the industry regulator ofgem (The Office of Gas and Electricity Markets) who will instead work on a regulatory framework (Hendry & Mogg, 2010). Information about energy use is expected to cut energy consumption and appears to be a pre-requisite to smart grid and distributed energy generation.

Demand side measures such as behavior change and adaptation contribute to the capacity to meet demand (Owen *et al.*, 2010), but these changes are not linear. For example, a rebound effect occurs when a technological improvement creates efficiency and a cost saving to the consumer. The consumer essentially compensates for the cost saving by using more energy (Berkhout *et al.*, 2000). If spending can be diverted to investment in domestic or community level generation, then consumers could become prosumers (producers and consumers) (Burr, 2010) and reduce demand on central generator capacity.

Understanding the behavior of energy consumers is critical. Behavior of different customer segments is likely to require different approaches. The desirable massive deployment of distributed energy resources requires that new products be designed for specific purposes for each type of customer (Encinas *et al.*, 2007). A feature of the new energy market is that products meet specific needs and at different scales. The opportunities are many and varied depending on the type of the property, e.g., space for pumps; ready access to renewable energy, e.g., coastal wind; and so on. There are on the other hand, risks and constraints: economic, social, political, and environmental, which may explain why distributed energy generation is minimal even though the market is the UK market is deregulated. A barrier greater even than the technological readiness of generating and storage devices at all scales, and appropriate information systems and networks, is the willingness for households and firms to invest and embrace a regime of distributed energy generation. Extremely attractive government subsidies have proliferated (see for example http://www.decc.gov.uk/en/content/cms/news/pn10_010/pn10_010.aspx) but still uptake stagnates.

Research Design

This paper proposes a two-stage research design to explain this resistance and to suggest interventions which may speed a transition to Smart Grid. Each stage uses well-known research methods, one from psychology (Kelly, 1963) and the other from modeling (Casti, 1997). The Repertory Grid Technique founded upon Personal Construct Theory (PCT) is used as the first stage in a mixed methods 'embedded design' (Creswell & Plano Clark, 2010). The outcome of this first stage is a classification of increasingly mature energy behaviors. The second stage is the creation of an agent-based model (Gilbert, 2007) using the classification in the first stage to differentiate different classes of agents. This method allows the current system to be modeled from the perspective of the individuals who collectively constitute the structure of the electricity system (Giddens, 1984) and bases the behaviors of the agents upon their constructs, and therefore the ways in which they know the electricity system. Demand growth, central generation capacity and technological improvement can be modelled in order to identify the rate of growth needed for Smart Grid to meet carbon targets, improve resilience and security, and reduce dependency on imports.

The model is then developed to overcome the constraints it has identified. This would be achieved by introducing global variables which could each be altered using appropriate scales. The variables themselves would represent the critical factors which inspired the constructs of the (small) class of consumers who are already prosumers to change their behaviors. These critical changes can be encouraged through global variables (e.g., economic, and otherwise) which would inspire some individual agents to transition from one class of agent to another which is more mature in terms of energy generation. Measures of overall system efficiency (energy losses), economic viability, environmental impact and energy security would be compared to the base-line model. The agent based model would demonstrate a successful transition pathway to a working smart grid with the desired outcomes.

The real-world problem of demand management and distributed energy generation uptake is an opportunity to connect the behavior of individuals to social structure within the operation of large-scale industrial systems. The method proposed here starts with the behavior of different types of player (or agent) in the system. Structured interviews take place to identify the rules by which different types of agent behave. The rules are the explanation of behavior based on the constructs which are elicited during the interview. For example, a middle-class property owner may not have economic barriers to adoption (or consumption), but may lack social constructions of environmental concern.

The domestic sector accounts for the largest number of independent decision-makers, who in 2008 spent 58% of electricity consumption on space heating, 24% on water heating, 16% on lighting and appliances, and 3% on cooking (DECC, 2010). Industrial and commercial (services, public administration and agriculture) users differ from domestic users, for example, in terms of size of buildings, and scale of machinery. Sectors are somewhat similar in that the owner (and bill payer) and user may be different: landlords/tenants, parents/children for example in the domestic sector, factory owners/production managers in industry, retail companies/store managers in services. The issue of control over use is exacerbated by the lack of consumption information, which is both aggregated (so not directly attributable to specific time/appliance usage) and untimely (bills arriving quarterly for example). Actual use is not easily controllable and knowledge of actual cost at time of use is not easily available. Intervention to reduce usage or generate using distributed devices requires the bill payer to perceive a benefit from changing behaviors.

This paper proposes an embedded design (Creswell & Plano Clark, 2010) in which the repertory grid data set provides a supportive, secondary role in a study based primarily on the agent based model data set. The agents and their rules are defined primarily from the literature but are supported by the repertory grid interviews data set in order to provide granularity, differentiation and real-life explanation. The steps in the proposed research design are:

1. Structured interviews with representative sample of agent-types to elicit constructs; mapping of constructs to interviewee behavior, creating agent rules;
2. Classification of agent types using similar types of interviewee; deciding on the ratios of agent types within the population;
3. Creation of a base-line agent based model using the agent types and agent rules, together with environmental information on technology capability, demand growth, resource depletion, etc., and;
4. Creation of a futures agent based model which includes new global variables to assist with transition (learning) for agents which would move them from one agent classification to a more desirable classification.

Personal Construct Theory (PCT)

The structured interview method proposed here is repertory grid technique (Rep Grid) from personal construct theory (Kelly, 1963). Kelly's fundamental postulate is that "a person's processes are psychologically channelized by the ways in which he anticipates events" (Kelly, 1963). A person's actions are determined by the ways in which the person expects events to turn out and so a person's behavior is explained by the constructs through which their expectations have been formulated. Sometimes the constructs are known only subconsciously to the person as they can rely on deeply embedded beliefs which have been developed over long periods of time.

Constructs are elicited in one of two ways: by providing elements for the interviewee or by asking the interviewee to name elements. These elements are compared in twos or threes (dyads or triads) through a structured interview technique which avoids interviewer bias. Elements will vary depending on the role of the agent. For the domestic prosumer, the events need to be related to electricity use, billing and generation. Critical events, such as the Buncefield Disaster (Buncefield Major Investigation Board, 2008), should be included.

There are a number of important reasons for using personal construct theory. Construing is by no means limited to those experiences a person can talk about or think about privately (Kelly, 1963). Construction systems draw in the experience of the interviewee from all aspects of life. Recurrent themes in the flow of a person's experience are drawn together and used as a basis for likeness and difference. The basis upon which a person finds it convenient to look at matters in a particular way rather than any other way is drawn out through the rep grid interview and is contained in the construction system which is elicited. A person's construction system evolves and embraces ordinal relationships between constructs. Ordinal relationships indicate that one construct subsumes another; it may itself be subsumed by another creating a complex super- and sub-ordinal structure. As events occur which lead to outcomes contrary to the predictions of their construct system, the person finds ways to transcend the construct system, revising and re-organizing as necessary (Kelly, 1963). Each construct is dichotomous. The contrasting end is both relevant and necessary to the meaning of the construct. A person places values upon each end of the construct. The constructs can be rated and ranked which can provide useful multi-grid comparative information.

Two people's psychological processes will be as similar as their constructions of experience. Cultural similarity therefore can be defined not only in terms of personal outlook but in terms of the impingement of social stimuli. A person's construct system anticipates not only what he/she will do but also what others will do (Kelly, 1963) and further, what he/she thinks they are expecting him/her to do! New wants can be acquired or learned from others (Witt, 2001) and have detrimental effects on consumption. This powerful connection of personal action to society's norms and expectations explains influences of social networks and particularly the ability to anticipate events given the experiences shared with various groups.

A construct is an abstraction or property associated with several events by which they can be differentiated into homogeneous groups (Kelly, 1963). Because of this, constructs are comparable across multiple respondents in order to find classes of respondents with similar understanding (constructions). An aid to further abstraction is the use of laddering (Reynolds & Gutman, 1988) which helps with the comparison of constructs in multiple grids. Nevertheless each construct system is personal and so functions as a whole. Permitting this pluralism admits that there is no single truth in a social system, or alternatively, that each person is limited in their ability to comprehend the full truth, even if exposed to all the experiences possible.

Agent-Based Modeling

The action of house holders is critical to the understanding of resistance to managing demand for electricity and uptake of distributed energy generation. Agents of various types (classes) may be defined together with their relative ratios within the model. The model can be coded so that individual agents occasionally act differently from the norm, or may learn more quickly than others, or be tempted by the action of their proximate neighbors. Agent-based models accommodate personal histories and change, such as migration, reproduction, work, expiration, etc.

Agent-based models are just one form of simulation (Dooley, 2002) and are appropriate when the system is modeled as a collection of agents who interpret the world around themselves and interact with one another via schema. The importance of change in agents' schema (individual agent construct systems) through learning and adaptation plus the exposure of emergent, self-organizing patterns of connectivity and interaction, mean that models can closely represent real-life systems.

The proposed approach is embedded in complexity thinking (Simon, 2001; Capra, 2005; Prigogine & Stengers, 1984). The energy generation and consumption system is a complex system (see for example, (Dobson *et al.*, 2007)). It is made up of many and varied agents (Cilliers, 1998). From the dynamics in the electricity system we can discern a dissipative structure, that is, a dynamical regime, not in equilibrium, because the agents and objects which constitute the system are coevolving. The dynamical regime makes use of resources (e.g., coal) and creates new resources (e.g. wind-farms) for the benefit of the agents in the system.

The PCT:ABM method will create an evolutionary model because although it makes assumptions about a boundary and classification, it does not require average types, average processes, nor average events (Allen, 1993). Averaging assumptions skew results, oversimplify reality and fail to shed insights into the system. Nevertheless even an evolutionary model is the real system (Gilbert, 2007) since not all aspects of any system can be modeled, otherwise it would be the real system! Note also that as in all complex systems, some constituent parts are not themselves complex (Holland, 1995), for example, a mechanical device or a power plant. These can be treated as agents governed by unchanging rules, although these too can have limited life-spans and require maintenance in order to be high-performing.

Conclusions And Further Research

Initial steps have started on both PCT and ABM data collection in order to shed light on a possible transition to smart grid. Demographics are needed for both the (potential) prosumer and their property in order to support the comparison of grids. These demographics are based in the literature and are the fundamental ways in which behavior is expected to differ, e.g. type of property, disposable income, owner/tenant. Preliminary constructs identified during pilot interviews of consumers (those who pay the bills) are time (lapse in smart meter installation; lapse in replacing electrical appliances); wealth; economics of new devices, risks of selecting the right companies for services, desire/motivation to use less electricity, control over usage (the consumer versus the user), barriers (lack of empowerment) to produce electricity for own use or community use.

Other than domestic and industrial prosumers and consumers, the learning agents for the electricity system are generators, suppliers, system operators (Transmission, Distribution and Network), legislators/regulators, environmental organizations and lobbyists. The objects (or agents which do not learn) in the current electricity system are power stations, National Grid transmission infrastructure, and electricity consuming devices. From literature review, the new objects in the future electricity system are: storage agents (at varying scale with varying technologies such as hydrogen cell, Electric Vehicles (EV), hydro-electric storage, smart appliances within households (automated active demand agents), and smart distributed generators at household and/or community scale (automated active demand agents). Some agents can operate at different time-scales from others. Infrastructure planning, approval, construction, etc. operates slowly with respect to new energy product creation and deployment, and much more slowly than electricity flow in existing operations.

System level measures include efficiency losses, demand growth, economics of investment, environmental impact and energy security. These measures will help to identify critical stages in the transition from a traditional power grid to a smart grid, in the context of increasing technological capability and socially connected world.

In conclusion, a method based in complex systems thinking is put forward here for explaining the transition to a smart grid which solves the challenges of diminishing non-renewable resources, power station emissions, demand growth, and security (and resilience) of supply needs. The system is constituted from various agents, each with associated behaviors, some of which can learn. PCT is used to capture existing behaviors through repertory grid interviews, using constructs to map agent rules. ABMs which attempt to model the real world are then created with the right diversity of agents and objects which interact over different time-scales. In combination, PCT:ABM connect the individual to the social group to system demand and supply. Interventions can be included in the model to demonstrate how the transition to smart grid could be speeded up. However, transitions in energy are revolutionary in nature and will take decades to achieve given the investment in existing infrastructure designed to handle non-renewable sources of energy (Economides & Wood, 2009).

Acknowledgements:

This work was supported by EPSRC grant: EP/G059969/1 "Complex Adaptive Agents, Cognitive Agents and Distributed Energy: A complexity science-based investigation into the smart grid concept".

References

Allen, P. (1993). "Evolution: Persistent ignorance from continual learning," in R.H. Day and P. Chen (eds.), *Nonlinear Dynamics and Evolutionary Economics*, ISBN 9780195078596, pp. 101-112.

Berkhout, P.H.G., Muskens, J.C. and Velthuijsen, J.W. (2000). "Defining the rebound effect," *Energy Policy*, ISSN 0301-4215, 28(6-7): 425-432.

BP (2010). Statistical Review of World Energy, http://www.bp.com/liveassets/bp_internet/globalbp/globalbp_uk_english/reports_and_publications/statistical_energy_ review_2008/STAGING/local_assets/2010_downloads/statistical_review_of_world_energy_full_report_2010.pdf.

Buncefield Major Investigation Board (2008). *The Buncefield Incident 11 December 2005: The Final Report of the Major Incident Investigation Board*, ISBN 9780717662708.

Burr, M.T. (2010). "Fill 'er up: Smart Grid as Quick-E Mart," Public Utilities Fortnightly, 148(4): 4-5.

Capra, F. (2005). "Complexity and life," *Theory, Culture and Society*, ISSN 1460-3616, 22(5): 33-44.

Casti, J.L. (1997). *Would-be Worlds: How Simulation is Changing the Frontiers of Science*, ISBN 9780471123088.

Cilliers, P. (1998). *Complexity and Postmodernism: Understanding Complex Systems*, ISBN 9780415152860.

Creswell, J.W. and Plano Clark, V.L. (2010). *Designing and Conducting Mixed Methods Research*, ISBN 9781412975179.

DECC (2011). British Electricity Trading and Transmission Arrangements (BETTA), http://www.nationalgrid.com/corporate/Our+Businesses/transmission/.

DECC (2010). Digest of United Kingdom energy statistics (DUKES), http://www.decc.gov.uk/en/content/cms/statistics/publications/dukes/dukes.aspx.

DECC (2009). Digest of United Kingdom energy statistics (DUKES), http://www.decc.gov.uk/assets/decc/Statistics/publications/dukes/1_20100208131106_e_@@_dukes09.pdf.

Dobson, I., Carreras, B.A., Lynch, V.E. and Newman, D.E. (2007). "Complex systems analysis of series of blackouts: Cascading failure, critical points, and self-organization," *Chaos*, ISSN 1089-7682, 17(2).

Dooley, K. J. (2002). "Simulation Research Methods," in J. Baum (ed.), *Companion to Organizations*, ISBN 9780631216940, pp. 829-848.

Economides, M.J. and Wood, D.A. (2009). "The state of natural gas," *Journal of Natural Gas Science and Engineering*, ISSN 1875-5100, 1: 1-13.

Encinas, N., Alfonso, D., Lvarez, C. A., Perez-Navarro, A. and Garcia-Franco, F. (2007). "Energy market segmentation for distributed energy resources implementation purposes," *IET Generation, Transmission & Distribution*, ISSN 1751-8687, 1(2): 324-330.

European Communities (2006). *European SmartGrids Technology Platform: Vision and Strategy for Europe's Electricty Networks of the Future*, http://www.smartgrids.eu/documents/vision.pdf.

Giddens, A. (1984). *The Constitution of Society*, ISBN 9780745600062.

Gilbert, N. (2007). *Agent-Based Models*, ISBN 9781412949644.

Hendry, C. and Mogg, L. (2010). "Smart electricity and gas meters," http://www.decc.gov.uk/en/content/cms/what_we_do/consumers/smart_meters/smart_meters.aspx.

Holland, J. (1995). *Hidden Order: How Adaptation Builds Complexity*, ISBN 9780201442304.

Holt, G. C. (2009). "Carbon Neutrality: What does it mean?" Nanotechnology Perceptions, ISSN 1660-6795, 5: 135-145.

IHS Global Insight (2009). "Cisco and Yello Strom launch energy-saving Smart Grid pilot in Germany," http://newsroom.cisco.com/dlls/2009/prod_100509e.html.

Jansen, J.C. and Seebregts, A. (2010). "Long-term energy services security: What is it and how can it be measured and valued?" *Energy Policy*, ISSN 0301-4215, 38: 1654-1664.

Jones, K.C. (2009). "IT gives smart grid initiatives a jolt," InformationWeek, ISSN 8750-6874, http://www.informationweek.com/news/government/policy/showArticle.jhtml?articleID=215901346.

Kelly, G.A. (1963). A Theory of Personality: The Psychology of Personal Constructs (originally published 1955), ISBN 9780393001525.

Nair, N. C. and Zhang, L. (2009). "SmartGrid: Future networks for New Zealand power systems incorporating distributed generation," *Energy Policy*, ISSN 0301-4215, 37: 3418-3427.

Ofgem (2010). "Action needed to ensure Britain's energy supplies remain secure," http://www.ofgem.gov.uk/Media/PressRel/Documents1/Ofgem%20-%20Discovery%20phase%20II%20Draft%20v15.pdf.

Owen, N.A., Inderwildi, O.R. and King, D.A. (2010). "The status of conventional world oil reserves: Hype or cause for concern?" *Energy Policy*, ISSN 0301-4215, 38(8): 4743-4749.

Prigogine, I. and Stengers, I. (1984). *Order out of Chaos: Man's New Dialogue with Nature*, ISBN 9780006541158.

Reynolds, T.J. and Gutman, J. (1988). "Laddering theory method, analysis, and interpretation," *Journal of Advertising Research*, ISSN 0021-8499, 28(1): 11-31.

Simon, H.A. (2001). "Complex systems: The interplay of organizations and markets in contemporary society," *Computational and Mathematical Organization Theory*, ISSN 1572-9346, 7(2): 79-85.

The Daily Telegraph (2009). "Blackout Britain: The biggest power plants," http://www.telegraph.co.uk/earth/energy/6122911/Blackout-Britain-The-biggest-power-plants.html.

The European Technology Platform SmartGrids (2010). "Strategic deployment document for Europe's electricity networks of the future," http://www.smartgrids.eu/

documents/SmartGrids_SDD_FINAL_APRIL2010.pdf.

The Independent (2007). "World oil supplies are set to run out faster than expected, warn scientists," http://www.independent.co.uk/news/science/world-oil-supplies-are-set-to-run-out-faster-than-expected-warn-scientists-453068.html.

The National Grid (2011). Transmission, http://www.nationalgrid.com/corporate/Our+Businesses/transmission/.

The National Grid (2010). "GB seven year statement," http://www.nationalgrid.com/NR/rdonlyres/A8DE6BAF-8465-4E25-8BB1-3FD376DA1E80/41460/NETSSYS2010Chapter3.pdf.

UK parliament (2008). Climate Change Act, http://www.england-legislation.hmso.gov.uk/acts/acts2008a.

Witt, U. (2001). "Learning to consume: A theory of wants and growth of demand," *Journal of Evolutionary Economics*, ISSN 1432-1386, 11: 23-36.

Zhu, F., Zheng, Y., Guo, X. and Wang, S. (2005). "Environmental impacts and benefits of regional power grid interconnections for China," *Energy Policy*, ISSN 0301-4215, 33(14): 1797-1805.

Liz Varga (PhD, Cranfield University, 2009; MBA, Cranfield University, 2002; BA (Honours) in Pure Maths and Data analysis/design, Open University, 1989) is Director of the Complex Systems Research Centre (CSRC) at Cranfield University. Although located in the School of Management the CSRC is a cross-school centre for research and teaching of complexity science applied to real world management problems. Liz's work attempts to explain socio-economic system behavior and adaptation in the context of infrastructure, technology and political environments. This focus has been developed through several grant-funded research projects mostly in collaboration with one or more other universities. These include various Innovative Manufacturing Research Council research projects, EU QosCosGrid (Grid Computing) project; and ESRC Modelling the Evolution of the Aerospace Supply Chain project. This legacy continues with two current projects: EPSRC CASCADE (a Complexity Science-Based Investigation into the Smart Grid Concept) and Technology Strategy Board funded ABIL project (Agent Based Intelligent Logistics—multi-modal freight transport systems modelling). Liz has secured funding for three new EPSRC projects starting in 2011: Transforming Utilities' Conversion Points (TUCP), All-in-One (Feasibility analysis of supplying all services through one utility product); and Land of the MUSCos (Multi-Utility Service Companies). All are inter-university collaborations. Liz is a member of the Complexity Society and a reviewer for the journal *Emergence: Complexity and Organization*. She has a growing schedule of teaching commitments, particularly in relation to research methods and designs incorporating agent-based modelling.

4. On Understanding Software Agility: A Social Complexity Point Of View

Joseph Pelrine
MetaProg GmbH, CHE

Over the last decade, the field of so-called Agile software development has grown to be a major force in the socio-economic arena of delivering quality software on time, on budget, and on spec. The acceleration in changing needs brought on by the rise in popularity of the Internet has helped push Agile practices far beyond their original boundaries, and possibly into domains where their application is not the optimal solution to the problems at hand. The question of where Agile software development practices and techniques make sense, and where are they out of place, is a valid one. It can be addressed by looking at software development as a complex endeavour, and using tools and techniques from the Cynefin method and other models of social complexity.

Introduction

Over the course of the last decade, a soft revolution has taken place in the field of software development. Experiences with projects delivering late and over budget have led people to question some of the basic tenets of software project development and management. Starting with a few provocative theses in Kent Beck's 1999 book *eXtreme Programming*, the field of so-called Agile software development has grown to be a major force in the socio-economic arena of delivering quality software to people on time, on budget, and on spec. The acceleration in changing needs brought on by the rise in popularity of the Internet has helped push Agile practices far beyond their original boundaries, and possibly into domains where their application is not the optimal solution to the problems at hand.

So, where do Agile software development practices and techniques make sense, and where are they out of place? To answer that question, it is necessary to first understand how, and more importantly why, Agile practices work. In the mid 1990s, it was maintained that practices such as *eXtreme Programming* could not possibly work, although even then dozens of projects successfully completed had proved otherwise. Later, after sufficient empirical evidence had accumulated to irrefutably prove that Agile practices do work, the question of why they worked still remained.

As a consequence of the increasing complexity and unpredictability of the world around us, Agile practice is increasingly seen as the solution. Agile represents a new paradigm in the truest sense, a complete abandonment of old methods that cannot be done in half measures. At its core, Agile addresses complexity in a manageable fashion, attuned to the needs of the human psyche. The Agile approach, though, constitutes a revolution in our modes of thinking, working and interacting. Agile processes have grown and developed as the body of knowledge acquires new ideas from its practitioners. Expansion is not always a positive force, however. The discipline that originally allowed Agile to explode linear, mechanistic development practices has often vanished, replaced by a cargo-cult "by-the-rules" interpretation of Agile based on checklists. It seems today that some 'Agile' teams are practicing nothing more than 'air guitar and attitude' (to quote Alan Kay).

Ultimately, models are only as good as the people applying them. Too many teams have come to regard Agile as something like a cookery recipe—follow this set of instructions and procedures for a tasty result. But in software development, as in cooking, what you get out is not simply the sum of what you put in. We need to develop an understanding of what makes Agile work, and indeed what makes it fail. This understanding is a prerequisite for sustaining and scaling Agile efforts. Acknowledging that Agile is not working in a particular situation is an inherent part of Agile practice, but it's one that's often ignored.

The purpose of this paper is to explore the following questions:

- Is software development (in whole or in part) a socially complex endeavour?
- What can be gained by treating it as complex, and using tools and techniques from social complexity science?
- Why do Agile practices work so well?

Why is Agile the best method around for meeting challenges in software development? As projects become more complex, and customer requirements ever more ambitious but ever less clearly defined, how does Agile help developers produce usable software on time and on budget? Despite the recent quantum shift in the field towards the adoption of Agile, many organizations have not yet made the conceptual adjustments necessary to apply it successfully. No method is without its detractors, and those who have yet to be convinced by Agile can point to the lack of hard evidence, of rigorous analysis, of a theoretical, scientific basis, in the literature. For Agile fans, rigorous does not mean rigid, and it is not 'anti-Agile' to question the assumptions on which we base our processes, quite the opposite is true!

The core realization inherent in Agile is that people build software, that team dynamics are fundamental to it, and that teams of people are complex and unpredictable. The need to factor psychological and cognitive concepts into project design and implementation is novel in a world where mechanization and uniformity are frequently encountered, and often admired, within corporate cultures. A more realistic model of corporate information sharing began with Knowledge Management and its applications, but Agile goes much farther.

Modern software development is performed by teams of motivated individuals. The prevailing attitude for much of the field's history, though, has been to treat software development as a predictable 'factory' process, where adding a given amount of money, time, programmers and managers will produce the desired result. Within this context, the development process is broken down into a sequential pathway, with deliverable outcomes predicted at set points. This 'traditional' approach is exemplified by the Waterfall model. It can work—if the requirements are known, in detail, right from the start, the product is straightforward, and nothing goes horribly wrong. But who has ever worked on a project like that? Successful Waterfall projects do exist, but they are few and far between, and on closer analysis may not be adopting 'pure' linear management models—some flexibility, the beginnings of Agility, has crept in!

Trying to establish computing as an engineering discipline led people to believe that managing projects in computing is also an engineering discipline. Engineering is for the most part based on Newtonian mechanics and physics, especially in terms of causality. Events can be predicted, responses can be known in advance and planning and optimize for certain attributes is possible from the outset. Effectively, this reductionist approach assumes that the whole is the sum of the parts, and that parts can be replaced wherever and whenever necessary to address problems. This machine paradigm necessitates planning everything in advance because the machine does not think. This approach is fundamentally incapable of dealing with the complexity and change that actually happens in projects.

Traditional	**Agile**
Sequential	Iterative
Defined	Empirical
Plan-driven	Result-driven
Big-bang	Incremental
Specialised teams	Cross-functional teams
Test at end	Test-first

Figure 1 *Traditional vs. Agile.*

Consider what happens if you manage a project like a production line. Developers are assigned tasks, code is pumped through, and the finished product rolls off at the end. There are two problems with this paradigm. Firstly, the production line approach is more suited to generating multiple repetitive units, something that is rarely entailed in software development. Secondly, as soon as the product stops working at the end, the entire production line must be analyzed and fixed to solve the fault. Usability is a good example—often, aspects of product usability are left until the end stages of a project, with the expectation that it can be fine-tuned as needed. Frequently, there are flaws deep within the software that are not trivial to fix. The whole process must be reworked, but the team has already given it massive investment in terms of resources, time and effort. For the author, adopting Agile becomes the task of increasing awareness of, and finding the best process for answering the question, 'when do you want to know you have a problem?' (That assumes, naturally, that you do want to know you have a problem, an equally important point!)

> *One of the most highly developed skills in contemporary Western civilization is dissection: the split-up of problems into their smallest possible components. We are good at it. So good, we often forget to put the pieces together again* (Toffler, 1984)

The reductionist approach described by Toffler has served us well in the past, but we need to move beyond it. Many people still regard building software as a complicated undertaking, but in fact it is a defining example of a complex or a 'wicked' problem. The concept of wicked problems was originally proposed by Horst Rittel and Marcus Webber (Rittel & Webber, 1973). Wicked problems have incomplete, contradictory, and changing requirements; and solutions to them are often difficult to recognize as such, because of complex interdependencies. Rittel and Webber stated that while attempting to solve a wicked problem, the solution of one of its aspects may reveal or create other, even more complex problems. Rittel expounded on the nature of ill-defined design and planning problems, which he termed 'wicked' (that is, messy, circular, aggressive) to contrast against the relatively 'tame' problems of mathematics, chess, or puzzle solving. In the author's experience, certain ground rules of Agile software development have emerged that address the limitations of the Waterfall and other linear development paradigms in tackling such problems.

Communication and team dynamics represent the other area where Agile differs fundamentally older development paradigms. The functioning of the team, and the contributions and roles of individuals within the team, are fundamental to productivity. Team roles are no longer fixed, but members are allowed to self-organise. Management takes on the role of facilitating and coaching the team, rather than issuing orders. Scrum (Schwaber & Beedle, 2001) sees self-organizing teams as a fundamental aspect of the process. In applying Scrum, there is an emphasis on skills, not knowledge, and there are few rules. The author has distilled three 'rules of thumb/rules of Scrum' from experience in practice: the first is 'we don't make mistakes, we learn,' i.e., set up a safe-fail work environment where it is OK to learn and to correct behavior, estimates etc., on the basis of that learning. Secondly 'whoever has the risk, makes the decision.' Increase awareness of roles, rights, and responsibilities of the various partners in the development process. And last 'if you're not having fun, we're doing something wrong.' Keeping people happy and motivated isn't easy over a long project, but there are techniques that can be used from the outset to promote good team practice.

Getting Comfortable With Complexity—Sense Making The Agile Way

What is a complex system? Complexity theory can be considered one of the most revolutionary products of 20th century thought. Theories of chaos, complexity and emergence have shattered the conceptual frameworks of science, technology and economics, and provide unifying themes across previously distant disciplines. Scientists, sociologists, economists and engineers are finding common ground that transcends the terms of reference of each particular field. We have gone from the assumption that everything can be modelled given enough time, intelligence or processing power, to the realization that not everything we experience can be drilled into predictable patterns that we can recognise and understand. The human mind does not readily grasp complexity. It is counterintuitive; we prefer to recognise patterns in mechanistic systems.

How can a complex system be defined? Ask ten or twenty people working on complexity and emergence to describe such a system and you will get as many answers. One of the best sets of criteria for complexity is provided by professor George Rzevski:

1. INTERACTION—A complex system consists of a large number of diverse components (Agents) engaged in rich interaction;
2. AUTONOMY—Agents are largely autonomous but subject to certain laws, rules or norms; there is no central control but agent behavior is not random;
3. EMERGENCE—Global behavior of a complex system "emerges" from the interaction of agents and is therefore unpredictable;
4. FAR FROM EQUILIBRIUM—Complex systems are "far from equilibrium" because frequent occurrences of disruptive events do not allow the system to return to the equilibrium;
5. NONLINEARITY—Nonlinearity occasionally causes an insignificant input to be amplified into an extreme event (butterfly effect);
6. SELF-ORGANIZATION—Complex systems are capable of self-organization in response to disruptive events, and;
7. COEVOLUTION—Complex systems irreversibly coevolve with their environments.

Is software development a complex domain, and if so, why? This is the key question. At one level, the software development process seems to fulfil all of Rzevski's criteria, but on another level there seem to be exceptions and questions. This question may not be able to be answered definitely, but as we will see, interesting things happen when we TREAT software development as complex. We might also question which other domains may benefit from this treatment.

Many customers and developers alike regard building software as a complicated undertaking, but in fact it is a prime example of a complex problem. In adopting Agile processes, the field is beginning to address this and to become more comfortable with complexity. Unfortunately, the typical Agilist perception of complexity is not quite aligned with any of the main scientific definitions of the term. Agile literature abounds with romanticized, subjective interpretations of terms such as complexity, self-organization, emergence, which can only be understood by remembering that 'a little knowledge is often a dangerous thing'.

If we even succeed in establishing that developing software is a complex endeavour, a wicked problem, how then do we address it effectively? Complexity is counterintuitive to many. This is one of the reasons that a mechanistic, Newtonian view of projects has persisted in management thought.

> *Even as it was toppled from its unassailable position in science, Newtonian mechanics remained firmly lodged as the mental model of management, from the first stirrings of the industrial revolution right through the advent of modern-day MBA studies* (Petzinger, 1999).

Complexity theory represents software development more realistically than the engineering model. Understanding the theory is only the first step. How can complex problems be tackled practically on a daily basis? How can one differentiate easily between the complex and, e.g., complicated aspects of a complex domain? The art of management and leadership is having an array of approaches and being aware of when to use which approach.

Thinking About Complex Problems

The challenges of a Wicked problem are manifold. At the outset, goals may be unclear, yet expectations are high. We are tempted to set out a grand plan, mapping the project from start to finish, with meticulous allocation of time and resources. Yet the chances of such a plan being followed are remote—even if the initial stages appear to be going well, reality will rapidly cause divergence from the pathway. New information, changing variables and requirements, external factors such as competitor activity, cannot be factored in to a plan made months before. However, how many times do managers insist on struggling forward with a battered, modified version of the original plan?

This tendency to cling to our initial assumptions and plotted course is down to the way our minds deal with new situations. The process of first-fit pattern matching evolved to make us capable of fast decisions in danger, based on previous experience. Once that 'fit' has been made it's hard for us to let go and consider alternatives within a complex problem. It also makes humans bad at cutting their losses and changing tack mid-project. Research shows we value things we already have more highly than things of equal or greater value that we don't possess, for example. We're also good at seeing patterns where none exist, and imputing causality in random chains of events. A classic example is cumulative winnings or losses from betting on heads or tails in a coin toss. These purely random outputs can be modelled by a first-order Markov chain, which as is well known, readily exhibits pseudo cycles and pseudo trends, with stationary mean and non-stationary variance.

The problem of how to figure out a solution to a complex problem goes further. It is another part of complexity science known as multi-ontological sense making. The sense making process says that there is not one fit solution. Sense making is looking at things pre-hypothetically, that is crossing the line between unknown and known. As G. Spencer Brown said in his book *The Laws of Form* (Spencer-Brown, 1979), the first thing to do is to draw a distinction, which is exactly drawing a line between unknown and known. What we can know is cause and effect, which is the basic observation we make, i.e. phenomenology. We see something happen along the temporal axis and we often input causality: the first caused the second. Because our level of resolution of perception allows us to perceive them as to separate events, we interpret a causal connection between them. I push that light switch and a light goes on, I do it again and again and the light always changes. So I assume that there has to be some repeatable cause of connection. In this way we can predict the future.

Dave Snowden says, "sense making is the way that humans choose between multiple possible explanations of sensory and other input as they seek to conform the phenomenological with the real in order to act in such a way as to determine or respond to the world around them."

A basic premise of sense making is that we need to understand our thought processes when we analyze things. Our opinion, our evaluation of something says as much about us as it does about the thing we are looking at. This is called cognitive bias, and it influences our interpretation of everything around us, for example, what we consider to be complex.

Agile As A Technique For Addressing Complexity

The basic science necessary to understand complex systems was just starting to be established when the first Agile literature was published. At that time, the works of Stacey, Nonaka, and others sufficed to provide ideas and impulses for some Agile pioneers, but lacked the full breadth and rigour necessary for providing a foundation for understanding the Agile process as a whole. Only with the publication of Dave Snowden's papers on the Cynefin model did a system emerge that finally allowed researchers and practitioners to understand social complexity science, and its position as the theoretical basis of software Agility.

This paper will discuss one of many aspects of social complexity science, the Cynefin approach, and one of its practical applications to Agile software development. Many aspects of software development fall into the complex domain.

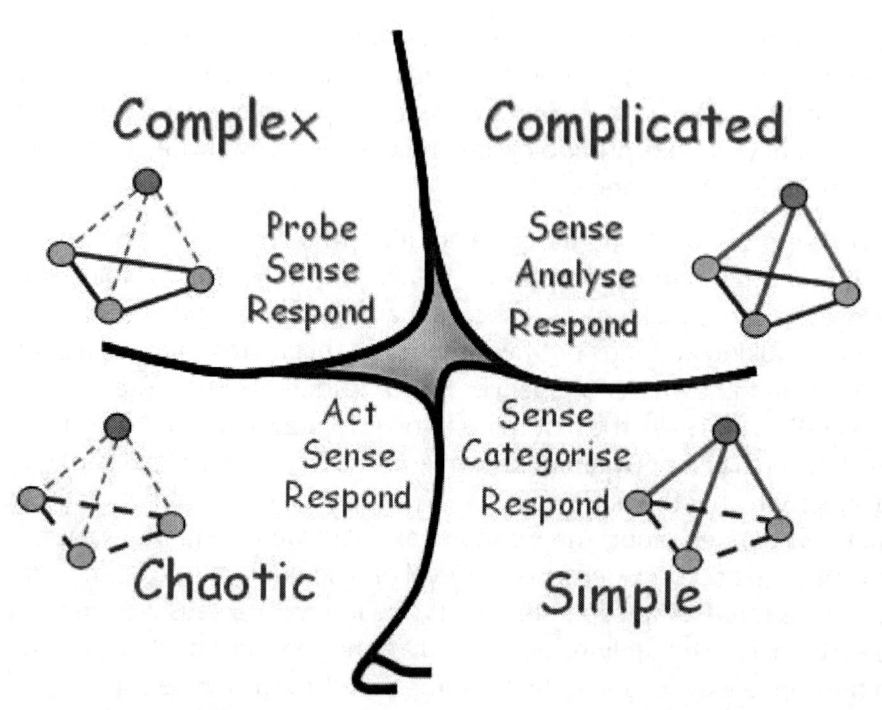

Figure 2 *The Cynefin Framework: Common Summary*

The Cynefin sense-making model has been described in a number of papers (Kurtz & Snowden, 2001; Snowden, 2005), and will not be covered in detail here. In addition to the sense-making model, though, the Cynefin method contains a number of techniques and exercises, which can be used to help groups make sense of their domain, helping them understand which methods and techniques can then be best applied.

In a study conducted over a number of years, the author has run the Cynefin 'butterfly stamping' exercise (see Cognitive Edge methods in the references) with over 300 people involved in Agile software development, with the goal of sensitizing them to scientific definitions of complexity and related concepts, of interpreting their cognitive biases related to software development, and of understanding whether software development as a whole could be considered a complex domain. During an introductory session, the participants are asked to brainstorm and collect topics they deal with and activities they engage in as part of their work. Later, after explaining the Cynefin model, definitions of the different domains, and the sense-making process, they do the exercise by assigning to the different Cynefin domains a set of situations, themes, and subjects provided by Dave Snowden, and which the participants were agnostic about. After "warming up" with these situations, themes, and subjects, and getting an active awareness for the meaning of the different domains, the participants then make sense of the activities and topics they identified and collected themselves.

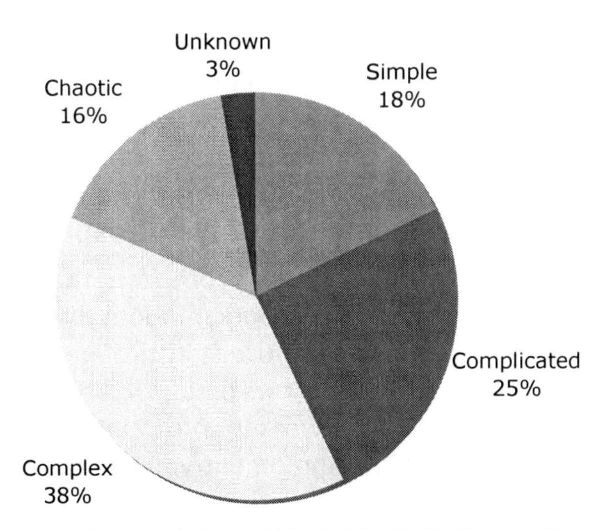

***Figure 3** Breakdown Of Typical Activities In Software Development*

Table 1 offers a sample of typical activities provided by participants, together with their sense-making results:

Simple	Complicated	Complex	Chaotic	Unordered
Knowing when a task is done	Ambitious (political) time-line	Changing requirements	Arguing about coding standards	No release deadline
Monitoring actual time spent	Fixing the build	Countering a belief in magic	Retrospectives without consequence	Resource shortage
Featuritis	Finding who to talk to	Task Estimation	Project volume too big	Lack of trust

Table 1 *Some Typical Software Development Activities*

Interpreting the results of the exercises led to the following realizations:

- Software development is a rich domain, with aspects and activities in all the different domains. The interactions between these aspects and activities are themselves often of a complex nature.
- Software development is a multi-level domain with self-similar characteristics, i.e., activities often tend to consist of sub-activities, each of which may be located in a different domain to the basic activity.
- The activities tend to be weighted more to the complicated and complex domains, with activities related to the coding aspect of software development landing in the complicated (or sometimes simple) domain, and activities associated with project management landing in the complex (sometimes chaotic) domain. Tasks dealing with interaction with a computer tended to be in the ordered domains, tasks dealing with interaction with other humans tended to be in the un-ordered, i.e., complex and chaotic, domains.

- The highest percentage of tasks and activities were in the complex domain. Although this is not sufficient to argue that software development as a whole is complex, it does suggest that many parts of it are amenable to analysis and treatment using complexity-based tools and techniques.

Success In Software Development Is Only Retrospectively Coherent

One thing that makes complex systems complex is their causality. As Dave Snowden says. 'If the system is chaotic/random then agent behavior is deterministic, which means I can use statistical instruments. If it's constrained, then the constraint structure allows predictability/repeatability. Strong constraints produce linear causality; weaker constraints provide repeatability that may be non-linear. However the moment you get the phase shift into a coevolutionary relationship between agents and system then there is no repeatability except by accident. In that context there is no meaningful causality, and any causality is only retrospectively coherent.'

In an ordered system, if you do something, you expect a specific result. Do it again, expect the same result. It's that simple. In a complex system, causality emerges as the system itself emerges, so that at the end, you can say how you got to where you are, but you can't guarantee that by doing exactly the same things, you'll get to the same place again—and you probably won't. In complex systems, we say the causality is retrospectively coherent. A classic example of retrospective coherence is task estimation. Before you do a task, it's almost impossible to estimate how long it will take. Afterwards, though, you can say exactly how long it took, and why it took that long. The same goes for projects. As a project goes on, the reasons for its success become established, not before. After it's over, you can say that the project was a success, and that certain things took place during the project—but you cannot say that the project was a success because these certain things took place!

Now shift the focus to a project. We tend to make the mistakes above when it comes to project planning. What worked last time? Why? Well, it must have been the people, the methodology, the meeting schedule—let's do it the exact same way again. Is it any surprise, then, that a project planned on this basis is likely to be a flop?

Contrary to Einstein's definition, in a socially complex system, insanity is doing the same thing over and over again and expecting the same result!

Complex Activities Require A Probe-Sense-Respond Model Of Action

Now that we have a reasonable basis for asserting that software development (as a whole) is a complex endeavour, or rather treating is as such, let's turn back to the Cynefin model. Since causality in the complex domain is retrospectively coherent, we'll always only know afterwards whether our efforts were crowned with success. To maximise flexibility in the face of this uncertainty, the Cynefin method suggests a probe-sense-respond technique. Set boundaries for the system to emerge. Employ numerous probes, which will provide feedback on what works and what doesn't. Apply sense-making to the results to the feedback. Then respond by continuing or intensifying the things that work, correcting or changing those that don't. Tighten this into a small iterative loop, observing emergent patterns, amplifying good ones, and disrupting bad ones, until you end up successful at your endeavour. In fact, it's often the case that by applying this process, you discover value at points along the way. Your final product can end up looking very different to the original plan, and being very much better. You could not have defined these benefits at the outset and aimed for them; these are an example of emergent properties of the complex system.

The 'apply-inspect-adapt' model of agile development is a probe-sense-respond model. The Scrum project management framework utilises an iterative, incremental model of development, with work divided into iterations (called 'Sprints'), and a review and reflection step at the end of each iteration. This technique, called 'inspect and adapt', should properly be called apply-inspect-adapt, otherwise the focus is not on actually doing anything. If we think about this, it becomes clear that the apply-inspect-adapt loop is nothing else then the probe-sense-response cycle used in the Cynefin method for dealing with issues in complex domains.

This insight brings us full circle. We used a social complexity method to gain understanding of the cognitive bias of Agilists towards the field of software development, and ended by noticing that the Scrum framework implements the exact method called for by the Cynefin method for managing work in the complex domain. This leads us to the following conclusions:

- The theoretical underpinnings of Agile methods need to be understood for such methods to become truly scalable and sustainable. Insights, and answers to many of the questions, can be found in social complexity methods such as Cynefin.
- Agile methods such as Scrum provide a lightweight, proven framework for managing work in the complex domain, be it software development or something else.

Software development is a rich domain, containing many aspects, a large percentage of which can be classified as complex. The interaction between these aspects is also complex. Just as we have benefited from treating software development as complex, and taking advantage of the toolbox of social complexity, namely the Cynefin method, so the field (as well as many other fields of human endeavour) would benefit from a multi-ontological approach, taking the best techniques for the various domains, and combining them in an appropriate and flexible manner. More work is needed to reach a deeper understanding of the inter-workings of agility and complexity, and it is the author's hope that the first (and following) workshops on complexity and real-world applications will not only provide insights, but also motivate other researchers to look into these fascinating fields.

Acknowledgements

Kent Beck, for being a strong source of motivation and inspiration; Dave Snowden and George Rzevski, for their input and insights into social complexity; Mark McKergow, for his excellent review and comments; Evelyn Harvey, for research and editorial assistance

References

Beck, K. (1999). *Extreme Programming Explained: Embrace Change*, ISBN 9780201616415.

Cognitive Edge methods: Butterfly Stamping, http://www.cognitive-edge.com/method.php?mid=45.

Kurtz, C. and Snowden, D. (2003). "The new dynamics of strategy: Sense-making in a complex and complicated world," *IBM Systems Journal*, ISSN 0018-8670, 42(3): 462-483, http://www.research.ibm.com/journal/sj/423/kurtz.html.

Petzinger, T. (1999). *The New Pioneers: Men and Women Who are Transforming the Workplace*, ISBN 9780684846361.

Rittel, H. and Webber, M. (1973). "Dilemmas in a general theory of planning," *Policy Sciences*, ISSN 0032-2687, 4: 155-169.

Schwaber, K. and Beedle, M. (2001). *Agile Software Development with Scrum*, ISBN 9780130676344.

Snowden, D. (2005). "Multi-ontological sense-making: A new simplicity in decision making," http://www.cognitive-edge.com/ceresources/articles/40_Multi-ontology_sense_makingv2_May05.pdf.

Snowden, D. and Boone, M. (2007). "A leader's framework for decision making," *Harvard Business Review*, ISSN 0017-8012, 85(11): 69-76.

Spencer-Brown, G. (1979). The Laws of Form, ISBN 9780525475446.

Joseph Pelrine is C*O of MetaProg, a company devoted to increasing the quality of software and its development process, and is one of Europe's leading experts on Agile software development. After studying philosophy, psychology, and music in Vienna, his interests led him to work in the field of artificial intelligence and software development. He worked as an assistant to Kent Beck in developing eXtreme Programming, and is Europe's first certified ScrumMaster Practitioner and Trainer. Joseph Pelrine is an accredited practitioner for the Cognitive Edge Network, and his work focus is on the field of social complexity science and its application to Agile processes.

5. TECS: A Browser-Based Test Environment For Complex Systems

Kurt A. Richardson
Exploratory Solutions, USA

The Test Environment for Complex Systems (TECS) is a comprehensive browser-based solution designed to support System Engineers, Test Engineers and Program Management through the design, execution and review of the Integration and Test (I&T) programs for complex engineered systems such as satellites. The aim of this paper is to introduce TECS and explore its potential role(s) in the important activity of I&T.

Introduction

The Integration and Testing (I&T) phase of any complex engineered system, such as a satellite, is one of the most challenging phases of system testing. Various components, often built by different manufacturers (using different test standards, processes and documentation), are brought together for the first time and tested as a complete system, rather than a collection of parts. Developing and maintaining the documentation and records to guide the I&T phase is a major undertaking and accounts for a substantial amount of the effort put into I&T. The creation and maintenance of test documentation is primarily the responsibility of Systems Engineers, but these are not the only 'users' of such documentation. For example, Test Engineers and Operators rely on detailed and accurate test documentation to allow them to plan and perform the necessary testing, whereas, Program Managers rely on detailed and accurate records to ensure program requirements have been met and tested thoroughly. Alongside the documentation required to perform a specific test, is the data generated from a particular test, which includes accounts of any anomalies that were observed during testing. Test documentation, test data, and anomaly records (which include both accounts of the anomalies and how they were overcome/solved) form a complex web of detail about the I&T process. As a result of this complexity, it takes considerable time and effort to navigate the 'test space' and understand exactly what the status is of any test program. The Test Environment for Complex Systems (TECS) is a set of tools that have been designed to define, maintain, record, and traverse this 'test space' with ease.

TECS is a comprehensive browser-based solution designed to support Systems Engineers, Test Engineers and Program Management through the design, execution and review of the Integration and Test (I&T) programs for complex engineered systems such as satellites. The aim of this paper is to introduce TECS and explore its potential role(s) in the important activity of I&T. The test environment provided through TECS can be divided into four major areas (in parentheses are the primary users for each area):

1. Test documentation creation, review and maintenance (System Engineers);
2. Test execution (Test Engineers, Test Conductors, and Test Operators);
3. Test anomaly management (System Engineers), and;
4. Monitoring of the overall testing program (Program Management).

Each of these areas will be discussed in the following sections. But before doing so I would like to comment on the connection between TECS and complexity thinking.

In a sense TECS would seem to be non-complex in that it is really nothing more than a tailored document management system—in many ways the tool itself is very much linear. However, the activity it attempts to support (I&T of complex engineered systems) is undoubtedly complex. In particular it is the changing and revisionary nature of an I&T program that TECS attempts to support. By making the process of creating and reviewing documentation and data as simple and as flexible as possible, it makes it far easier for users embedded in the complex I&T process to change and revise the ongoing record and plan of the process itself, as well to synthesize an accurate account of current status. In this sense it is argued that such linear tools can have an important, even central, role to play in managing complex systems if designed and utilized appropriately (see Richardson, 2008, for a discussion of the use of linear tools within a nonlinear modeling culture). The tool itself might be linear, but it is intended for use within a nonlinear system—rocket science needn't always be rocket science!

Indeed, there seems to be a general principle involved here: both the understanding and most efficacious utilization of systems imbued with complex dynamics can benefit from a host of very different tools or methods, whether linear or nonlinear, mechanical or dynamical, freeze-framed or processual, functional or algorithmic. Because the specific tool or method may prove helpful, though, doesn't warrant jumping to the conclusion that the properties characterizing that tool or method can be projected wholesale into an adequate explanation of the complex system under study. This is especially pertinent when it comes to linearization techniques used in dealing with nonlinear systems since there is the strong temptation to reduce the system's dynamics to linear functions which, of course, are much easier to handle mathematically.

A related case in point is how fractal generating functions, although they may display graphical and/or statistical properties that may seem remarkable in their capacity for simulating the apparently fractal nature of some natural object, say, the branching of a tree, do not necessarily translate into an adequate explanation of how in fact the natural object itself was generated. Intelligent working scientists (and engineers), accordingly, always need to keep in mind that methods and tools do not by themselves explain anything outside of the specific context in which they were found useful, and that these contexts have to be incorporated into theorizing beyond the particular use of the tool.

Furthermore, a good design is a good design whether it emerges naturally in a natural complex system like the mammalian brain or it is intentionally designed into a complex engineered system such as a the production of a satellite. That is one reason that complexity science has prompted a burgeoning of research and publications involving applications of naturally-occurring emergent design into intentional design (e.g., see Peng & Gero, 2009) as well as applications of intentional design into the study of organisms (e.g., see Lewens, 2005).

Test Documentation Creation, Review and Maintenance

For every test performed in the I&T process a procedure is needed to direct the test, and for a comprehensive performance test (CPT), for example, these can be very complicated documents. TECS was designed to ease the creation of such documents substantially. Regardless of how much effort goes into preparing a procedure, as soon as testing starts, changes to the procedure will undoubtedly follow as more is learnt about how the spacecraft bus and payload (for example) interact with each other. It is this dynamism that TECS is particular good for capturing and managing.

The TECS Procedure Development Tool (PDT) implements a bottom-up approach to procedure development that helps avoid inconsistencies, errors, and the time required to implement procedural revisions, as well as to develop new procedures or sub-tests. It also provides a clear record of when changes were made, why, and by who, which (among other things) means users can go back to any prior date and reconstruct a procedure exactly how it appeared for a particular past test, i.e., it provides a very robust audit trail (without any additional work). Digital versions of the issue/error documentation that are created when something 'off-procedure' occurs can also be referenced, via the TECS Anomaly Management (AM) tool, making it much easier to trace the development/evolution of a particular test, or an entire procedure. The output of TECS-PDT is HTML-based which means that procedures can be developed and read via conventional browser software (such as Internet Explorer or Firefox) by persons with the appropriate access profiles, and navigated with ease. The browser-based approach means that no software (other than the standard browser that comes with most operating systems) needs to be maintained on users' machines, making TECS distribution a relatively trivial exercise; the whole system can be hosted on one central Windows server (behind a firewall). Although TECS has been written for a Windows-based server, it can be accessed from any OS because of the cross-platform nature of HTML. TECS is written using ASP.NET 4.0, AJAX and LINQ, state of the art programming frameworks with extensive support.

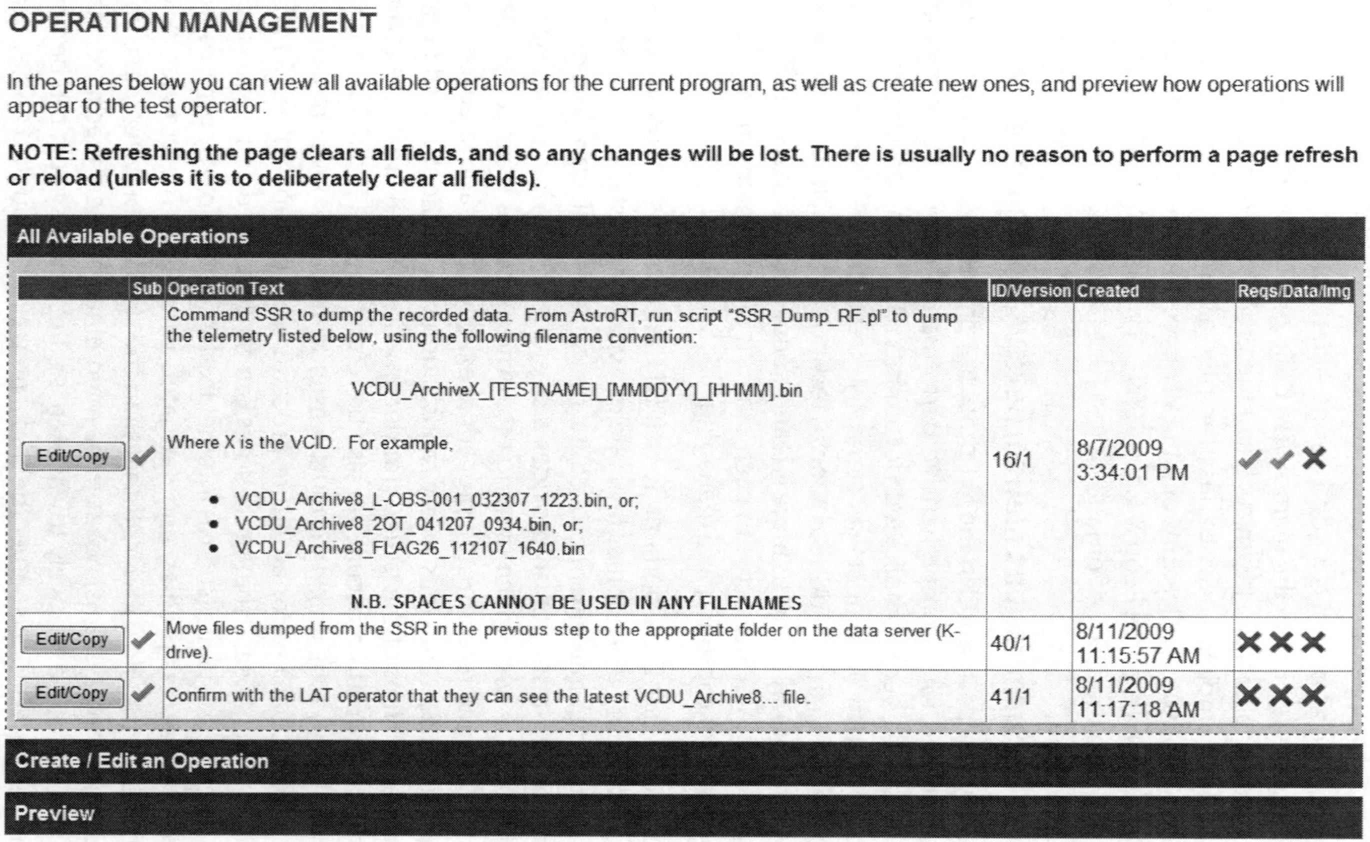

OPERATION MANAGEMENT

In the panes below you can view all available operations for the current program, as well as create new ones, and preview how operations will appear to the test operator.

NOTE: Refreshing the page clears all fields, and so any changes will be lost. There is usually no reason to perform a page refresh or reload (unless it is to deliberately clear all fields).

All Available Operations

	Sub	Operation Text	ID/Version	Created	Reqs/Data/Img
Edit/Copy	✓	Command SSR to dump the recorded data. From AstroRT, run script "SSR_Dump_RF.pl" to dump the telemetry listed below, using the following filename convention: VCDU_ArchiveX_[TESTNAME]_[MMDDYY]_[HHMM].bin Where X is the VCID. For example, • VCDU_Archive8_L-OBS-001_032307_1223.bin, or; • VCDU_Archive8_2OT_041207_0934.bin, or; • VCDU_Archive8_FLAG26_112107_1640.bin **N.B. SPACES CANNOT BE USED IN ANY FILENAMES**	16/1	8/7/2009 3:34:01 PM	✓ ✓ ✗
Edit/Copy	✓	Move files dumped from the SSR in the previous step to the appropriate folder on the data server (K-drive).	40/1	8/11/2009 11:15:57 AM	✗ ✗ ✗
Edit/Copy	✓	Confirm with the LAT operator that they can see the latest VCDU_Archive8... file.	41/1	8/11/2009 11:17:18 AM	✗ ✗ ✗

Create / Edit an Operation

Preview

Figure 1 *A Screenshot of the Operations Management part of TECS-PDT – ' All Available Operations'.*

Operations, Sequences, Tests and Procedures

The basic building blocks of TECS-PDT are operations, or steps. These are individual indivisible steps, such as, "select script XXXXX.pl". These operations in themselves do not achieve much. These are then grouped into sequences, which do have some functional value. For example, for the Gamma Ray Large Area Space Telescope (GLAST—now Fermi) there were three operations involved in performing a solid state data recorder (SSDR) dump. In the TECS-PDT terminology, these three operations comprise a sequence. Operations and (short) sequences are the most important elements in PDT. From these units, whole tests and entire procedures can be constructed with relative ease. Operations and sequences are subsystem specific, and operations for different subsystems cannot be combined into the same sequence.

Even at the level of sequences there is a lot of commonality. The details of specific sequences tend to vary between different hardware configurations, but these differences are often small. TECS-PDT's ability to easily create duplicates of operations and sequences and adjust them to specific hardware configurations saves time as well as ensures a common 'language' and 'grammar' throughout different tests in different hardware configurations. One of the advantages of this bottom-up approach is that changes made to particular operations or sequences will diffuse automatically through other procedural levels. This can also lead to inadvertent changes, and so TECS-PDT contains tools (such as the Operation Mapper) that enable the procedure developer to see how changes made at one level will proliferate through other levels.

Figure 1 is a screenshot of the Operation Management part of TECS-PDT. When you first open this page you are shown the three tabs: 'All Available Operations', 'Create / Edit an Operation', and 'Preview'. The 'All Available Operations' tab shows the full list of operations that have already been created for the current Test Program. It should be noted that users can only view this page is they have the appropriate access profile (which is managed via the TECS Security Layer (TECS-SL), and only operations for the current program are shown, which means that no unauthorized user can accidently see operations created for a different test program. In this example, only three operations have been created, and basic information is shown, such as the operation text, ID/Version, date created, etc. The user can choose to create a new operation from scratch, create a new operation based on an existing operations (which is useful, for example, for when only a script name needs to be changed), or edit an existing operation (editing an existing operation essentially results in the creation of a new operation, i.e., a full version history is maintained). For a program such as GLAST, despite the complexity of the 'test space', there were only ~250 distinctly different operations that were used to construct all test documentation. In this case, all the operations can be grouped, filtered, searched against, so that the page is not filled with a list of all operations, and relevant operations can be found efficiently.

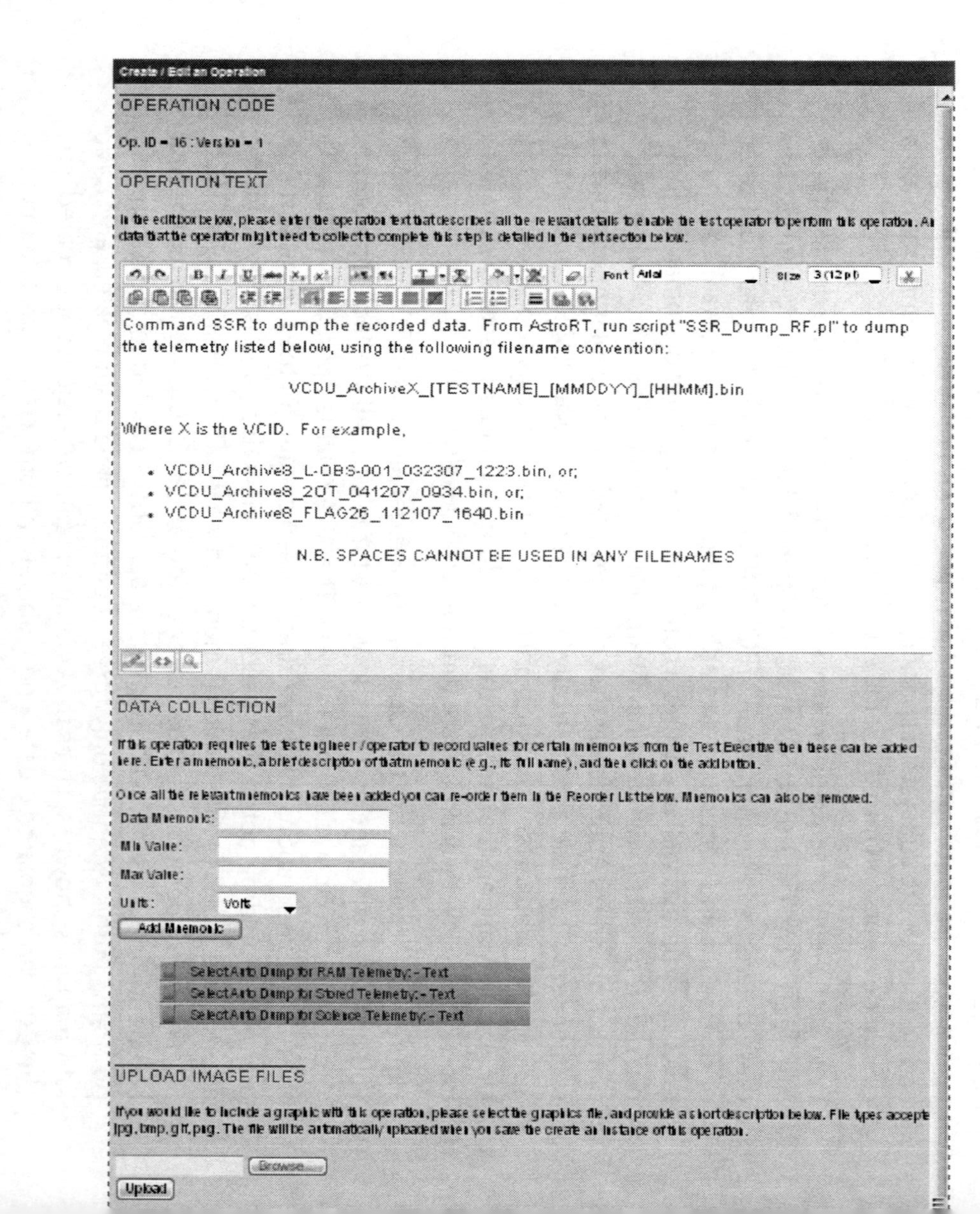

Create / Edit an Operation
OPERATION CODE
Op. ID = 16 : Version = 1
OPERATION TEXT
In the editbox below, please enter the operation text that describes all the relevant details to enable the test operator to perform this operation. An data that the operator might need to collect to complete this step is detailed in the next section below.
Font Arial
Size 3 (12 pt)
Command SSR to dump the recorded data. From AstroRT, run script "SSR_Dump_RF.pl" to dump the telemetry listed below, using the following filename convention:
VCDU_ArchiveX_[TESTNAME]_[MMDDYY]_[HHMM].bin
Where X is the VCID. For example,
VCDU_Archive8_L-OBS-001_032307_1223.bin, or;
VCDU_Archive8_2OT_041207_0934.bin, or;
VCDU_Archive8_FLAG26_112107_1640.bin
N.B. SPACES CANNOT BE USED IN ANY FILENAMES
DATA COLLECTION
If this operation requires the test engineer / operator to record values for certain mnemonics from the Test Executive then these can be added here. Enter a mnemonic, a brief description of that mnemonic (e.g., its full name), and then click on the add button.
Once all the relevant mnemonics have been added you can re-order them in the Reorder List below. Mnemonics can also be removed.
Data Mnemonic:
Min Value:
Max Value:
Units:
Volts
Add Mnemonic
Select Auto Dump for RAM Telemetry: - Text
Select Auto Dump for Stored Telemetry: - Text
Select Auto Dump for Science Telemetry: - Text
UPLOAD IMAGE FILES
If you would like to include a graphic with this operation, please select the graphics file, and provide a short description below. File types accepte jpg, bmp, gif, png. The file will be automatically uploaded when you save the create an instance of this operation.
Browse...
Upload

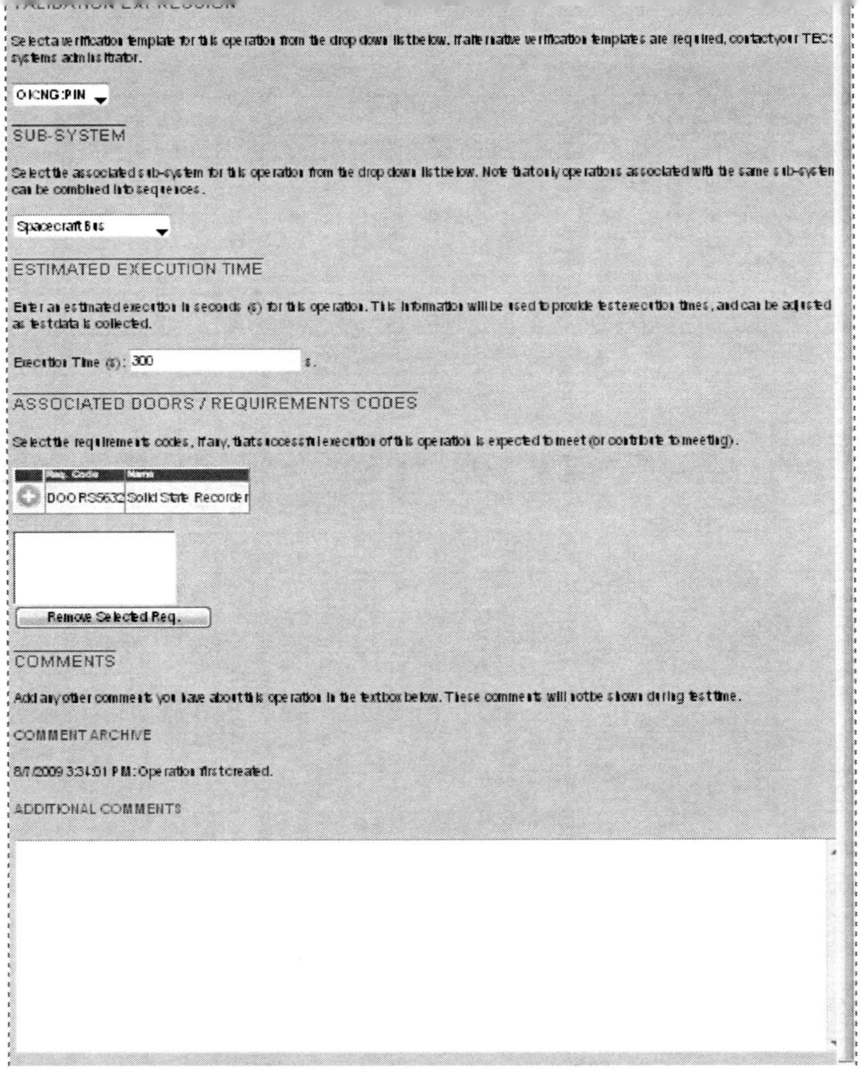

Select a verification template for this operation from the drop down list below. If alternative verification templates are required, contact your TECS systems administrator.

OICNG:PIN

SUB-SYSTEM

Select the associated sub-system for this operation from the drop down list below. Note that only operations associated with the same sub-system can be combined into sequences.

Spacecraft Bus

ESTIMATED EXECUTION TIME

Enter an estimated execution in seconds (s) for this operation. This information will be used to provide test execution times, and can be adjusted as test data is collected.

Execution Time (s): 300 s.

ASSOCIATED DOORS / REQUIREMENTS CODES

Select the requirements codes, if any, that successful execution of this operation is expected to meet (or contribute to meeting).

	Req. Code	Name
+	DOORS5632	Solid State Recorder

Remove Selected Req.

COMMENTS

Add any other comments you have about this operation in the textbox below. These comments will not be shown during test time.

COMMENT ARCHIVE

8/7/2009 3:34:01 PM: Operation first created.

ADDITIONAL COMMENTS

Figure 2 *A Screenshot of the Operations Management part of TECS-PDT – 'Create / Edit an Operation'.*

Once the user has decided to either create, duplicate, or modify an operation they move on to the 'Create / Edit an Operation' tab, which is shown in Figure 2. Here is where all the details that define a particular operation are entered. The main 'fields' are:

- *Operation Text* (required): The direct guidance to the test operator. A sophisticated text editor is used to allow the clearest possible guidance to be designed;
- *Data Collection* (optional): A list of the data mnemonic values that need to be recorded as the time the operation is executed;
- *Image* (optional): The filename of the graphic that should appear;
- *Validation Expression* (required): What form of validation is required to allow the test operator to move onto the next operation. In the example shown in Figure 2 the verification template (for which there are a variety of choices), required the test operator to indicate whether the operation was executed successfully (OK), or failed (NG, no go), and then enter his or her unique pin. If NG is selected then the Test Operator will automatically be given the opportunity to create a Test Anomaly Report, or provide a reason as to why, given a failure, the test is to continue;
- *Sub-System* (required): The sub-system for which the operation is designated;
- *Estimated Execution Time* (required): An estimate of how long it takes to execute the operation. This estimate can be revised with real test data. This is very useful when it comes to planning, as it becomes a trivial task to generate time predictions for much longer tests such as a CPT;
- *Requirements Codes* (optional): If the program uses a standard requirements system such as DOORS then requirements codes relevant to the operation can be selected from the requirements database. This enables System Engineers and Program Managers to easily monitor which requirements have been met, which tests contributed to their having been met, etc.;
- *Comments* (optional): Any comments that the user feels are important to record, e.g., if the operation is valid only for a particular hardware configuration. When changes are made to existing operations, there are a number of automated comments that are generated, such as which fields where changed, what the previous operation ID was, when the change was made and by whom. This results in a comprehensive audit trail for each operation. This same trail is created at all levels of the TECS hierarchy.

Once all details have been entered, a preview can be generated (Figure 3) which shows how the operation will appear to the Test Operator at test time. Note that some standard elements are automatically generated, such as navigation buttons, email, anomaly generation, etc. (not all of which are shown in the preview).

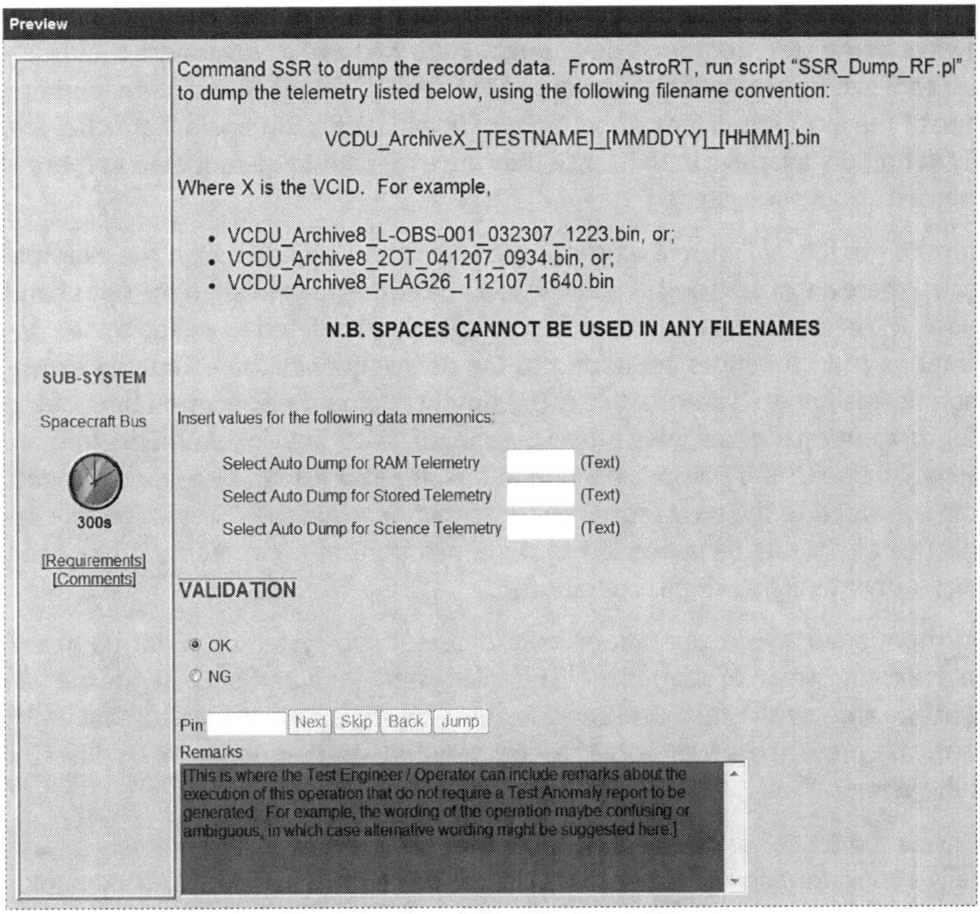

Preview

Command SSR to dump the recorded data. From AstroRT, run script "SSR_Dump_RF.pl" to dump the telemetry listed below, using the following filename convention:

VCDU_ArchiveX_[TESTNAME]_[MMDDYY]_[HHMM].bin

Where X is the VCID. For example,

- VCDU_Archive8_L-OBS-001_032307_1223.bin, or;
- VCDU_Archive8_2OT_041207_0934.bin, or;
- VCDU_Archive8_FLAG26_112107_1640.bin

N.B. SPACES CANNOT BE USED IN ANY FILENAMES

SUB-SYSTEM

Spacecraft Bus

300s

[Requirements]
[Comments]

Insert values for the following data mnemonics:

Select Auto Dump for RAM Telemetry (Text)
Select Auto Dump for Stored Telemetry (Text)
Select Auto Dump for Science Telemetry (Text)

VALIDATION

OK
NG

Pin: Next Skip Back Jump

Remarks

[This is where the Test Engineer / Operator can include remarks about the execution of this operation that do not require a Test Anomaly report to be generated. For example, the wording of the operation maybe confusing or ambiguous, in which case alternative wording might be suggested here.]

Figure 3 *A Screenshot of the Operations Management part of TECS-PDT – 'Preview'.*

What cannot be shown in the figures is that each object in the screenshot has hidden 'Intellisense' that pops-up whenever the mouse cursor hovers over that object. This provides the user with built-in instructions on how to use the various tools, further minimizing errors, and encouraging consistent usage.

After a number of fundamental operations have been created and saved, the user can then move onto develop meaningful 'sequences' of operations to achieve a sub-system specific outcome. Figure 4 shows the Sequence Management page from TECS-PDT. Here, existing operations and sequences are combined to produce new sequences. As with the Operation Management page, sequences can be created, duplicated, or modified, and a full auditable record is generated along the way. A sequence may be defined to, for example, detail how the command window is passed from one sub-system team to another, or how a solid state recorder dump is performed. It is likely that such sequences could be reused with a variety of hardware configurations without change. There are no hard and fast rules about how long or short a sequence should be. Through continued use the experienced Systems Engineer will figure out what the appropriate compartmentalization should be, and this will undoubtedly evolve as the test program is developed. Here we have a case where a fairly linear tool can be employed to develop heuristics of a nonlinear system, perhaps one of the most important advantages of linearization. But again, heuristics are, by definition, adaptive in the sense they are expected to be modified as more is learned about the system.

On the top left of Figure 4 existing operations and sequences (in the example given there are no existing sequences) can be added to the list on the right hand side. These entries can be moved up or down and deleted as needs by. As operations and sequences are added to the new sequence the estimated execution time is updated automatically (by simply adding the execution time of the component operations, plus a 10% overhead). Also, any requirements that are associated with the comprising operations are also listed. New requirements can be added to the new sequence for scenarios when the completion of a sequence of operations is needed to demonstrate that a requirement has been met, rather than just single operations.

As mentioned above, operations from different sub-systems cannot be mixed in the composition of sequences. The sub-system for a particular sequence will be the same as the first operation in the sequence, and the editor will automatically prevent the user from mixing operations and sequences from different sub-systems.

Once a sequence has been defined, the user can preview it. The preview is basically a concatenation of screenshots like the one shown in Figure 3. This enables the user to step through the sequence as it will appear to the Test Operator.

In reality, there will be far fewer sequences than operations, and more often than not the relatively fewer sequences will be used to build combinations higher up the TECS hierarchy, such as tests and procedures. For emphasis, the real test detail is captured at the operation level. Sequences are really no more than a collection of operation and sequence codes, so although the definition of an operation can be quite involved, a sequence might be expressed as simply as, for example, op_16, op_40, op_41, sq_3. What the TECS-PDT various editors do is introduce a bottom-up structure for test procedure development, and a simple scripting language to build more complex test documentation from low level elements.

The next level in the TECS-PDT hierarchy is the test level, where full self-contained tests can be created by bringing together various operations, sequences, and even other tests. An example of a 'test' is the full functional test of one particular hardware configuration, or perhaps a flight software upload test, which would include the upload itself as one test, and the subsequent regression testing as a separate test or set of tests. Fine tuning a test so that it is valid for a range of hardware configurations is a very simple matter. The Test Management editor is

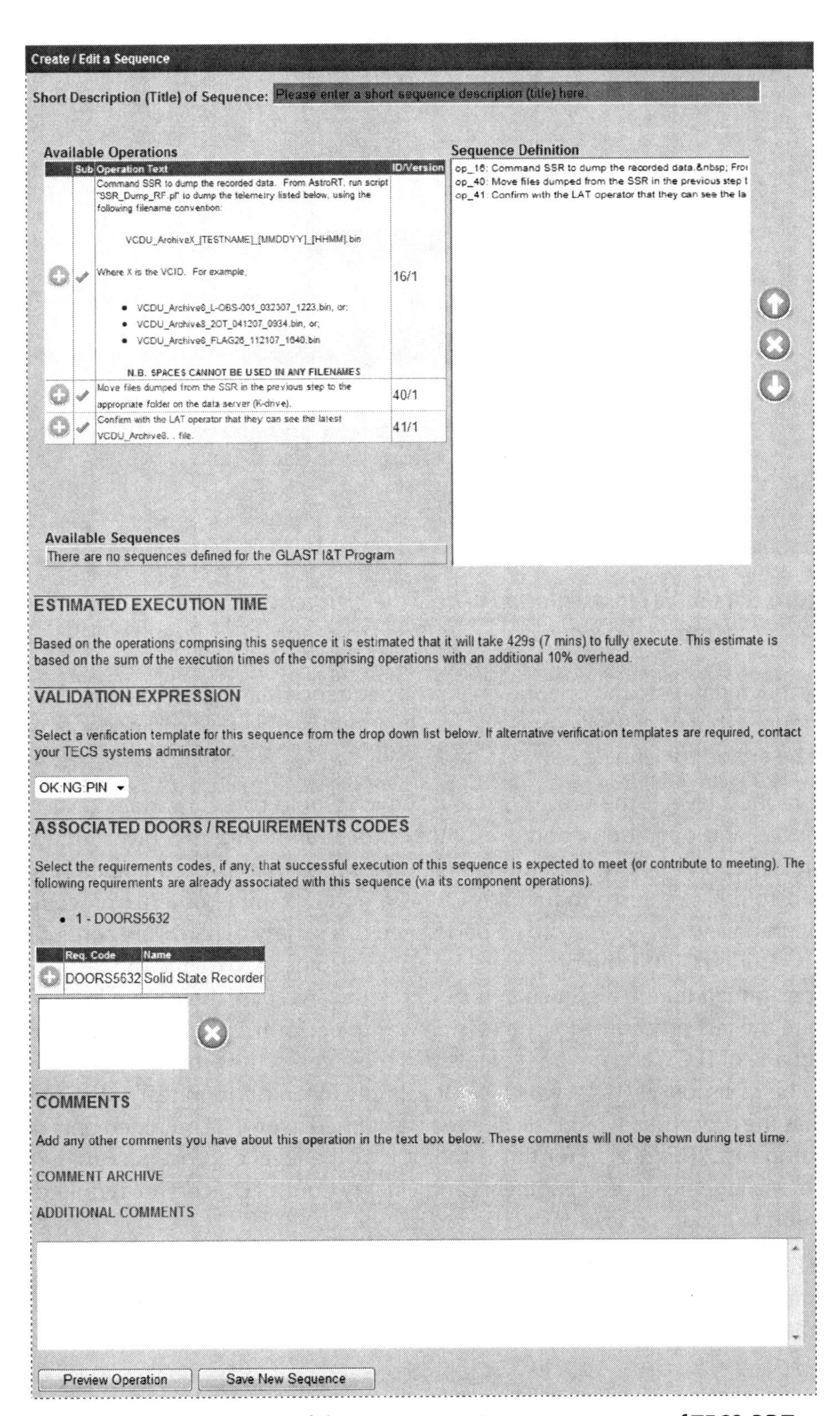

Figure 4 *A Screenshot of the Sequence Management part of TECS-PDT – 'Create / Edit a Sequence'.*

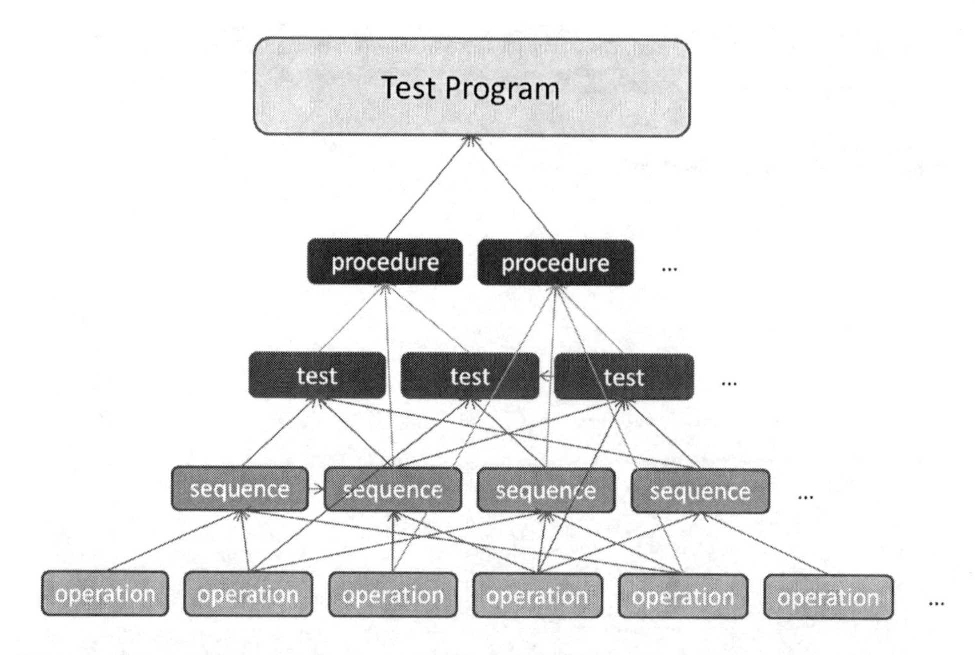

Figure 5 *A simple representation of how the hierarchy of operation, sequences, tests and procedures are combined to define a Test Program.*

very much similar to the one shown for the Sequence Management page shown in Figure 4, but of course existing tests is included in the list of components that can be added to define a new test.

The highest level is the procedural level which would contain various tests, sequences, and operations, and even other procedures. Figure 5 gives a simple representation of how operation, sequences, tests and procedures are combined to define a Test Program. A CPT is a good example of a full procedure containing a variety of tests to be performed in a variety of hardware configurations. During GLAST the procedure-level was also used to define the testing to be performed during a particular testing session. As such, there were instances where the test procedure for the session was to perform a particular 'test' (in the language of TECS) and not a set of tests. In these cases, the procedure definition may be as simple as "ts_4", which would simply mean perform test 4. The point is that the procedure would still need to be defined even if it contained only one existing test. The reason for this is that it is at the level of procedure that Program Management, Lead Engineers and Quality Control Officers are required to sign-off to allow any sessions' testing to go ahead.

The hierarchy of the TECS is engineered-in, so to speak, but this same designed hierarchy gives inklings of how nature typically employs "self-organizationally" hierarchical design because of the latter's many expediencies.

TECS & Authentication

At each level the test designer can state who has the authority and responsibility to sign-off on the execution of each test phase. An individual with the correct role and designation must be logged in and authenticated before s/he is authorized by TECS to sign-off on the successful execution of an operation, sequence, test, or procedure. If say a Program Manager, a Systems Engineer, and a Quality Engineer are specified as having the authority to sign-off on a particular procedure for example, then that procedure won't be recorded as complete until those individuals log in and digitally 'sign' the document. All authorized personnel are maintained in a TECS program-specific roster. This provides an effective and simple way to monitor I&T progress and identify which tasks are complete and which ones need further investigation.

TECS & Requirements

As has already been discussed briefly above, during the test program design process, requirements (such as DOORS codes) can be associated with the successful completion of individual operations, sequences, tests and procedures, and even entire test programs. With such data made available, combined with as-run test results (also maintained in the TECS central SQL database), it becomes a trivial matter to see which requirements have been met at any stage during the test program. Furthermore, given the database centered approach adopted by TECS, generating output to be imported into other management systems is a simple case of defining the appropriate search query, and the output template.

TECS & Flexibility

The toolset developed in TECS-PDT allows for the creation of robust and consistent procedures via a simple browser-based graphical user interface (GUI). Another valuable attribute becomes apparent during testing itself. CPT dry runs, for example, rarely run smoothly and often procedural changes are needed quickly to allow testing to continue in a smooth and timely manner. Using the database of operations, sequences, etc. that is created as part of the procedure development process, 'new' procedures and tests can be rapidly prototyped, reviewed and run. For example, for the last CPT the Large Area space Telescope (LAT, the primary instrument onboard Fermi) team put together a 300+ step procedure that hadn't been run before. The resulting procedure required only minor changes on review as it was constructed from operations and sequences that had already be reviewed and executed a number of times in previous tests.

An important part of this flexibility is the quality assurance (QA) process. QA is essential if testing is to meet requirements efficiently without damaging hardware, and/or wasting test time. Often the review of test documentation occurs at the procedural level. As full procedures can be long complex documents, this review process can be time consuming, and not always thorough. Who can honestly say that they have reviewed a 500-page test procedure thoroughly from beginning to end? With TECS-PDT the QA process can occur at a lower level, e.g., if a certain configuration independent sequence has been QAed, then it can be reused in other procedures, tests and sequences without revisiting the review process. When larger tests or procedures are created in the dynamic pre-CPT environment, for example, then the review/QA process need only be concerned with how the various sequences and tests interact with each other, rather than revisiting every individual operation. This makes the QA process far more efficient, without losing its effectiveness (the document development process basically becomes self-QAed). Whenever a procedure, test, sequence or operation is opened in the TECS-PDT environment a full history for that element is provided which represents a complete audit trail. Integration and testing generates an enormous amount of paperwork and having a tool to easily and successfully navigate this 'document space' is a genuine asset.

Once a procedure or test is ready for release and execution, TECS-PDT generates an instance of that procedure to be run separate from the current version stored in the TECS database. This way all the data for a particular test can be captured and isolated from other tests, to facilitate further analysis and the ability to easily access the details of tests perhaps performed months or years in the past. A paper-based version can also be easily generated, and various user-defined templates designed and utilized.

Test Execution

During actual execution of a test, usually performed by a Test Conductor accompanied by a number of Test Operators, the operators (who must first log in and be authenticated) are presented with a simple GUI (via test console's browser) showing details of the test about to be performed, and some high level details about the test procedure. Once the required sign-offs are confirmed (this is easily incorporated by simply defining an operation that requires a number of sign-offs and placing this operation at the beginning of the procedure) the test proceeds by displaying one operation at a time. From this display the operator can confirm that the operation has been performed, create a test anomaly case in the event that the execution of the operation fails (via TECS-AM), move onto the next operation, access roster details, send an email directly from the test console to anyone in the program roster, and move to the next step. Whenever the test operator interacts with TECS an internal log is updated. The operator must sign-off on an operation before being permitted to move onto the next operation in the procedure.

The test team can also submit change requests for any part of the procedure. These are often not associated with a test anomaly, but are simpler changes that might reduce ambiguity or improve the flow of the procedure. Such effective communication between Systems Engineers and Test Engineers is another essential dimension in the effective execution of an I&T program.

Where TECS Fits In

In the case of test documentation itself, TECS is designed to replace the paper-based documentation systems of the many engineering firms that comprise the aerospace industry, and provide a standard way to generate such documentation (which alone would make for a smoother more efficient and effective I&T process, especially where teams coming from different companies are working together). In the simplest terms, TECS is a flexible document management system specifically designed for the I&T phase of manufacturing of complex engineering products, such as satellite systems.

A typical test set-up for a multiple instrument satellite system, for example, would be for separate test consoles (comprising separate computer systems and monitors) for each of the instrument teams. A Test Conductor would wield overall control of how the test proceeds, but execution of instrument-specific aspects of the test is delegated to the respective instrument teams. A bound paper version of the procedure to be performed is passed from team to team as their turn is taken, or copies are available to each team. During actual testing, this paper-based procedure is replaced with a browser-based system that can simply be realized as another window on the instrument team's console. TECS does not communicate directly with the hardware (which is done through a Test Executive), but simply provides a digital representation of the procedure, along with digital recording mechanisms for collecting certain test data. Although anyone can log in and view the status of a particular test at any point, only the individual or role associated with each operation is able to make changes. This is achieved via the two viewing modes:

- Passive view: This mode allows any authenticated user to step through any procedure, but without the ability to make any changes. Once a procedure has been run, only passive view is available;
- Active view: This mode allows an authenticated user who is also listed as an authorized sign-off agent for the current operation to update and sign-off on the current operation. The view only becomes active if the individual is signed-on, and all previous operations have been completed and signed-off.

TECS Anomaly Management

In the real world testing does not always go to plan, and integrated systems do not always behave as simply the sum of their respective parts. Issues emerge at the integrated level that were not necessarily foreseen at the sub-system level. This fact, in combination with the fact that documentation can often contain errors, and Test Operators sometimes make mistakes, means that problems that might occur during testing, generally do occur during testing, especially early in the I&T process as Engineers learn more about how the new integrated system behaves. As Test Operators execute particular operations, TECS provides tools for them to report anomalous behavior via a Test Anomaly Report (TAR). If an operation fails for whatever reason, TECS automatically requires that a TAR is completed. Similarly to how operations are designed and maintained, TECS provides a TAR Management System (TMS) that keeps a full record of each TAR (including the test session and operation that was being undertaken at the time of the observed anomalous, or 'off-procedure', event), allows Engineers to add further documentation, evidence, comments, resolution, etc. to the original record, and, importantly, provides a status of each TAR as 'open' or 'closed'. A list of required sign-offs can be associated with each TAR, which determines who are the responsible agents for evaluating the closure resolution. At any time, a full account of any TAR can be viewed, as well as the status of all TARs for a particular procedure or program. Although an as-run procedure may be considered 'completed', even when there exist TARs associated with that procedure, a Program cannot even begin to be considered completed before all TARs are satisfactorily resolved. Whenever the resolution of a TAR results in a change to the test documentation, the ID of the relevant TAR is captured in the change history of the operation, sequence, test or procedure that was amended. This aspect of TECS again contributes to the comprehensive audit trail that using TECS for I&T results in; a trail that can be readily summarized, searched, and critiqued at any point during the I&T program.

I&T Program Monitoring

The benefits of the digitization process that TECS brings to test procedure development, maintenance and execution is that a very detailed audit trail is created, so essential to test robustness and quality. As such Program Management can remotely log into the system and easily generate status reports that detail how a particular test is proceeding, how long any part of the test program takes to execute (which can also be used to generate predictions regarding test execution durations), the progress of the entire test program, details of which design requirements the testing has already verified, overviews of the number and nature of outstanding test/design anomalies, etc. There exist many opportunities for generating pertinent reports to facilitate Program Management planning and budgeting as a direct result of digitizing a major proportion of the integration and test process. Although TECS provides a number of standard reporting tools, it is a very simple manner to add further reporting capabilities derived from the captured test data, and even extend the database fields when gaps in the collected data are identified.

Summary

TECS has been designed to facilitate the robust development, maintenance, review, and execution of testing procedures for complex engineered systems. A cut-down development version (desktop focused rather than browser based) was used with considerable effect by the LAT team during observatory I&T for NASA's recently launched Fermi telescope. For LAT, as an example, although there were hundreds of pages of test procedure, there were only around 250 distinctly different test operations, many of these being variations on only a few themes. TECS, then, was able to reuse many operations, sequences, and tests (with perhaps only minor adjustments) during all phases of the test program. It is asserted that such a bottom-up structured approach to I&T can lead to significant time and, therefore, cost savings for any complex I&T program. As a result engineers can spend more time focusing on engineering issues, rather than document management, which would undoubtedly result in higher quality final products and, in the case of satellite design and construction, reduced on-orbit failures and errors.

Acknowledgement

I'd like to thank Jeffrey A. Goldstein for insightful additions to my original manuscript in relation to clarifying the ongoing relevance of linear systems in understanding and guiding nonlinear systems.

References

Lewens, T. (2005). *Organisms and Artifacts: Design in Nature and Elsewhere*, ISBN 9780262621991.

Peng, W. and Gero, J. (2009). *A Design Interaction Tool that Adapts*, ISBN 9783639135893.

Richardson, K.A. (2008). "On the limits of bottom-up computer simulation: Towards a nonlinear modeling culture," in L. Dennard, K.A. Richardson and G. Morçöl (eds.), *Complexity and Policy Analysis: Tools and Methods for Designing Robust Policies in a Complex World*, ISBN 9780981703220, pp. 37-53.

Kurt A. Richardson, PhD is currently the owner of Emergent Publications—a small publishing house specializing in the field of complex systems thinking; the owner and chief technical offer for Exploratory Solutions—a startup company focusing on developing hardware and software for a range of high tech companies (including NASA and General Dynamics); the lead software developer for Inergy Systems—a new startup company developing software and hardware for household and commercial smart grid solutions; and is also the lead FPGA (field programmable gate arrays) engineer for Orbital Network Engineering, who builds satellite telemetry systems for the likes of Raytheon and Lockheed Martin. His research focus is on the development of fundamental theory for complex systems, and their epistemological implications. This work has lead to 15 books, and more than 40 articles on complexity. On the commercial front his interests lie in applying principles of complexity thinking to both software and hardware design in a wide range of fields from publishing technology to smart grid products.

6. Using Principles Of 'Holistic Business Science' Facilitated Through A 'Process And Emergence Tool' To Support Conventional Businesses Deal With Complexity

Claudius Peter van Wyk
Transformation Strategies, ZAF

The on-going global financial crisis demonstrates the inadequacy of conventional business models to deal with complexity. This calls for the transformation of business practice in accordance with a holistic business science. Complexity science provides insights to flesh out a holistic model and praxis of business. A Process and Emergence Tool 'PET' offers the opportunity to operationalize complexity theory providing a new business lens to address the paradigm blindness of the conventional approach and thus enhancing the coevolutionary nature of complexity thinking. The ethical shift from a mechanistic to a holistic model of business requires the organization to change from its current nature as a separate-from-society money-making machine, with ethics ensured by its being populated by legally responsible persons, to something different. Since the whole is greater than the sum of the parts business requires a focused collective responsibility that transcends shareholder interest. It should transform to an integrated dynamic 'whole' with a moral conscience defined by its concern for the well being of society and nature.

Introduction

The World Economic Forum (Davos 2009) attributed the causes of the current global financial difficulties to inappropriate macro-economic policies. These included unsustainable models of development, prolonged low savings, high consumption, and the excessive expansion of financial institutions in blind pursuit of profit. It was also attributed to the lack of self-discipline on the part of financial institutions that included distortions of risk information and asset pricing, the failure of financial supervision and regulation to keep up with financial innovation and the increasing risk of financial derivatives. The suggested solution from Davos was the institution of some further world financial control, through a World Financial Organization (WFO) to supplement the WTO and other existing regulatory bodies. In line with this approach former World Bank chief economist Joseph Stiglitz (2010) came out in support of the South

African government's 'New Growth Path'. A core element of the program was positioning government at the centre of economic reconstruction. Stiglitz contrasted this with the so-called short-termism of unregulated markets which, he said, had been so evident in the economic crisis.

Stiglitz's view is seen against two contrasting opinions on how to respond to the economic crisis. One is that the crisis was caused by the failing economic system of capitalism and consequently that old fashioned socialist planning was the only coherent alternative to a collapsing capitalist economy. The other argument was that capitalism per se is not the problem; rather it is the way it has been employed. This especially relates to the simplistic premise that the principle of a market driven exchange of perceived value according to demand as the underpinning economic principle is the only way of managing business. Pro free-market commentators criticize the South African economic plan for essentially ignoring the vital role in the country's economic development of the private sector. They point out that governments generally have a dismal record of positive economic intervention, and in South Africa particularly most government enterprises have essentially failed. Both the Davos proposals and the SA economic development plan it is argued must be viewed against constraints of what is termed institutional saturation and especially the dangers of centralized control over complex systems. Furthermore neither the assumptions of socialism or capitalism, it is shown, adequately address the core issue 'demand' of economic exchange. A core question here to be addressed is on what principles that demand is to be based. Socialism, requiring as it does the state to be the essential player in economic affairs, attempts to define citizen needs, still of necessity encounters the same problems of satisfying demand, even when that demand is not intelligent. Examples of the inefficiencies of central planning in failed socialist states are legend. Therefore a core issue to be addressed in an approach to a new business model is the inculcation of intelligent demand, in turn implying an evolution of economic intelligence, or an enhanced consciousness in business.

Some position needs to be established about the current free-market versus socialist debate. At least in South Africa the tension between collectivism and individualism, socialistically orientated versus free-market economics, is not going to go away—such is the vexed nature of South African society. What is clear however is that the classic neo-liberal business model will have to redefine itself in respect of national goals—certainly a core outcome of the economic development plan. Mill's old capitalist 'invisible hand' patently does not serve the universal requirements of South African society. A business model thus needs to be developed capable of dealing with both the growing phenomenon of complexity and its accompanying cohort, unpredictability. It must also be capable of addressing deeper social needs. This will surely require that business be relegitimized through a reformed business ethics based on its relationship and responsibility to the whole of society and also the natural ecology that sustains that society. Consequently it is suggested that what is required is a transformed system of economic and business science that recognizes that true sustainability is a function of a genuine exchange of value. The corollary notion in a Holistic Business Science is that a defining feature will be self-regulation.

Holism's Potential Contribution

Jan Smuts (1926) coined the term 'holism' and described its natural manifestation in nature and its application to human affairs. For him it was in the generative engagement of disparate entities that new vital wholes were formed. It is thus contended that in this generative and sustainable exchange of value the greater whole is constituted. Therefore in terms of economic management a holistic economic and business science would see economic health and life in terms of the robustness, vitality and sustainability of exchange. The vital importance of trade, and consequently of general business, is thus established and the issue of intelligent demand is again surfaced. Holistic business science requires the consideration of the impact of such exchange activity on the whole as a primary ethical consideration. This, in turn, informs the apportionment of value, genuine or other. Consequently Stiglitz's problem of short-termism could be addressed whilst the free enterprise system's generative power of individual and corporate achievement would be coupled to social and ecological good. Consequently economics could be looked at the though the lens of ecological sustainability. In a certain way biological principles would apply. Smuts said:

> *Plants, like human beings, should be studied in their social relationships. Ecology is in reality nothing but the sociology of plants, a study of their loves and friendships and antagonisms.* (Private letter, 1937)

Holistic Business Science demands a certain transformation of the prevailing individual materialistic mindset of business players. The prevailing tendency is still one in which ego-based self-recognition is satisfied by often showy symbols of wealth and overt or covert manifestations of power. Hence profit is exploited to that end. Rather, holistic business practice would require a transformed recognition of self-worth that would manifest in generative participation in, and service to, the holistic order. Profit would be optimized to providing for own creative needs whilst also strengthening the corporation's capacity to engage meaningfully with the whole. For Smuts it was in the human being's engagement and identification with and consequent responsibility for the full milieu of existence that the true 'spiritual life' was to be found. He warned:

> *The denial of what is deepest in our spiritual nature must lead to a materialist, mechanical civilization where economic goods take the place of spiritual values and where mankind can at best only achieve a distorted and stunted growth*[1].

---

1. In respect of Smuts' notion of spiritual life it is interesting to note that for Polanyi life's ir-

Something like this 'distorted and stunted growth' was Marx's concern with the legacy of capitalism, but it is here contended, that has also been a function of modernism. Merely chaining capitalism in a restrictive legal framework or taking over its natural function by bureaucracy does not address the issue raised by Smuts, namely, the need for a vital and freely creative relationship with the enabling milieu. Whilst in a holistic approach a transformed free enterprise system would still deliver its intrinsic economic value certain legislative safeguards probably would still remain a necessary evil until the anticipated paradigmatic transformation has taken place. If free enterprise were finally to lose its credibility, and hence its capacity of offering the creative factor of self-motivated individuals operating in a free but responsible way, it will be catastrophic to the global economic future. This leaves the question of how the implied dangers of the unstructured nature of free-market economics are to be addressed. The enormous damage to vulnerable communities in economic downswings has, as has been shown, been attributed to the free-market system, with increased centralized management seen as the only alternative. The dangers of institutional saturation and central control over complex systems have been raised. Consequently the aforementioned argument is a simplistic assumption based on linear mechanistic thinking that ignores the reality, as shown in complexity science, of the phenomenon of unpredictability and the nature of emergence[2]. This is thus where complexity theory has to be applied to business science. It presents a daunting challenge to complexity theorists to come down from philosophical abstractions and more actively address the real world of practical economics. This discussion of the desirability of a model of holistic business science is pre-framed by the observation that any paradigmatic change is facilitated by stress. This statement anticipates the approach to be offered here with the PET instrument. In this way the normal, albeit covert change process will be modeled, and by rendering it overt, could enable the fast tracking of transformation. Its application will be offered to support the anticipated paradigm shift in business thinking to Bertalanffy's (1971) 'whole systems' and Polanyi's (1997) 'emergence' information processing that characterizes complexity thinking. It is shown to provide a mechanism to address the practical challenges of applying complexity thinking to support the transformation of business methodology. The complexity notion of 'coevolution' in which complexity thinking enables adaptability is thus specifically related to the business world.

---

reducible structure, for example the information contained in the DNA molecule, is irreducible to physics and chemistry. Although the DNA molecule could not exist in the absence of physical properties, its physical possibilities are constrained by higher level ordering principles.

2. Polanyi criticizes the mechanistic worldview and advocates 'emergence' and claims that there are several levels of reality and causality. Boundary conditions supply degrees of freedom that, instead of being random, are determined by higher-level realities whose properties are dependent but distinct from the lower level from which they emerge.

Holistic Business Science—Fantasy Or Reality

A number of interrelated issues are dealt with in presenting this case for holistic business science and the application of the PET. The issues are preliminarily identified and therefore not exhaustive. Probably the most significant of these are that firstly the nature of the organization as a complex adaptive system is recognized. Secondly the vexing question of ethics is dealt with against a natural context (Prigogine *et al.*, 1985) of chaos and order. Thirdly there it is an attempt to identify the structures of the most complex adaptive system of all, namely that of human subjective experience. Given the challenge that these issues are enfolded with each other, on what fundamental presuppositions can a Holistic Business Science be founded? Following Smuts, a new point of departure would be the recognition of the tendency for whole making in evolution. His notion of whole making also included the notion of self-perfection later echoed by Szent-Gyorgyi (1977). In this respect holism it is suggested could be seen as the strange attractor operating in tandem with natural selection in the evolutionary process. Such a proposition could, for example, be supported by the identified phenomenon, Weinberg/van Wyk (1992), that the neurological integration in the brain underpinning and enabling the phenomenon of consciousness, shows additional brain cells being recruited to integrate neurological parallel processing—hence the generation of a greater holistic network. Secondly, and highly significant for Smuts as implicitly echoed by Polanyi (*ibid.*), in the human person, matter, life, mind, society values, ethics and aesthetics are all integrated. This resulted in him suggesting that the developed human Personality was the apex of the evolutionary process. Thirdly there is the notion that subjective experience can no longer be separated from 'real' phenomena for scientific investigation. This then would suggest that the business of 'business' could be reinterpreted through Holistic Business Science in that greater context of the full human experience.

For Prigogine (*ibid.*) complexity science is seen as relating to the structures that are found between the conditions of chaos and order. This is thus where 'man's new dialogue with nature' needs to take place. Furthermore, for this argument, in holism complex systems are seen as being organized. From that perspective this paper is an attempt to address a few of the more practical implications of complexity science in respect of Prigogine's 'dialogue' as it affects the organization. Davies and Gribbin (1992) assert that when non-linearity becomes important it is no longer possible to proceed by analysis, because the whole is now greater than the sum of its parts. Non-linear systems, they say, can display a rich and complex repertoire of behavior and do unexpected things—they can, for example, go chaotic. This effort then, in describing the PET, is also an attempt to show how human behaviors go chaotic. Ethics is introduced to measure behavior and the correlations between ethics and complexity are examined. The more recent notion of corporate 'environmental fit' suggests that for business it is not only about profit but also about a form of praxis that embraces the well being of society and nature. Since business life is such a fundamental part of modern living, its ethics is appropriately the subject of on-going investigation. Holistic Business Science must address this question of ethics directly whilst also heeding warnings that philosophy is not like science, and moral knowledge is not like scientific knowledge. Philosophers claim to be a little more attuned to the complexities of moral questions and the difficulties involved in knowing how to answer them. Beukes (1989) contends that Smuts' problem with classical evolution was that living creatures were regarded as materially and mechanically conditioned and life regarded as a struggle and a survival of the fittest. In business, in politics and international affairs this led to incalculable harm for the validity of moral values were denied. This assertion both validates the role of philosophers and calls for an implementation of Smuts reformed scientific view. Growing complexity in the business world presents an increasingly complex moral issue and, as such, ethics becomes an issue of business sustainability subject to scientific scrutiny. Regarding such a potential 'science' of business a further significant complexifying feature in the application of business ethics is that of 'cultural relativism'. This is an expression of habituated ways of thinking in different communities, and it is inevitably exacerbated in the process of globalization. Verhane (2002) argues that we generally filter all experience through one point of view, she terms this the 'conceptual scheme'. It is argued here that this 'scheme', hereafter to be described as the 'inner mental context', is also a dynamical system subject to the same laws of complexity and applies both to individuals and communities, and thus, it will be shown, can be tracked with the PET. Before proceeding with the opening question of whether a model or definition of Holistic Business Science is feasible Franses (2010) offers a timely warning:

> *...in seeking to give the whole expression, holistic science is addressing a potential that is also by definition yet to be complete.*

He warns about trying to catch it in some halfway definition of what it appears to be, and says it is probably still in its infancy and needs greater incubation. Whilst Franses's point is salutary, in describing the PET it will be suggested that the attitude and behavior that takes us to a new level of competency—a new psychological 'whole' of responsibility—is the capacity to formulate a vision, an ideal, and then to experiment. Consequently in this spirit of experimentation a provisional set of assumptions is offered, not intended to negate current business models but to supplement them, in order to define Holistic Business Science:

1. People and organizations are complex adaptive systems (CAS)—consequently so is a business corporation;
2. It is in the adaptation/collaboration where those CAS's come together as part of a greater order that creativity emerges. Therefore this principle points to a gradual social transformation from conflict to cooperation, and in business

to a transformation from simple competition in the market place to creative collaboration;

3. Organisms, whether they are biological entities or economic systems, gain their sustainability through the genuine, mutually enriching, exchange of value. Consequently this exchange must be seen as the engine of development and evolutionary sustainable growth. The inculcation of intelligent demand can be seen within this context;
4. Mazlow (1943) argues that human biological and psychological needs form the lower rungs of such an exchange of value. Self-actualization is seen as the higher rung, herein supporting Smuts, and in turn defining a genuine exchange of value. Holistic Business Science then of necessity pitches itself at supporting organizations at strategically positioning themselves in the vital mix of where their enterprise satisfies the 'real' requirements of the persons constituting the market;
5. When corporations are prepared to engage in this qualitative transformation their purpose for existence will transcend the simple requirement of providing shareholder and stakeholder value. They willingly embrace a greater responsibility to society. The 'triple bottom line' (shareholder/society/ecology) of the King III report on corporate governance is an example of promoting responsible corporate citizenship. However it is here added that what defines Holistic Business Science is its requirement for the genuine life-enrichment of all those participating and also those being influenced by the business activities, and;
6. It follows that business ethics from such a vantage point is not to be seen as a regulatory or controlling mechanism, leading to temptations of a tick-box response to compliance with a code, but rather ethics is the overview perspective that becomes intrinsic to business operations—giving it validity, creativity and sustainability.

In short, therefore, Holistic Business Science can be seen as being directed by the ethics of a sustainable and genuinely life enriching exchange of value. Given the uncertainty of an increasingly dynamic business milieu the question would be how are corporations to do this?

Complexity And Business Practice

In the face of this complexity Fitzgerald (1996) asserts that Prigogine's chaos science provides a new set of lenses for seeing the universe that is gaining space in the psyche of modern management. In this science of complex, non-linear dynamical systems, she confirms, the enterprise is seen as a complex dynamical whole that is not reducible into parts but to a few certain properties. According to her these properties are: consciousness, connectivity, indeterminacy, dissipation and emergence. In respect of consciousness she identifies the limitations of scientific management in the habit of seeing organizations as passive machines. This is replaced with the notion that the enterprise is a living and thinking organism that learns, grows, creates, destroys and dies like any other form of existence. Fitzgerald declares:

> *Organizations are complex, dynamical patterns of active intelligence that involve every human participant in an integral and inextricable way.*

She concludes that to change the system profoundly the mind must be changed profoundly. As long as the organizational mind fails to hold profound systemic change as both possible and desirable any effort to affect a change strategy will be futile[3]. It furthermore recognizes that the complexity notion of 'dissipation' is also relevant for such change to happen. Fitzgerald shows that it is indispensable for chaotic systems to fall apart (destruction) before they can reconfigure (creation) into a unique new form (stabilization)—an emergence that is ideally more complex and coherent. She insists that every enterprise will reach its 'limits' where it will face this choice. Barrett (2006) shows that corporations follow an emergence that can be correlated to the Mazlow hierarchy of needs but that there is a critical point of fundamental transformation related to a collective internal psychological shift[4]. Still Fitzgerald affirms that chaotic systems are able to transfigure themselves to ever greater levels of complexity and in so doing shape their relationships with their local milieu in order to sustain themselves. Notwithstanding that, however, the issue of timeliness is critical and subjectivity remains an impediment—particularly when it is informed by an outmoded way of thinking. Consciousness thus remains the critical issue—and ethics, as shall be shown, is integral to consciousness. Before proceeding to the issue of ethics a few correlations from complexity science to a holistic model of economics are offered:

Polanyi's (*ibid.*) critique from complexity theory investigates where boundaries of the 'whole' are to be found. It is suggested that the effect of attractors might define boundaries—especially in competing with other attractors where qualities of dissipation might emerge. If wholes are indeed to be defined by attractors (including strange attractors) then in holistic systems there must be multiple attractors—including competing attractors. This would account for the dynamism of polarity including its potential stabilizing effect. However central to the attractor of that competition is the attractor of holism itself—it must thus mediate all other attractors. These dynamics must then underlie those conditions described as far from equilibrium (FFE) and near to equilibrium (NTE).

---

3. It is against this statement that the PET offered here is designed specifically to address the possibilities of an organizational mind change.

4. The PET identifies that this process is constantly occurring in the human psyche. Consequently this tool is able to address Fitzgerald's further challenge, namely, that only when a system acquires accurate and timely information telling of its location and speed with respect to its limits can it change strategically.

Consequently the vitality of Complex Adaptive Systems must lie somewhere between conditions of FFE and NTE—FFE providing the conditions of openness to engage with the dynamic conditions of the context of the whole—the greater whole—NTE providing the coherence of the entity within that whole. CAS (complex adaptive systems) emerge as such through the degree of autonomy in the various parts. Agency theory thus investigates the effect of lower or higher degrees of autonomy of agents as contributing factors to FFE or NTE conditions.

The implication of the above is that the socialist model of economy (high regulation) inevitably tends to centrism, hence lower agency autonomy, and hence greater danger of NTE conditions (statism) and death. Conversely the free enterprise economy (low regulation) tends to decentralization with higher agency autonomy and hence greater dangers of FFE (chaos) and dramatic emergence (economic fluctuations) etc. Holistic economics thus respects the potential autonomy of the individual or agent commensurate with the degree of responsibility capable of being displayed to the 'greater order' holistic principle and therefore requires managing agents according to that maturity. This certainly points to the need for some highly efficient and intelligent regulatory structure as identified at Davos—but challenges the notion that to avoid accusations of imperialism they need to be fully democratic. This notion presents further ethical challenges that fall outside of the immediate scope of this enquiry.

Complexity And Ethics

A simplistic approach to organizational behavior would be some sort of algorithm, some cost benefit analysis, but it would still represent the mechanistic view. Lozano (2002) suggests that any model of moral choice affirms the primary importance of conscious deliberation. Donaldson (2004) continues that what is needed is a general moral theory working in tandem with an analysis of the foundations of corporate existence. Practically needed, he asserts, is an interpretative mechanism, or algorithm, to determine the implications of own moral views, particularly in the more complex multinational situations. For example Letiche (2005) quotes Jeurissen on moral complexity in organizations. The latter mentions four applications of ethics that apply—rules, values, stakeholder interests and social dialogue. Three dynamics emerge in that quadrant, namely, complimentarity, contrariety and contradiction. Letiche queries Jeurissen as advocating the resolution of these dynamics through social dialogue, and this being directed to achieving organizational and economic efficiency. Letiche cautions against what he terms identifying an overriding criterion of organizational 'performativity' over the social ethical criteria of 'justice' or 'human flourishing'. Consequently for him values should be chosen above rules because values as a criterion 'involve more consciousness of action and self-reflexivity'.

Solomon (1993) points out that Aristotle already required us to think of ourselves as members of a larger community and to strive to excel, and bring out what is best in our shared enterprise. Accordingly the larger community defines what is best in us, our virtues. Ethicists assert that Aristotle was attempting to derive an ethical ideal from experiential reality. Consequently what is needed in business ethics is a theory of practice and account of business as a fully human activity in which ethics provides not just an abstract set of principles of side constraints but the very framework of business activity.

The aforementioned issue provides the background to the challenge of engaging in ethics with the real world applications of complexity theory. The merit of that intention is endorsed by Spinoza *et al.* (1997) who argue that human beings are at their best not when they are engaged in abstract reflection, but when they are intensely involved in changing taken-for-granted every day practices in some domain of their culture. How better, for our purposes then, than address the problematic thinking behind business praxis that ignores the negatives realities of business' growing influence on society and ecology? This accepts that thinking then will be enriched through Donaldson's (2004)contractual approach that there is a responsibility for increasing prosperity for the whole of society. Solomon (1993) argues that it was this schism between business and the rest of life that so infuriated Aristotle. Life was supposed to fit together in a coherent whole. For Solomon it is this same holistic idea that business people and corporations are first of all part of a larger community that also drives business ethics. It is thus this loss of connectivity, a core element in complexity science, which accounts for the detrimental impact of common business practice on human lives and ecosystems. It is the lack of holistic information processing systems in organizational mind that in turn accounts for this failure of business morality. This information processing needs to be applied at various dimensions. Enderle (1999) who analyzed business ethics at the macro, meso and micro levels, argued that the micro level corresponds to the interpersonal dimension, the meso level corresponds to the organizational level and the macro level corresponds to the social dimension. Consequently it is here asserted that getting in touch with the subtlety of information that constitutes the fuller human experience requires a greater connectedness with our own physical and emotional responses, as well as that of others. Emotional quotient (EQ) thus becomes a feature of greater sensitivity to complexity.

Holistic Business Science And The PET

A feature of Holistic Science that distinguishes it from conventional science already mentioned is that it embraces subjectivity in its notion of reality. It is now suggested that the largest part of subjective experience lies in the unconscious. The unconscious then becomes a significant feature of holistic business science, especially with reference to corporate culture. Bateson says (1972):

I shall argue that... a problem of integration and what is to be integrated is the diverse parts of the mind—especially those multiple levels of which one extreme is called 'consciousness' and the other the 'unconscious".

An outcome of the application of the PET is to focus on reestablishing a sensitive and empowering relationship with the emotional self thereby bypassing the perceptual filters and hence accessing early signals of growing systemic disorder. Dealing with the unexpected requires heightened perception, this is both a way of conscious thinking and also enabling the phenomenon of intuition, namely, the early discernment of patterns of emergence and being alerted to them by emotional disturbance. The transformation and creative emergence thus facilitated can have significance on global business practice. Correlations of the PET thus to complexity science are both obvious and nuanced. The to-be-described experience of conscious and unconscious competence can be viewed as complex, albeit stable systems, with unconscious and conscious Incompetence characterized as increasingly chaotic systems. They are the sensory mechanisms of consciousness and generate the experience of competence or incompetence, whether genuine or perceived. This applies, from boundary perspective, at the micro, meso and macro levels as previously identified. The PET, it is also suggested, is offered for leveraging organizational mind from a mechanistic to a holistic business worldview. The mismatch between the mechanistic and linear materialistic business epistemology and that of an emerging complex dynamical and non-linear context, with emergence being a function of interplay of chaos and order is thus addressed. The intended outcome is that the business organization becomes integral and creative to sustainable global society. One feature that complexity science is able to contribute to the development of more holistic business practice is in stressing the creative role of connectivity in whole systems. The PET is an instrument to support such connectivity.

Productivizing The Process And Emergence Tool (PET)

It is stressed that the PET is not intended to provide methodologies for corporate reengineering but rather to offer a way of reevaluating corporate experience of creative adaptation to change. The PET is an instrument to become aware where a person or group is cognitively and emotionally situated in this process. It enables the leader to be able to discern in any arena where the follower or colleague is situated so that the appropriate support or intervention can take place. In order to move to a new place of comfort, that is a place where the inner mental context is effectively a more coherent representation of the real external environment, new learning needs to take place. The PET is consequently introduced into the corporate environment as a management and HR instrument specifically directed towards managing transformational leadership. In corporate training and facilitation the interactive process of engagement with learners with PET is pre-framed with an introduction to the neurolinguistic programming-based (NLP) 'communication model'. This emphasizes the processing of the externally based sensory input through value-based perceptual filters into a subjective internal cognitive representation. It is further experientially demonstrated that the perceptual filters are malleable and also that emotional responses are affected by cognitive interpretation as much as by the original experience itself. Physiological responses are then experientially tracked to these information-processing sources. This is important in respect of engaging with the PET where emotional responses are then identified as serving as potent guides.

In the facilitation of the PET the dynamics of transformation, through process and emergence, are explored and delegates are invited to identify their own personal experience of the process as leaders. In the course of the facilitation delegates are required to acknowledge and integrate the need for continuous learning, learn to personally reframe the uncertainty and anxiety experienced in the phase of 'Conscious Incompetence' and learn how to nurture and guide their own teams and colleagues through the learning phases. This initial process is facilitated in a minimum of a full morning—ideally it needs a day.[5] The four distinct, albeit, overlapping phases in the transformation and learning process are defined in the following sub-sections.

PHASE 1: Conscious Incompetence

People generally tend to enter the conscious learning process when they ultimately become aware of their deficiency of knowledge (a prerequisite for meaningful engagement with complexity science) or through negative feedback to their own activities. Feedback may be in the form of verbal feedback, or awareness that things are just not working out. Because of the negative feedback, it becomes apparent that it is what they are doing that does not work. This negative feedback is inevitably accompanied by discomfort. This phase of finding new responses in the face of the uncomfortable experience of negative feedback is thus entitled 'Conscious Incompetence'. When it is realized that coping mechanisms are inadequate to addressing the challenges the experience is found to be characterized by stress, confusion, fear and some disorientation. The resultant likely behavior is characterized by initial uncertainty, nevertheless being willing to ask for help, a growing sense of determination and the willingness to experimentation.

PHASE 2: Conscious Competence

If learning is pursued through experimentation, in this phase there is awareness that certain newfound behaviors are more suitable to the circumstances or environment and therefore there is a growing ease in carrying them out. At this level there will be an increasing willingness to experiment with these behaviors

5. The source of the concepts of 'consciousness' and 'competence' are acknowledged as being derived from the conventional 'Learning Hierarchy'.

and transfer them into other circumstances or environments. This represents a change of inner mental context. To the extent that the behavior 'works' in new circumstances and environments, the experimental explorative phase is continued and the experience remains pleasurable. However at this stage carrying out those activities still requires conscious processing—the task has to be consciously concentrated on. The core experience of positive engagement is likely to be that of celebrating success, relating to others engaged and co-creating

PHASE 3: Unconscious Competence

The newfound capacities soon become unconscious habits—they can then be performed whilst conscious attention is engaged with other matters. The brain will quickly associate the acquired habit to the appropriate circumstances. The inner mental context has changed and allows the new competence to become unconscious, or habituated. This enables the conscious mind to be freed up to deal with other, more relevant issues. This phase is called Unconscious Competence and is often the foundation of what is regard as 'expertise'. This unconscious habit only remains a resource if there is no significant further change in the circumstances or in inner mental context. If the circumstances or inner mental context does change the resourceful habit acquired may now again become an unresourceful habit—it is no longer suited to the new circumstances or environment. This will then be the start of the next learning phase, which is unconscious incompetence. Meanwhile the experience of easy competence will be characterized by enhanced sense of presence, heightened perception and access to intuition. Behaviors would be characterized by a capacity to stay in process, to flex, flow and adapt

PHASE 4: Unconscious Incompetence

Here firstly the inner mental context might have changed and the old habits are still being performed. However, whilst they worked in the old inner mental context, they do not work in this new inner mental context. For example some situation has changed in the environment that has not been detected, or consciously acknowledged, and the response is still carried out in the old way. At the business level the culture (inner collective mental context) of the old company is carried into the new company. At the primary level instincts can be seen as inherited habits that normally work in the environment. Negative feedback might be experienced but the subject may be unable at this stage to associate the negative feedback to own responses and behaviors. It is then natural to project the cause of the negative experience into the environment. The problem is 'out there'. In the workplace a new manager of a division might have been tasked with different outcomes for the division he/she is to lead. The employees of the division will continue with their old behaviors, until now unconscious competence, and will want to project their discomfort at the adjustments needed to meet the new objectives on to the manager. He might even be seen as being incompetent. They will be unaware that in terms of the new objectives they are now unconsciously incompetent. The experience of unconscious incompetence is likely to be characterized by experience ranging from arrogance to anger to despair to a sense of helplessness and the accompanying fear and an experience of victim hood. Behaviors would tend to be characterized by denial, resistance, projection of the problem to the context and a further tendency to specific blame. Attitude or orientation would be characterized by a tendency to focus on the future or the past, a need to reestablish control, striving for fixed objective and even a tendency to superstition, and ultimate self-pity

Context

There are an additional four elements, related to boundary conditions in complexity science (related to attractors), to consider in coaching followers to new behaviors. These are always relevant to specific operational arenas. These arenas would in turn normally be categorized under the general term of 'context'. In order to facilitate the subtlety of leadership required, this notion is unpacked further.

- **Specific environments**: This is the physical space in which activities take place—activities are ultimately governed by the space in which the participants find themselves. (All behavior can be seen as being geared to adaptation).
- **Inner mental contexts**: These are the assumptions, agreements, habits, culture, defined outcomes for activities that take place - i.e. the frames in peoples minds, that allows attention to be paid to one thing over another.
- **Situations/circumstances**: These are the specifics of events/ dynamics/ actions prevailing at any moment in the operational space responsible for sparking the response. (It is the inner mental context that determines the significance of the circumstances and hence the requirement for response.
- **Ecology:** This term is generally ascribed to nature but in the human experience it refers to responses in relation to the interaction of all the above dynamics—whether these responses work sustainably, productively, harmoniously or otherwise—ecology itself can thus be seen as a sort of holistic inner mental context.

The PET—Practical Application

The PET is applied in most of the change work facilitated by Transformation Strategies Ltd. It is currently being applied at the executive leadership level in a SA-based multinational construction group that is intended to promote its transformation to the responsible corporate citizenship described. The chief executive, Bruce[6], describes the intention with the following phrases:

---

6. Brian Bruce, 'Robust, Murray & Roberts Group Magazine, Johannesburg March 2010

...developing new styles of leadership for unlocking rather than exploiting the developmental potential of the organization.

...accepting responsibilities not for personal benefit but for its value to ourselves and to the organization...our own personal enhancement naturally followed without being the primary motivation.

...the overriding purpose...is to serve the developmental needs of societies and nations.

...the world abounds with opportunity which may not be accessible in the conventional manner, so there is no point seeking it according to the old paradigm.

This intention, it is concluded, characterizes the attitudinal shift to Holistic Business Science.

The PET is also being applied in Johannesburg, London and Stockholm to facilitate culture change in a multinational gaming corporation focused on a transformed management style to promote personal engagement, development, operational efficiency, pro-activity and innovation.

The PET thus enables leaders in their function of capacity/skills building to engage with and experiment with it, apply the successes, re-adjust the failures, and integrate the entire process into their management repertoire.

Conclusion

Holistic awareness, it has been argued, is the subtle quality for proactive and innovative leadership. From the perspective of complexity science where organizations are seen as complex adaptive systems it requires taking the subjective experience of the whole organization into account and incorporating this insight into the various phases of the learning process. As such holistic awareness that proactively embraces notions of complexity and emergence is essential for the practice of Holistic Business Science.

References

Barrett, R. (2006). *Building a Values-Based Organization*, ISBN 9780750679749.

Bateson, G. (1972). *Steps to an Ecology of Mind*, ISBN 9780226039053.

Bertalanffy, L. von (1971). *General System Theory: Foundations Development Applications*, Allen Lane (1968).

Beukes, P. (1989). *The Holistic Smuts: A Study in Personality*, ISBN 9780798125376.

Davies, P. and Gribbin, J. (1992). *The Matter Myth: Beyond Chaos and Complexity*, ISBN 9780671728410.

Donaldson, T. (2004). "Moral minimums for multinationals," in S. Collins-Chobanian (ed.), *Ethical Challenges to Business as Usual*, ISBN 9780130487636, pp. 129-141.

Enderle, G. (ed.) (1999). *International Business Ethics: Challenges and Approaches*, ISBN 9780268012137.

Fitzgerald, L.A. (1996). *Organizations and Other Things Fractal: A Primer on Chaos for Agents of Change*, The Consultancy Inc., Denver.

Franses, P. (2010). "Editorial," *Holistic Science Journal*, ISSN 2044-4370, 1(2).

Institute for International Business (2010). "Davos-Trade Ministers Downbeat on WTO Prospects", Reuters, http://news.alibaba.com/article/detail/markets/100242062-1-update-1-davos-trade-ministers-downbeat-wto.html.

Letiche (2005). "Comments on Jeurissen: Organization and moral complexity," in M. Korthals and R.J. Bogers (eds.), *Ethics for Life Scientists*, ISBN 9781402031793, pp. 21-25.

Lozano, J.M. (2001). *Ethics and Organizations: Understanding Business Ethics as a Learning Process*, ISBN 9781402003622.

Maslow, A.H. (1943). "A theory of human motivation," *Psychological Review*, ISSN 0033-295X, 50(4): 370-96.

Polanyi, M. (1997). *Science, Economics and Philosophy: Selected Papers of Michael Polanyi*, ISBN 9781560002789.

Prigogine, I. and Stengers I. (1985). *Order out of Chaos: Man's New Dialogue with Nature*, ISBN 9780006541158.

Smuts, J.C. (1926). *Holism and Evolution*, ISBN 9781428623286.

Solomon, R.C. (1993). *Ethics and Excellence: Cooperation and Integrity in Business*, ISBN 9780195064308.

Spinosa, C., Flores, F. and Dreyfus, H.L. (1997). *Disclosing New Worlds: Entrepreneurship, Democratic Action, and the Cultivation of Solidarity*, ISBN 9780262193818.

Stiglitz, J. (2010). "Nobel Laureate backs government's New Growth Path," http://www.info.gov.za/speech/DynamicAction?pageid=461&sid=15202&tid=25643.

Szent-Gyorgyi, A. (1977). "Drive in living matter to perfect itself," *Synthesis*, ISSN 1570-2693, 1(1): 14-26.

Verhane, P.H. (2002). "The very idea of a conceptual scheme," in T. Donaldson, P.H. Verhane and M. Cording, *Ethical Issues in Business*, ISBN 9780130923875.

Weinberg, I.R. and van Wyk, C.P. (1992). "An integrated neurological model of consciousness: The case for quantum determinism," www.claudiusvanwyk.com/articles/quantum-determinism-weinbergvan-wyk/.

World Economic Forum (2009). Annual Report, Davos, http://www3.weforum.org/en/events/AnnualMeeting2009/index.htm.

Claudius Peter van Wyk is a Master Practitioner of Neuro Linguistic Programming (NLP). He works in the business environment within the framework of Holistic Business Science in the arena of corporate transformation, value-management and leadership. He facilitates corporate culture change in Southern Africa and Europe. Claudius holds a M.Sc. degree in Organizational Behavior and a M.Phil. degree in Applied Ethics (Business specialization). Claudius has also practiced in the field of Psychoneuro-immunology (PNI)—working with mind to enhance healing. He has been featured in the press, on radio and television in South Africa and Zimbabwe on this subject. He has lectured to the Psychological Association of South Africa (PASA), selected Academics and the Wits Medical School, the Family Life Association of South Africa (FAMSA), the Zimbabwe Independent Medical Association (ZIMA), etc. He has a particular interest in the potential application of PNI to the corporate work place. Claudius was awarded a D.Sc. in Alternative Medicine (Open International University Complimentary Medicines) his thesis being 'A Holistic Revision of the Epistemological Framework of the Biomedical Model with Reference to the Role of Consciousness as Subjective Experience.' Claudius is a registered member of the Natural Healer's Association. With his active interest in sustainability he supported the scientific movement, 'Towards Gondwana Alive, Stemming the Sixth Extinction', in achieving a visible profile during the Johannesburg World Summit on Sustainability, 2002. He organized a conference on biosphere reserves with attendance from UNESCO, the IUCN, MAB, and other key players and with Princess Irene of the Netherlands as the keynote speaker. He facilitated UNESCO recognition for Gondwana Alive and participated in the official opening of the People's Earth Summit. The main thrust for the initiative was the early expression and sketching of a program of Eco-literate Capacity Building. Subsequently Claudius was given ten minutes on the platform with Deepak Chopra on this subject. Claudius has contributed articles to the Green-build Journal on the role of consciousness in sustainability. In respect of complexity science in November 2004 Claudius presented a paper in Rio de Janeiro to an International Workshop on Complexity and Emergence – Philosophical Implications. The paper was entitled: "Complexity-based Ethics and the Challenges of Morality in the Emerging South Africa". He was subsequently invited to present at Liverpool University and University of Havana. Claudius was a contributor to the book 'Quantum Leap' by Ian Weinberg, and coauthored with him the paper: 'An Integrative Neurological Model Of Consciousness: The Case for Quantum-determinism', by Ian R. Weinberg, M.B., Bch. F.C.S. (Neuro) and Claudius P. van Wyk. He also contributed a chapter on transformed economics to 'Diamonds in the Dust' edited by Fritz Holscher. Recently Claudius contributed an article on the contribution of Smuts to holistic science in the Journal of Holistic Science.

7. Assessing Capacity And Maturity For Change In Organizations: A Patterns-Based Tool Derived From Complexity And Archetypes

Stefanos Michiotis[1] *& Bruce Cronin*[2]
1 TETRAS Consultants, GRC
2 University of Greenwich Business School, ENG

In transition times leaders should be aware of the hidden and intangible assets of their organizations or communities. While linear—analytical assessment tools face major difficulties in meeting this challenge, we suggest that the use of archetypal models as knowledge systems can be of help. A new tool, based on the use of geometrical patterns and aiming to reveal and assess a system's capacity and maturity for change, is presented here.

Introduction

Our era is characterized by increasing complexity, interdependence, fuzziness and instability. Yet, although change is inherent in every living organization, is often perceived by stakeholders as 'going too fast' and beyond their consensus. Strategic goals are interpreted variously, according to their interests or needs and when this multiplicity of meaning cannot be bridged or synthesized centrifugal forces emerge and create polarization and resistance. The mainstream attitude of dealing with such polarities and contrasting worldviews through imposition of power leads almost unavoidably to intense conflict and then, very often, to policy failures. The latter magnify the existing intractable problems instead of resolving them. Indeed, in many cases the problem occurs because of the wrong way a certain difficulty is dealt with; it is the attempted 'solution' that creates the real problem (Watzlawick *et al.*, 1974; Tsoukas, 2005). Eventually, this results in a growing sense of powerlessness and anxiety among governments, organizations and individuals (Peat, 2008).

In many change initiatives at this point there is a crossroad for the leaders: a) to abort the initiative, 'downsizing' the need that created it until it returns more aggressive and uglier or b) to 'declare war on the organization' (Holder, 2003), to attempt to eliminate diversity and polyphony for 'emergency reasons' and to establish a new reductionist order. As Kahane (2004) indicates, 'tough problems' either get stuck or solved by force; but the 'solution' imposed cannot last for long. Meeting such challenges is not just a matter of intellect or authority but of talking, thinking and acting together (Senge *et al.*, 2004).Yet, there is a third option, which will be examined further below.

As an organization (or community) moves away from stability and order , it becomes crucial for its leaders and change agents to adopt a more inclusive picture about the starting pointThis depends on one's viewpoint and the transition path, which is almost never the same as before or elsewhere. This is especially so in turbulent times and in cases of higher order or large scale changes (Tsoukas & Papoulias, 2004; Pelagidis, 2005), when the dominant interpretive scheme of the organization is challenged and change involves a shift from one dominant archetype to another (Greenwood & Hinings, 1993). In such cases, people usually sense a threat to the system's autonomy and values and retreat to their deeper beliefs to seek certainty. This kind of reaction, commonly known as 'resistance to change', is natural and results from the impact another attractor, which already exists and is different from the desired or the emerging one (Goldstein, 1992).

However, the structure of this attractor is not always known to leaders or change agents. Being accustomed to the established way of doing things or empowered by their own vision, they generally neglect those aspects of the collective reality that do not fit with their assumptions or vision. Thus, they cannot make sense of the whole picture, especially its emerging elements; on the other hand, they cannot 'see' their own blind spots. Most of the times leaders aim to 'change the other's mind' (manipulation) and 'make them understand' (propaganda), in order to fall in line with the leadership decisions (conscription). But, fortunately, meaning imposition is impossible and synergy cannot be conscripted (Snowden, 2002). In fact, this 'preacher's (or missioner's) attitude' normally results in even greater resistance on behalf of the population.

The third option mentioned earlier has to do with the leaders' choice to dispute and disrupt the dominant patterns that may be of no use anymore, in order to allow the emergence of new ones. For that, leaders should estimate: a) whether the implicit qualities and collective priorities of the system are compatible with their vision and b) the level of readiness for, or resistance to, change within the system. This knowledge will help them choose and prioritize among contradictory ideas and plans on a safer basis. It will also enable them to decide which of the existing patterns should be strengthened, which should be reduced and which new patterns should be introduced. In this way, they can probably avoid some crucial, recurrent and sometimes irreversible mistakes that usually activate the system's 'shadow', which is its collective non-conscious and often neglected or rejected side. On the other hand, they might find and ignite the proper catalyst for the transformation process.

Within the second half of the twentieth century, System Dynamics and Systems Engineering aspired to understand the constructing components of a system; describe the relationships between its parts; and improve their functionality. However, what works with artificial and ordered systems cannot always work with living ones, for humans most of the time are complex and self-adapting, while in the next moment or in another context can be simplistic or chaotic

(Kurtz & Snowden, 2003). Moreover, whatever is tacit or subliminal (i.e., hopes, dreams, ideas, talents, emotions, tensions, jealousy, power struggles) is usually ambiguous too and at the same time very real and powerful (Mindell, 2000). However, the linear—analytic assessment tools have been proved quite poor to work with the non-measurable aspects of a system, because of their fundamental assumptions, among which one could mention the perception of a fragmented world (Dimitrov, 2005), the belief in the rational causality and choice (Snowden, 2002), the inability to handle ambiguity and paradox (Lane, 1998), and the loss of balance between what can be measured and what cannot (Senge *et al.*, 1994). Moreover, experts are often unable to assess the human factor in an 'objective and accurate' way, beyond power or conformism, and beyond their own 'lens and authority' because of insufficient attention to informal relations and the pattern-based character of human behavior(Michiotis, Cronin & Devletoglou, 2010).

The Contribution Of Archetypes

To explore and respond to the dynamics of collective behavior, it is useful to consider the role of archetypes. Archetypes are both properties of the collective unconscious and at the same time inherent patterns common to all humans that can be expressed in different ways, according to the given personality or context. Archetypes have an emerging and dynamic nature, expressed through evolving images over time. They possess both positive and negative aspects, light and darkness; they are able to unite opposites within themselves and thus, to incorporate contradictions and integrate ambiguity (Jung, 1940, 1968). This is something that linear models (simple or algorithm-based) fail to do. The official indicators for measurement and comparative evaluation of innovation could be a good example for this case. They calculate and sum tangible assets, such as R&D expenses, number of patents, PhDs, new end products etc, but they cannot catch the emergent character of innovation. They fail to estimate 'accurately' the dynamics of synergy or competition that evolve, beyond rational choice or cost-benefit criteria, among the parts involved.

Archetypes are only relatively isolated from one another; they often mix with each other in a network of relations (von Frantz, 1975) and strive for balance in a given situation or system (Pearson, 1998). When, in a given context or personality, dominant archetypes leave no space for others to manifest, the latter turn to a *shadow* mode, which corresponds to unconsciousness. Shadow includes whatever is too painful to be consciously recognized as one's own and therefore is rejected or discharged on others, as well as the undeveloped parts or skills (in potentia) that seek to manifest. However, most of the time, encountering with a shadow issue generates a negative reaction, as it 'reminds' us that this issue is still unsolved. In this way, archetypes can express and represent different facets of a collective life, such as the organizational culture, elements of a community's tradition, trends of a society or a market or a successful brand mixture, for example.

The significance of archetypes lies in the fact that human experience is structured on and around these preexisting principles (McDowell, 2000) which profoundly influence how we perceive and experience the world and underpin our behavior. Indeed, the more deeply one understands the archetypal elements present and their meaning, relation and influence to oneself, the freer can one be in dealing with them and use their power in one's own favor (Stevens, 1982; Pearson, 1998).

Several theorists have referred to archetypes as psychic or strange attractors, basins of attraction, and energy patterns of potential that can create unpredictability and raise entropy (Jacobi, 1974; Rossi, 1989; Van Eenwyck, 1997; Matthews, 2002). Such key characteristics have been also identified in the complex and chaotic systems. Adopting Jung's concept of archetypes, Matthews considers them as loosely defined rules of a game that vary according to circumstances of time and place; they operate as a ordering or organizing principles and probability rules in the landscape of management and social life, indicating feasible journeys of human behavior in a given context.

Being abstract, archetypes cannot be observed directly or understood in a rational descriptive manner but can only be experienced and recognized through their effects that are imprinted in diverse images and patterns. These are encountered mainly in narratives (Edinger, 1972; Jacobi, 1962; von Frantz, 1975). An archetypal image can have an abstract or geometrical form (square, circle, wheel, etc or their combinations in symbols) or possess a figure of real or fantastic creature, plant, natural element or planet (e.g., mother, father, child, hero, god, fair lady, dwarf, giant, lion, dragon, tree, bush, fire, sea, river, sun, moon etc). Each archetypal pattern has certain elements (the goal, the characters, the turning points or thresholds and the treasure earned through trials), which constitute an archetypal story, myth or metaphor. Each time such a story is told or thought by someone, it adds meaning to the exact data (facts) of a specific event. It is like a theatrical show that is performed in different places; each time the local actors dress up in local clothes and perform the same play known in advance to everyone (Campbell, 1989). Although the story seems particular, its evidence is general. Thus, allowing different interpretations and accepting deeply the individuals' right for free will and choice, archetypes enable a 'holistic' perspective and facilitate the expression of a system's complexity.

As we have indicated elsewhere (Michiotis *et al.*, 2010), the term 'archetype' has been used in a different meaning within the organizational context. Senge (1990) introduced *system archetypes* as generic guiding structures and resulting behavior patterns and suggested ten exemplary cases. These were meant to help leaders recognize the cycles that systems go through and predict what is about to come. Yet, in practice, they have been used by practitioners and managers with the (hidden) anticipation of controlling events. The concept has not delivered the expected results, however, mainly due to the inability of system thinkers to deal with ambiguity and diversity of interpretations; system archetypes could deal most readily with explicit problems rather implicit ones. Recently, a number of models and tools employing archetypal figures have been developed, mainly in the areas of counseling, marketing and organizational behavior (Pearson, 2003; Rooke & Torbert, 2005; Neville & Dalmau, 2006). They try to provide a comprehensive view of the overall culture of the brands and organizations and to facilitate transitions. While Senge was interested in procedural and situational archetypes, these models focus mainly on figures that express the different spirits that are mostly encountered in organizations. For that, they have incorporated many years of experience in their structural components and some of them have used archetypal and mythological assets. However, the preset classification of the qualities of these 'archetypes' can turn to be non-contextual and in need of specially trained and well experienced facilitators, in order for the qualities of the figures to be properly clarified.

On the other hand, Snowden (2001) insists that organizational archetypes should be contextual and not universal (non-Jungian). He emphasizes that the archetypes he deals with are emergent properties of the discourse within an organization at all levels of it and refer not only to characters but also to situations, values and themes. They can be used as a means for understanding customers, tapping tacit knowledge, designing role-plays or lessons-learned programs, representing the existing culture or introducing different ones, as well as in cases of merger and acquisition. The main limitations of Snowden's method for archetype extraction is that it can get stuck in tedious rounds of process and that it is difficult to relate the findings of one case to another, as they are purely contextual; especially when comparing the findings of different groups or settings is meaningful (e.g., within the same system).

We suggest in this paper that a number of the above mentioned limitations can be addressed by the use of archetypal models. An archetypal model can be either a typology for the structure and content of a non-linear system or an attempt to model the dynamics of its behavior. In the first case, it uncovers a system's basic elements and the relationships between them. In a social or organizational context this representation could take the form of key players and the oppositional or collaborative forces among them. Some of the most widely known examples of such models are the *Four Elements* and the *Olympian gods*. The former have been used in Greek philosophy, Hinduism, Buddhism, Tibetan and Japanese tradition as a *structure* to represent the primal energy in nature, the life-sustaining force. Empedocles, Heraclitus, Pythagoras, Aristotle and others have considered the Four Elements as *principles* and '*roots*', while Jung viewed them as a major symbol of the archetype of quaternity (Jung, 1940, 1968). They can represent different aspects of reality, without reducing the latter to a simplistic description, as each of them can symbolize many different qualities. On the other hand, the widely known model of *Olympian Gods* consists of twelve ambiguous characters that represent fundamental characteristics recurrently encountered among people, inherent *qualities* of a system that are active at a given time or waiting to get activated. In the second case, an archetypal model can refer to the life stages and the initiation rituals at the thresholds between these. It is about the maturation process of individuals through society and resembles an inner road map, made by people who have already traveled in those areas (Campbell, 1989). Typical examples of this case are the Hero's Journey template (Campbell, 1949) and the labors of Hercules (Pitsouli, 2010), at the thresholds of which new perception and behavior patterns are shaped as the old role fades away or is shaken off and a new one emerges through pain and turbulence.

Geometry And Meaning

According to Jung (1968, 1940) and von Frantz (1974), geometrical schemes are considered as images of the deepest archetypes. It is also true that all the highly developed cultures of the world have used some geometric constructions as their symbols (e.g., the triangular pattern for trinity, the cycle or square mandala, the cross pointing towards the four directions, the interlacing triangles, the sacred hoop, the world tree, the snake that swallows its own tail etc). Indeed, across the ages, philosophers and scientists approached the powerful relationship between geometry and meaning. In his work *Timaeus*), Plato presented his cosmology model based on five solids related to the four natural elements (fire, water, air and earth) plus quintessence. His concept was adapted and further researched by Johannes Kepler (Caspar, 1994). Since then, Buckminster Fuller through his work *Synergetics* (1975) and Arthur Young in his book *Geometry of Meaning* (1976) have, among others, contributed to this relationship. Recently, the geometry of thinking and meaning has been introduced into the organizational and business context. The geometrical metaphor has been extensively used in the articulation of identity and strategy; simple geometric forms seem to help organizational leaders and strategists structure their thinking and planning (Judge, 2009; Keidel, 1994; 2010).

When patterns are expressed through simple schemes or solids, making sense of complex behaviors and (archetypal) dynamics within a non-linear system is easier. This easiness enables a more participatory sense-making, corresponding more to the common sense, even sometimes at the expense of accuracy. For example, a threefold concept or a three-fold of choices, can be represented by the nodes of a triangle, while the relations among the alternatives or the components can be on the sides. Likewise, a four-fold concept or model can be represented by a square or a four-domain scheme. It should be particularly noted that the latest version of the *SenseMaker*, which is a pattern-imprinting tool, follows a three-fold (Cognitive Edge, 2009), while the Cynefin model follows a four-fold one (Kurtz & Snowden, 2003). In its third version of SenseMaker tools, the conventional dipoles of choice (Cognitive Edge, 2006) have been substituted by triangles, within which one's opinion, estimation or viewpoint can be marked; the location is meaningful. Therefore, our research has been led towards relating patterns, meaning and geometry, for the latter can enable conceptualization, visualize emergent properties and explore relationships.

Introducing A New Tool And Process

The tool described here has been developed as part of a research project by the University of Greenwich Business School in collaboration with a Greek consulting company. It aims to help organizations and communities make sense of the qualities and skills yet needed for the challenge(s) they encounter, as well as the degree of difficulty for obtaining them. More specifically, it aims to enable an organization or a community to:

Reveal its hidden potential and the patterns though which this can be conveyed and thus understand how the organization collectively reacts when facing significant challenges.

Assess the similarities and differentiations among the perception patterns of different control groups and thus assess their capacity to set and accomplish common goals.

Estimate the level of difficulty for the system to bridge the current status with the desired one and indicate possible traps and breakdowns.

Extract the organizational archetypes in the form of contextual figures, situations etc and indicate their qualities, along with their 'shadow' aspects.

The tool is based on the theory of archetypes and uses self-organization techniques and simple geometrical schemes, in order to reveal and assess the collective capacity and maturity for change. It has been built on the assumption that complex situations and problems are not assessable by an outside expert. These situations exist because people are complex and the dynamics (hidden or not) are difficult to grasp and understand before any change can take place. Moreover, the experts' rationalistic logic is not always the same as common sense, which is much stronger and (occasionally) much wiser and has to do with shared patterns in a given context. With this tool, all one needs to do is participate and observe. The patterns that emerge make it clear and easy for every interested party involved to understand these dynamics and their meaning for the organization. Additionally, since the process and the end results of the assessment are based on archetypal images, words, phrases etc, it is easier for the participants to understand, relate and 'connect' to the result and eventually derive meaning from it. The tool imprints not only the active but the shadow archetypes as well, revealing the 'blind spots' of the organization, which may be active or need to be activated in the system. Thus, people can easily see the 'whole picture' (where they stand in relation to others, how they affect results, how they affect the whole organizational scheme) and create the consensus necessary for change.

Making sense of complexity becomes then both a personal and collective thing, conceptualization being the core issue that will lead not only to personal awareness but to a collective one from within the system. It helps them see how they engage with an organizational scheme; or not (Michiotis *et al.*, 2010).

The term *capacity* is used here to include the sum of the qualities and inclinations that are inherent in an entity (individual, group, organization or community), along with the skills and knowledge that have been obtained throughout this entity's evolution. So, capacity is actually the sum of the intangible assets that are used as a collective gear by an organization or community to face inner or outer stimuli. Sometimes the system is aware of its patterns and the impact of the stimulus. In other cases the collective awareness is very poor (if any) and, as a result, the symptoms are neglected or the responsibility is transferred to others. Indeed, a system is usually aware only of some of its collective qualities, skills and knowledge, which tries to control consciously; for example to include them in operational procedures, training programs, vision-mission statements etc. The rest of the available assets, of which the system is not consciously aware, will become activated and emerge only when and if will be really needed. What really occurs in every organization is that its conscious knowledge about itself and environment exists only to a certain degree. Challenges and intangible assets are interrelated and interacting. Actually, it is the challenge that activates the system's capacity, which until then exists in-potentia. Some challenges and needs suit more to certain knowledge, skills and qualities available by the system; others do not. While the former can be faced easier by the system, without spending valuable resources, it is the latter that can become more fruitful, leading to valuable experience and enabling the actualization of hidden potential.

Mapping the system's intangible assets shows in which way the forces and needs that operate on an archetypal level resonate within the system. This is considered to be of great importance, as the role and impact of the tangibles are more or less denoted in most cases, while the intangibles, being neglected, interpreted differently or even rejected, remain usually in the 'shadow'. Together, they can depict a larger picture of the current reality and the desired future of the organization. This can provide more meaning and motivation to the organization as a whole. In other words, the system's collective capacity reveals its ability to make sense of itself and its environment and to adapt to it. However, this map should be contextually expressed, meaning in real personas, real problems and mainly in a language easily understood by everyone in the system.

On the other hand, by the term *maturity* is meant here whatever permits individuals, groups and societies to recognize their problems before trying to confront them. It is *not* the capability to operate according preset keys, rules and procedures, like traditional assessment tools claim (Curtis, Hefley & Miller, 2009). It is rather the ability and knowledge to discover the 'keys' needed for the problems and the will to use them. It is the ability for self-learning and at the same time the commitment to make the steps required towards this. It informs about the 'lessons learned' so far (by the organization or the community as a whole), as well as about the ones yet to be learned. Actually, maturity is the way in which a person or an organization behaves when encountering a potential lesson. It has to do with the space that one makes (or leaves not) for the proper (but 'unusual') skill to be used when dealing with a challenge, which has not faced before or remains unsurpassed so far. This tool defines four levels against which the collective maturity is assessed.

The 'lesson' is a pattern of meaning that emerges through experience, reflection and conversation. For those who think of evolution in a mainly developmental way, in each stage there are specific understandings to become aware of, missions to be accomplished or lessons to be learned, *before* moving to the next step; indeed, this is the common sense and belief for the 'true' meaning or the word maturity. However, in non-linear systems, lessons are available anytime and in any occasion, according to the existing level of self-awareness and maturity.

When such factors are sensitized enough while struggling with a challenge, they could bring a person or a system to a certain energy level that corresponds to a threshold of linearity. Beyond that threshold, a bifurcation can occur in the existing mental patterns and a new understanding (different from the initially aimed) is possible to emerge. Actually, the more experienced and ready one is for such 'discovery' opportunities, the more fruitful the process will be. Some potential categories of the lessons can refer to: a) operational skills or qualities, which can be related to targets, relationships, knowledge etc, b) developmental stages in programs of awareness building, personal development or transformation processes, c) challenges or traps that corresponding to the 'shadow' of archetypes (like the ones used in some of the models mentioned earlier) or d) specific capabilities required by particular operational models.

With regards to the scope of this paper, the discussion of the tool will be focused mainly on its structure and the process of the capacity assessment.

Structure Of The Tool

It should be mentioned from the beginning that the tool-and-process that has been developed can be adapted to various archetypal models with different structural elements, different relations between them and of course different content; this tool-and-process is rather a meta-methodology. However, for testing it, he had to employ a specific model, which is the Measure Formula model, introduced by Arthur Young (1976), a mathematician, engineer and inventor of the Bell helicopter. It is a model that consists of *twelve elements*, which can be classified in three sets of four (called *tetrads*) or in four sets of three (called *triads*) (Figures 1 and 2). Each tetrad informs and *describes* all possible modes of structure or operation, such as four psychological operations (thinking, feeling,

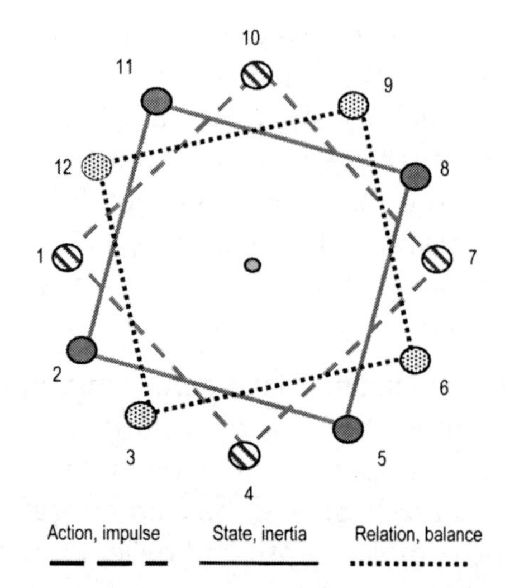

Figure 1 *Three Sets Of Four (Tetrads)*

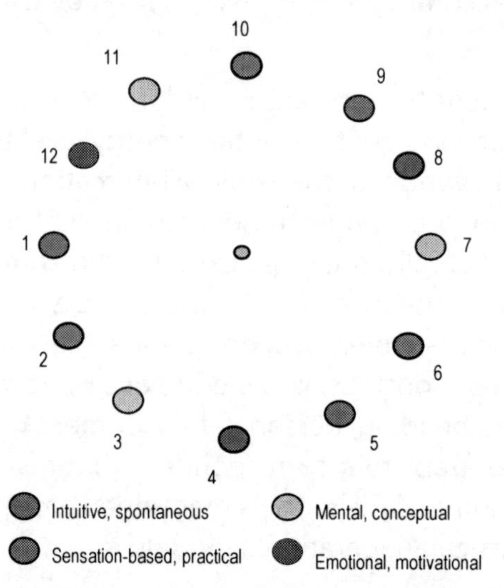

Figure 2 *Four Sets Of Three (Triads)*

intuition, sensation), four personality types or temperaments (sanguine, choleric phlegmatic, melancholic) and four natural elements (air, water, fire, earth). It is used to analyze concepts and forms in their fundamental components. On the other hand, the triad indicates a pathway of *transformation* (i.e., birth–maturity–death; creation–sustaining-destroy; impulse-inertia–balance; stimulus–reaction–result; or even the transcendence of a polarity or a conflict (i.e., black–white-rainbow).

According to Young, if combined, these two ways of classification can describe all possible situations that correspond to the ways that the fundamental components can express or evolve. For example, an emotional activity or situation can be expressed in an impulsive, controlled or balanced way, while an impulsive and intuition-based (risky) activity can lead to a practical status. Although this concept seems deterministic, it is not. It is the people, not the experts, who choose among existing possibilities and through their choices reveal the existing patterns. Based on the above, the structure of the tool consists of:

1. Twelve elements that constitute the nodes of a regular dodecagon inscribed in a cycle and, at the same time, the content of a 3 × 4 matrix. These elements represent:

   The intangible assets of the system's potential (qualities, skills, intentions, holdbacks etc.), corresponding to the potential ways a system responds when challenged and

   The acquired issues of the collective maturity that correspond to the lessons learned or yet to be learned.

2. A number of stimuli representing fundamental (archetypal) needs, forces or challenges that are encountered in the organizational context (e.g., identity, creativity, learning, risk, success, communication, stability, expansion, competition, etc.). These needs etc endure in time (e.g., the need to survive or to learn or to feel free etc is universal) but in parallel are expressed in various ways, following a typical example within a given cultural context (e.g., regarding survival, the 'eye for an eye' concept and practice is alternative to the trickster's, which is followed in different settings). These forces, needs and challenges stimulate the energy fields of the system, the collective personality of which is shaped through their confrontation.
3. A databank of archetypal images, phrases, situations, patterns etc that will be used during the process.

Capacity Assessment Process

The process assumes that when people are attracted to a certain *archetypal* image, phrase, pattern, story or situation among others, they indirectly indicate an influential archetype (dominant or shadow) in their context and informs the collective patterns of behavior within the group they belong to. This occurs because values are attached to *symbolic* images that attract or repel our attention through chaotic dynamics. The meaning of these symbolic images vastly transcends their content; the meaning of a symbol is synonymous to its capacity to generate a dynamic relationship between the one who interprets and that which is discovered (Van Eenwyk, 1997). As Pearson (2003) indicates, they act like meaning magnets for the psyche and provide a bridge between the deepest motivations and felt experience that fulfills (or promises to fulfills) basic human needs. Due to this fact, they are widely used in marketing in order to place and promote a firm among others, either by expressing more tangible stereotypes and caricatures or by referring to more abstract meanings or needs. For example, with regard to describe a certain situation, making a choice among a crown (that usually symbolizes power), an owl (that is generally related to knowledge and wisdom) and a fist on the air (that indicates struggle or imposition) informs of one's perception of that reality. Therefore, regarding the process described, such *archetypal* images, if properly selected by the facilitator(s), can be used as a means for the participants to depict, beyond rational descriptions, sides of their current or desired reality and bring up some unconscious facets, needs, intentions or feelings generated by it. Moreover, by spontaneously expressing an archetypal image or phrase in contextual terms (of their own reality), people provide the elements for a meaningful language, through which messages can be communicated effectively within the specific context.

Through this combination of archetypes with emergent techniques, different interpretations of reality can be expressed in a spontaneous and unaffected way and the main factors of system's complexity can be imprinted. As choices are made and added (over time or location) in this guiding map, one can see a) how the system collectively perceives reality and therefore what is capable for and b) which action is feasible or possible and which is out of the beaten track or difficult. At this point geometrical schemes and patterns are used to organize the emergent properties, visualize their relationships and enable the conceptualization of the findings of the process.

The process for the assessment of the organizational capacity is carried out in steps (Figure 3) with control groups derived from different staff subgroups, management, stakeholders etc.

At first (Step 1), each participant chooses one or two images that in his/her opinion resemble to the current organizational reality and a desired future. Then, each one expresses the main characteristics of these states in terms of dominant values or skills or a meaningful phrase; he/she also indicates the main obstacle for moving from current reality to a desired one. All these get written close to the relevant images or the 'present-future' lines. A quantitative analysis of these properties (Step 2.1), which is based on the aforementioned 3-fold, 4-fold and 12-fold categorizations, provides some initial indicators that refer: a) to the influence (strength) of each archetypal element or their sets (tetrads and triads) in the given context and b) to the dominant qualities (the most frequently indicated). Comparing these indicators with the ones derived from other groups(Step 2.2) aspects of hidden potential and blind spots can be revealed.

Then (Step 3), the participants proceed on a collective basis to discover some fundamental relationships that exist among the emerged qualities. They do this by relating any three of the emerged properties they assume that are encountered in combination within their context. The outcome of this step informs of which qualities can really work together, even if they don't necessarily 'fit' from a first view or even if this is not consciously accepted or stated.

These fundamental relationships of qualities serve as the core of the organizational figures (contextual protagonists) that will be shaped next (Step 4.1); at least as viewed by the specific group, which, if coherent, delivers in this way its

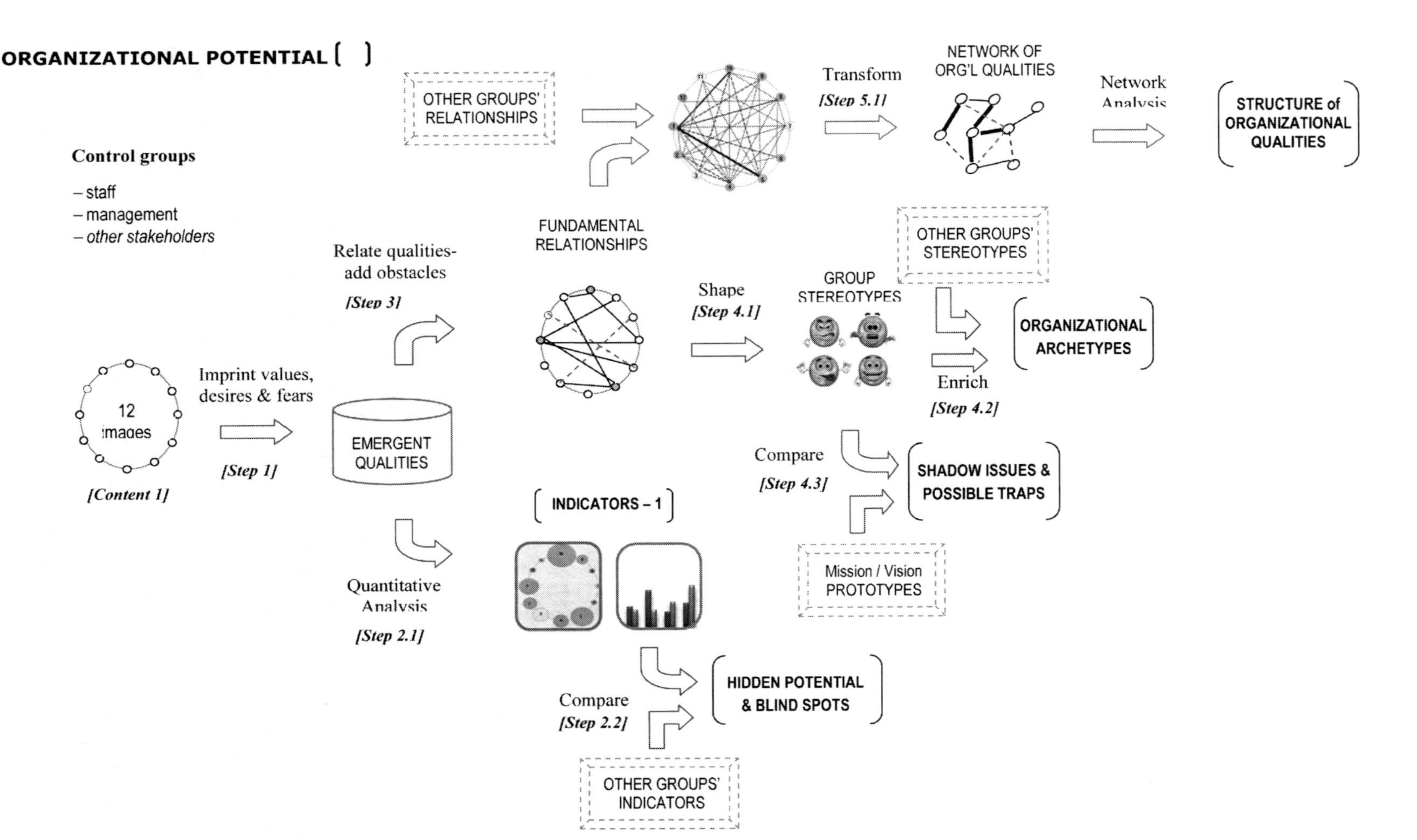

Figure 3 *Overview Of The Process For The Assessment Of The Organizational Capacity*

stereotypes. These figures can be then compared either to other groups' stereotypes (Step 4.2) or to the mission-vision prototypes (Step 4.3), in order to deliver either the organizational archetypes or some shadow issues and possible traps regarding corporate plans and initiatives.

Finally, by summing (overlaying) the outputs of Step 3 from different control groups, a total graph of the fundamental relationships within the specific context can be created. By transforming this pattern of lines inscribed in the dodecagon to a pattern used by the social network analysis (Step 5.1), we could have a "network of qualities" that could inform in which way (pathway) a certain *quality*, *knowledge* or *skill* can be obtained or cultivated within this particular context. Through iterative runs of the process for different stimuli and by using network analysis theory, we could be led to the structure of the existing (or the emerging) organizational attractor and from there to a safer answer regarding the question which of the existing patterns should be strengthened, which should be reduced and which new should be introduced.

After each step and being aided by some triggering questions, the participants discuss the findings and note whatever they consider unexpected or worthy to mention. Furthermore, comparing their data against the ones from other groups, they may suggest some initial points that could relate to organization's blind spots. Sometimes, a preparatory narrative gathering can enable the contextualization of sets of the archetypal phrases, images, stages, situations, etc, which are derived from the Databank and correspond to the twelve elements of the tool.

The process of the maturity assessment is more or less similar to the one previously described; it focus on: a) irregularities of the distribution of 'lessons learned' that indicate absence of experience or obstacles and b) situations that are dealt by using specific qualities that have been previously identified as dominant.

This tool has been tested as a prototype in different settings. The initial estimation from these pilot tests is that it is quite functional as it enables bias elimination, makes the process dynamic and more intriguing, encourages participation and personal expression, facilitates understanding, and seems to strengthen motivation and commitment. With appropriate clinical guidance, it could be potentially used in a counseling setting as a process facilitation tool, since all its components are based on psychological archetypal principles of personal growth, awareness and limitations, energy localization, shadow archetypes and their expression, etc. Specifically, it may aid individuals in the realization of dominant or shadow, positive or negative, current or past expression of archetypal figures in their life and motives / patterns of behavior. Another significant advantage of its concept is that it can be easily adapted to models fundamentally different, which follow a different structure than the 12-fold. These models can be simply represented by a different polygon, as long the analogies are kept in the process and the content of the databank. As for the challenges, their number is open to the facilitator's choice or the context's requirements (Michiotis *et al.*, 2010).

However, the facilitator requires some skills in employing the process:

- Beginner's mind: avoid 'judgment as usual', notice the bare bones and stay open,
- Moments of truth: challenge conventionality, embrace paradox and exploit tension
- Critical relations: enable what can be spread fast and easily and lead to a critical mass
- Trust the process: let the participants lead, don't worry about the outcome, be present
- Let go of your "armor", in order to evolve it

Epilogue

It seems that the intangible assets of an organization can be approached, mapped and assessed in a more comprehensive and participatory way by using pattern-based tools. For this, archetypes can be of help, especially when combined with participatory and self-organized techniques, as well as with geometry. Together they can enable the emergence, contextual expression and visualization of contradictions and complex aspects of individual behavior and organizational culture. The latter, being collectively visible, can be more easily comprehended and therefore be made more manageable.

References

Campbell, J. (1949). *The Hero with a Thousand Faces*, ISBN 9780586085714 (1993).

Campbell, J. with Moyers, B. (1989). *The Power of Myth*, ISBN 9780385247740.

Caspar, M. (1993). *Kepler*, ISBN 9780486676050.

Cognitive Edge (2006). "Pre-hypothesis research," (working paper, not available anymore online).

Cognitive Edge (2009). http://www.cognitive-edge.com/video-sensemaker-suite.php.

Curtis, B., Hefley, B. and Miller, S. (2009). *The People CMM: A Framework for Human Capital Management*, ISBN 9780321553904.

Dimitrov, V. (2005). *A New Kind of Social Science: Study of Self-Organization in Human Dynamics*, ISBN 9781411601420.

Edinger, E. (1972). *Ego and Archetype*, ISBN 9780877735762 (1991).

Fuller, B. with Applewhite, E. (1975). *Synergetics I: Explorations in the Geometry of Thinking*, ISBN 9780020653202 (1982).

Goldstein, J. (1994). *The Unshackled Organization: Facing the Challenge of Unpredictability through Spontaneous Reorganization*, ISBN 9781563270482.

Greenwood, R. and Hinings, C (1993). "Understanding strategic change: The contribution of archetypes," *Academy of Management Journal*, ISSN 0001-4273, 36(5): 1052-1081

Holder, B. (2003). "Why change efforts fail," The CEO Refresher Archives on Leading Change, December, http://www.refresher.com/!holder7.html.

Jacobi, J. (1962). *The Psychology of C.G.Jung*, ISBN 9780300016741 (1977).

Jacobi, J. (1974). *Complex, Archetype and Symbol*, ISBN 9780691017747 (1992).

Judge, A. (2009). "Geometry, topology and dynamics of identity," http://www.laetusinpraesens.org/musings/idengeom.php.

Jung, C.G. (1940). *The Integration of Personality*, London, UK: Routledge & Kegan Paul.

Jung, C.G. (1953-78). *The Collected Works of C.G. Jung*, London, UK: Routledge & Kegan Paul.

Jung, C.G. (1968). *The Archetypes and the Collective Unconscious*, ISBN 9780691018331 (1992)

Kahane, A. (2004). Solving Tough Problems: An Open Way of Talking, Listening, and Creating New Realities, ISBN 9781576754641 (2007).

Keidel, R. (1994). "Rethinking organizational design," *Academy of Management Executive*, ISSN 0896-3789, 8(4): 12-30.

Keidel, R. (2010). *The Geometry of Strategy: Concepts for Strategic Management*, ISBN 9780415999243.

Kurtz, C. and Snowden, D. (2003). "The new dynamics of strategy: sense-making in a complex and complicated world," *IBM Systems Journal*, ISSN 0018-8670, 43(2): 462-483.

Lane, D.C. (1998). "Can we have confidence in generic structures?" *Journal of the Operational Research Society*, ISSN 0160-5682, 49: 936-947.

Matthews R. (2002). "Competition archetype and creative imagination," *Journal of Organizational Change Management*, ISSN 0953-4814, 15(5): 461-476.

McDowell, M. (2000). "The landscape of possibility: A dynamic systems perspective on archetype and change," http://cogprints.org/1084/1/Jap_9.html.

Michiotis, S., Cronin, B. and Devletoglou, H. (2010). "Revealing hidden issues and assessing intangible assets in organizations and communities," *International Journal of Decision Sciences, Risk and Management*, ISSN 1753-7169, 2(3/4): 308-326.

Mindell, A. (2000). *The Leader as Martial Artist: Techniques and Strategies for Revealing Conflict and Creating Community*, ISBN 9781887078658.

Neville, B. and Dalmau, T. (2006). "Managing the polytheistic organization," *International Journal of Knowledge, Culture and Change Management*, ISSN 1447-9524, 5(8): 45-60.

Pearson, C. (1998). *The Hero Within: Six Archetypes We Live By*, ISBN 9780062515551.

Pearson, C. (2003). *Understanding Archetypes in Your Organization: An Introduction to the OTCI Basic Report*, ISBN 9780935652734.

Peat, D. (2008). *Gentle Action: Bringing Creative Change to a Turbulent World*, ISBN 9788895604039.

Pelagidis (2005). *The Tangle of Reformations in Greece* (in Greek) Papazissis editions 2005 Athens

Pitsouli, J. (2010). *Hercules He Hero within Us* (in Greek), Esoptron, Athens.

Rooke, D. and Torbert, W. (2005). "Seven transformations of leadership," *Harvard Business Review*, ISSN 0017-8012, April 2005

Rossi, E. (1989). "Archetypes as strange attractors," *Psychological Perspectives*, ISSN 0033-2925, 20: 4-14.

Senge, P. (1990). *The Fifth Discipline*, ISBN 9781905211203.

Senge, P., Kleiner, A., Roberts, C., Ross, R.B. and Smith B.J. (1994). *The Fifth Discipline Fieldbook*, ISBN 9781857880601.

Senge, P., Scharmer, C.O., Jaworski, J. and Flowers, B.S. (2004). *Presence: An Exploration of Profound Change in People, Organizations, and Society*, ISBN 9780974239019.

Snowden, D. (2001). "Simple but not simplistic: the art and science of story," *Strategic Communication Management*, ISSN 1363-9064, April.

Snowden, D. (2002), "Complex acts of knowing: Paradox and descriptive self-awareness," *Journal of Knowledge Management*, ISSN 1367-3270, 6(2): 100-111.

Stanford Encyclopedia of Philosophy (2005/2009), http://plato.stanford.edu/entries/plato-timaeus/.

Stevens, A. (1982). *Archetype: A Natural History of the Self*, ISBN 9780415052207 (1990).

Suchman, A. (2006). "A new theoretical foundation for relationship-centered care: Complex responsive processes of relating," *Journal of General Internal Medicine*, ISSN 0884-8734, 21, S40-S44.

Tsoukas, H. (2005). "The reformation as psychotherapy," (in Greek), in Pelagidis, *The Tangle of Reformations in Greece*, Papazissis ed., Athens.

Tsoukas, H. and Papoulias, D.B. (2005). "Managing third-order change: The case of the public power corporation in Greece," *Long Range Planning*, ISSN 0024-6301, 38: 79-95.

Van Eenwyck, J.R. (1997). *Archetypes and Strange Attractors: The Chaotic World of Symbols*, ISBN 9780919123762.

Von Franz, M.L. (1974). *Number and Time: Reflections Leading Toward a Unification of Depth Psychology and Physics*, ISBN 9780810105324 (1986).

Von Franz, M.L. (1975). *C.G. Jung: His Myth in Our Time*, ISBN 9780919123786 (1998).

Watzlawick, P., Weakland, J.H., Fisch, R. and Erikson, M.H. (1974). *Change: Principles of Problem Formation and Problem Resolution*, ISBN 9780393011043.

Young, A. (1976). *The Geometry of Meaning*, ISBN 9780960985050 (1984).

Stefanos Michiotis is a PhD researcher in Organizational Change (University of Greenwich Business School) and holds an MSc in Management Engineering (NTUA, Athens). He is founder and scientific director of TETRAS Consultants, where he works as adult trainer and organizational consultant. Corporate clients in Greece have included the National Centre for Public Administration, the Greek Developmental Agency for Local Government, the National Technical University of Athens and the University of Patras, the National Foundation for Youth, the Sivitanidios Foundation for Technical Education, various municipal / regional organizations and training centers, as well as a number of small and medium sized business firms. He is an accredited practitioner for the Cognitive Edge network and a trained member of the Presencing Institute community.

Bruce Cronin MA, MSc, PhD (Auck) specializes in the role of business networks in strategy and innovation within and across organizational and national boundaries. He convenes the University's Business Network Research group, works with the Knowledge Management Group and is an accredited consultant for the UK Cynefin Centre for Organizational Complexity. Corporate clients have included Unilever, The Pensions Regulator, Pearson, McGraw-Hill, Oxford University Press, The Work Foundation, The Institute of Management Consultants, Universities UK and the London Knowledge Network. He is editor of Global Knowledge Review and serves on the international advisory boards of the *Journal of Knowledge, Culture and Change in Organizations*, the *European Knowledge Management Conference* and the *World Political Economy Society*. He is a member of the European Group on Organization Studies and the International Network for Social Network Research.

8. The Application Of Complexity Thinking To Leaders' Boundary Work

Alice E. MacGillivray
Fielding Graduate University, CAN

This chapter—for the *1st International Workshop on Complexity and Real World Applications*—describes the early development stages of tools and processes to help leaders with boundary work. These tools and processes are tentatively branded as The Edge Effect™ Kit for leaders. The chapter outlines shifting perceptions of complexity, leadership and boundaries for work in complex systems. It notes the liminal nature of this workshop and describes the author's research, tools and processes as presented during the workshop. The presentation included leadership concepts developed by the author, such as "ecosystem-informed boundary critique" and the related concepts of "edge" and "intellectual estuary."

Context of the Chapter

This chapter was written for the *1st International Workshop on Complexity and Real World Applications*. The workshop organizers are accustomed to dealing with conceptual and theoretical papers. For this event, they provided specific directions about applied content. One specification was clearly atypical for scholarly papers:

> *In both your poster and paper please include a section that explicitly addresses the productization of your tool or service... who would use it, what form would it take (manual, software, etc.), how would it be distributed, what is need to make that a reality, time-frame, etc.* (May 10, 2010 e-mail correspondence from workshop organizers).

The workshop and its products were clearly positioned at the potentially productive edge between scholarship and practice.

The Nature of Complexity Thinking

Many researchers have written about the need for new ways of thinking, support systems, and tools for work with complex systems. There are no guarantees that new approaches will be successful, and even in the face of serious, long-term intractable issues, many leaders are hesitant to give up processes ingrained in their education and intermittent career successes.

Many complexity scientists and related researchers are interested in how their work might benefit organizations: the 1st International Workshop on Complexity and Real World Applications is one illustration of this interest. For practitioners new to the field, Kurt Richardson and Paul Cilliers's (2001) categorization of three complexity research communities may be helpful.

1. The *reductionist* school is a quest for general principles and overarching theories. This research has value, particularly in specific disciplines, but may not be useful for practitioners. They cite "Horgan's seductive syllogism" (p.6), which claims there are simple rules and patterns underlying complicated phenomena.
2. *Soft complexity science* is positioned as metaphorical, based on an underlying belief that the natural world and social constructs are fundamentally different. This contrasts with some neo-reductionists' views that reality can be represented by our constructs.
3. *Complexity thinking* is grounded in the premise that organizations are complex systems, which highlights "the limited and provisional nature of all understanding" (p. 8) and the need for a seachange in our ways of thinking about management and leadership. For example, if management is about maintenance of order and stability (Kotter, 1996, 2001), it may be losing relevance in comparison with new forms of leadership.

Of course these categories are separated by socially constructed boundaries, which some practitioners choose to span.

The Nature of Leadership

After WWII, there was a burst of leadership research. At that time, there was little conversation about knowledge workers: manufacturing in the Industrial Era remained a compelling context for study. The war undoubtedly reinforced heroic stereotypes, the importance of chains of command, and the perception that leaders must have formal authority. Henry Mintzberg's famous study of what [male] Chief Executive Officers really did in their day-to-day work reflected the common assumption that information in an organization should flow upwards to the top of a pyramid, so "the leader" could set the vision, oversee alignment, and make the best-informed decisions, which would flow down through the organization.

Leadership paradigms and related research have broadened, with mixed reactions. In a recent webinar announcement, the International Leadership Association (2010) described the presenter's work as follows:

> *George Graen has worked for forty years in the leadership field and written over two-hundred research articles and books. Best known for LMX theory, George has recently developed two alternative protocols that describe real world leadership and has been presenting seminars and writing about his latest attempts to "tidy up an extremely messy field."*

As a scholar, I share some of Graen's discomfort; as a practitioner, I am constantly amazed by how messy leadership is. Is someone who is masterful at ensuring compliance a good leader? If a "leader" sits at the top of a hugely profitable company, is that evidence of effective leadership? What if that same company is so lean and brittle that it falls apart in a crisis? If a leader encourages reflection, systems thinking and adherence to values, and several good employees leave the company because of disconnects with the values behind shareholder-driven practices, is that evidence of [in]effective leadership? If a leader convenes a meeting, and half the people leave inspired while the other half leave thinking the leader has walked over the backs of others to build personal power, what does that say about leadership? Is a leader good if she does not know details of what is going on in her vibrant and innovative community? Are people who think there are correct answers to these questions...correct?

People who welcome elements of messiness in leadership studies may not understand rigorous research. Or they may work with different epistemological assumptions. I believe both explanations have merit. In this chapter, I suggest that new assumptions are driving emerging research about leadership; resultant leadership theories and approaches have been described with terms such as *Leaderful, Relational, Complex, Shared* and *Distributed*. Some of the contributing authors are Mary Uhl-Bien, Russ Marion, Joseph Raelin, Peter Gronn and Jim Hazy.

These newer approaches are useful for some contexts, yet there is a huge amount of inertia behind the assumptions of formal, heroic, individual, position-based, controlling, masculine leadership. The assumptions have been built into theories, texts, models, competency frameworks, training programs and—of course—language. At a recent international conference, I asked a question of a new-paradigm leadership scholar. She replied by explaining how she must often compromise with language in her published work, including terms such as *leader, follower* and *subordinate*, even if they don't fit well with her findings. These challenges are intensified by a low tolerance for ambiguity and complexity in many western cultures. Despite new research that uncertainty in a certain world can correlate with persuasiveness (Tormala, 2011) we expect our politicians—for example—to sound confident and clear, even to the point of black and white statements such as George W. Bush's declaration: "You're either with us or against us" about the war on terror. We may be seeing reinforcing loops: leaders must sound confident and in-charge; subordinates are comforted and feel less pressure to be informed leaders; leaders gain more power; subordinates lose capacity and capability to lead effectively; centralized control grows; and the inevitable dissension is punished by central power.

Nonetheless, there is potential for progress. Several popular authors have captured public imagination with insights about new leadership models related to systems and complexity. And fiction can be an interesting place to look for trends. The call for papers for this workshop began with a quote from Canadian novelist Alice Munroe: "The complexity of things—the things within things—just seems to be endless. I mean nothing is easy, nothing is simple." A paper written for the event by Reynolds quoted Canadian poet Leonard Cohen. Complexity permeates our thinking even in situations requiring command and control. In a new Canadian television program *Flashpoint*, the commander of a SWAT team stated this is a "who and where situation" to which the female SWAT team member responded "Bonus points for why?" (Aired June 25, 2010). Police departments might seem the last place to find complexity-grounded models of leadership, but Police Chief Todd Wuestewald from Broken Arrow Oklahoma created an environment in which a shared model with complex system characteristics has thrived for several years (MacGillivray, 2009).

Aspects of our work worlds are becoming much more complex. We need a kind of meta-leadership for complexity: shifting how we think about leadership in order to lead effectively in complex systems. This chapter outlines one scholar-practitioner's current efforts to help with that shift, by supporting leaders' boundary work.

The Nature of Boundaries

In systems literature, boundaries are described as social constructs, whereas in organizations they are often treated as very real. This begs questions about the nature of reality, and about implications of different views of reality. The nature of reality with respect to boundaries is explored by Kurt Richardson and Michael Lissack, where they describe a domain as "an apparently autonomous (critically organized) structure that differentiates itself from the whole" (2001: 36). From this, one could infer boundaries are real and important. But the authors emphasize the importance of time, as domains may appear spontaneously and disappear unpredictably. Even if practitioners acknowledge such fluctuations, the acceptance of boundaries as real can compromise decision-making, ethics, resilience or adaptability.

Authors including Werner Ulrich and Gerald Midgley have suggested the boundary concept is central to systems thinking. Richardson has regularly emphasized the importance of boundaries in complexity thinking. Given that leaders are becoming more aware of the increased complexities with which they work, and given that boundary concepts may be central to that work, one would expect boundary concepts and strategies to be well integrated into leadership tools and competencies. However, this is not yet the case.

Boundary Critique

One particular aspect of boundary work that has overarching implications is that of boundary critique. Ulrich and Midgley have developed somewhat different approaches to the concept of boundary critique. Ulrich (2000) deals with the reflection of boundary judgments of systems in continual juxtaposition with judgments of fact and value judgments. Midgley (2000) has built on the work of Hegel, Churchman, Ulrich and others to develop a normative theory of boundary critique to emphasize the importance of reflection in the reification and use of boundaries in any sort of intervention. In essence, he describes how groups often operate within narrow boundaries where their values and norms reflect that narrow scope. If others have broader boundary judgments, there will be a marginal area between the two. As shown in Figure 1, the two boundaries can be labeled as primary (the narrower of the two) and secondary (including the broader, marginal group).

Using this lens of boundary critique, the core group will pass value judgments about the marginalized area within the secondary boundary. Often these judgments are mixed and there is related tension. We see this at many scales with nation states (Israel and Egypt are in the news as this chapter is being prepared) sections of organizations (such as research and development and finance) fields or disciplines (such as scholarship and practice) or even family members. Midgley uses the terms sacred and profane to describe valued and devalued. Stabilization can occur through firmly valuing or de-valuing the perspectives of the marginal group.

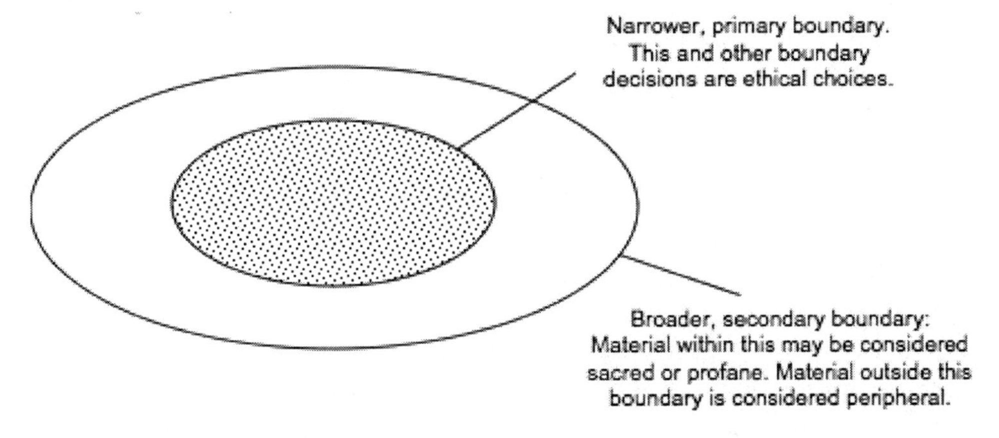

Figure 1 *Midgley's Normative Theory of Boundary Critique*

How does the theory of boundary critique apply to scholarly work? As one example, consider the questionable status of action research—research to effect practical improvements (Bentz & Shapiro, 1998)—in the Academy. Herr and Anderson (2005: 52) explore opposition to action research and write implicitly about primary and secondary boundaries.

> *For many academics* within the primary boundary, *the acceptance of action research is given only on the condition that a separate category of knowledge be created for it, i.e.,* forming a marginalized space. *This is usually expressed as some variation of formal (created in universities* and therefore core) *knowledge versus practical (created in practice settings* and therefore

marginal) *knowledge and a strict separation from research to practice* (delineated by the primary boundary).

How does boundary critique apply to work by practitioners? The Gulf of Mexico oil spill provides fertile ground for exploration. Blogger Avivah Wittenberg-Cox (2010) writes about values and gender at play in the valuing/de-valuing dynamics of BP. The theory of boundary critique is clear in her statement about a shift in attitude about environmental values in the margin:

> *In a public address at Stanford University in 2009,* [CEO] *Hayward said that BP had too many people "trying to save the world"—a reference to the 'Beyond Petroleum' green vision developed in the John Browne era.*

Wittenberg-Cox (2010) also hypothesizes about links between those values and gender:

> *One year ago, BP's most senior woman left the company. Vivienne Cox was the head of the company's renewable energy business. A lifelong proponent and pusher of sustainability issues, she was one of the many women to leave the company after the current CEO Tony Hayward took over from Lord Browne... one wonders had she and the other women stayed, would BP be in its current mess?'*

The theory of boundary critique is a simple way of framing system elements and planning actions. It can be seen as an umbrella under which other boundary concepts sit.

Boundary-Related Research Behind the Edge-Effect™ Tools

Since 2005, the author has researched boundaries—and related concepts such as edges—in organizational and community contexts. The most recent study explored how leaders respected for their work in complex environments understood and worked with boundaries. These leaders were identified through a nomination process. Each nominator described a leader's work in complex, boundary-spanning environments, as well as specific reasons behind their respect. In these complex environments, the leaders had little or no positional authority and the work was often voluntary.

In addition, nominated individuals were required to have experience as leaders with positional authority in formal organizations, so they could talk about transitions from one context to the other. Many participants were working simultaneously as formal leaders in a hierarchy and informal leaders in a boundary-spanning environment (such as a trans-organizational community of practice). Geographically, they were situated in Canada, the United States and the United Kingdom. There were no restrictions on sector or profession.

There was a certain paradox in use of the term "leader." The author has previously emphasized leadership as a process, not necessarily associated with positional authority or even with individuals. Because leadership is so strongly associated with individuals in our society, this compromise facilitated the nomination process.

In this most recent study, a modified form of phenomenography called systemic phenomenography (MacGillivray, 2009) was used to surface qualitatively different ways of understanding boundary-related concepts and strategies. Phenomenography is a qualitative method used to explore different ways of understanding a phenomenon. But in its original form, its focus tends to be at the micro level. For example, researchers might interview students to see how they understood a concept in a textbook paragraph. Systemic phenomenography has several modifications, including very lightly structured interviews coupled with direct observation for enriched data gathering.

Because the study participants were nominated as respected leaders, each way of understanding has demonstrated value in a particular context. The research participants moved beyond ways of understanding, and shared various strategies they used for boundary work. It is unlikely that there is a single way of understanding—or strategy—of value in every context. The strategies could also be seen as applied complexity processes, in that they were used in relatively unpredictable environments with many interacting entities, and with a focus on connections and relationships.

In addition to talking about boundaries and boundary work, participants frequently contrasted the nature of their vertical and horizontal environments. Their experiences of working in the two types of environments had been very different. This was not deliberately explored in the research and this finding was unexpected. Their stories led to my incorporation of Midgley's normative theory of boundary critique in the study's discussion. Using Midgley's language, participants placed vertical hierarchies within the primary boundary and complex, boundary-spanning environments in the marginalized zone within the secondary boundary. Participants invested considerable effort in boundary work between the two. This effort could often be described as enhancing options for a *space of possibilities*, a concept used in many fields including biology, design science and psychology. Participants—consciously or unconsciously—drew on ecosystem principles when they described these boundaries and how they worked with them. For example, "Lloyd" talked about the importance of periodically stepping aside from the rush of mainstream thinking. He likened this to paddling into an eddy at the side of a river where you can rest, look ahead, invite others into conversation, and plan new approaches. A number of these ecosystem-related concepts have been woven into the theory of boundary critique to develop *ecosystem-informed boundary critique*.

How does this depiction of vertical and horizontal environments (Figure 1) relate to Eileen Conn's model (in this volume) of external/public agencies vis-a-vis community residents? Conn has presented the public agencies and communities as two distinct systems. She considers the boundary around each system a primary boundary, with a space of possibilities between systems. I suggest that I could have presented my participants' vertical and horizontal environments in exactly the same way, as separate systems with a space of possibility between them. This would have been legitimate.

However, in my study, participant stories were so loaded with evidence of power in the vertical that I chose to use Midgley's boundary critique model to emphasize that power dynamic. Another way of representing elements of both would be to tip Midgley's model on its side, have the horizontal (for example, community of practice) layer become the bottom layer and the smaller but more powerful graphic element of the vertical core sit on top, with a space of possibilities between them. Or one could simply envision the boundary between the vertical and the horizontal as the space of possibilities. Without exception, leaders in my study were treating that intersection as a space of possibilities.

I cannot say whether it would be appropriate for Conn to present her public sector and community entities using Midgley's model, but I expect Midgley would consider it quite appropriate. Public sector and community groups are quite separate (even though some community members would work for government). Arguably, people in public service decision-making roles have more power and can choose to value or devalue the citizenry. And it could be productive to engage in dialogue around those issues and become aware of the impacts of the real or perceived boundaries.

The common ground between Conn's model and mine may be fueled by the public sector nature of some of our participants' work. A good part of my participants' work as complex leaders involved negotiating the boundaries between the vertical and horizontal and developing options within the space of possibilities. It is important to note that the leaders in my study were interested in knowledge flow across this space, but were not attempting to make the two very different environments homogeneous. As a matter of fact, they sometimes increased adaptive tensions by highlighting differences and thereby "shocking the system".

Many researchers, practitioners and authors are wrestling with boundary issues. The Edge-Effect™ processes and tools now under development are designed to help leaders work more effectively with boundaries, as a central concept in complex systems.

The Edge Effect™ Kit for leaders

Context

Formalized systems and complexity thinking are relatively new fields. Interest in systems and complexity is growing as evidenced by book sales, practitioner networks, select academic programs and conferences and the widespread workplace adoption and [mis]use of terms such as *emergence*, *synergy*, *complex*, *organic* and *edge of chaos*. Simultaneously, emphasis is shifting from a dominant focus on tangible assets to intangibles. While some forms of work, learning and metrics are strongly pre-structured—such as lean Six Sigma and Six Sigma black belts—many new learning forums are deliberately understructured, such as twitter chats, un-conferences and open space events. Yet leadership development resources still tend to focus on the individual leader with formal authority, setting a vision for a seemingly predictable future and motivating subordinates to align with the vision, achieving interim objectives as efficiently as possible.

This mechanistic view of leadership for a controlled world is appropriate in some industries and contexts where change is minimal and efficiency is optimal. It stands in contrast to an organic view of leadership that brings benefits for complex environments where change is ubiquitous and many interacting factors make the future unpredictable. Instead of efficiency, people and organizations are likely to benefit from great observation skills, adaptability, innovation, or resilience. Instead of the machine metaphor—where we talk about replacing people, leverage points, or greasing the wheels of progress--we can look to our natural world for metaphors that exemplify different types of strategies and strengths. Metaphors such as cross-pollination and osmosis are seeping into our workplace vocabularies to fill this void. The Edge Effect Kit™ has a rich variety of images, principles and metaphors from nature with which to work.

As stated in the initial planning document for this conference, the concept of "boundary" is central to systems and complexity thinking. Nature can teach us a lot about boundaries. Some organizationally-focused authors such as Susan Cohen, Andrew Sturdy with Karen Handley, Miguel Nicolelis, Matt Smith and Suzanne Werner with David Davis have written explicitly about boundaries, but with little reference to nature. Furthermore, the boundary theme has not yet permeated mainstream leadership literature and related development material. If leaders need to work with increasingly complex problems--or if leadership needs to emerge and be supported in new ways for complex work—we need readily available tools and processes to help leaders understand and work with boundaries. The Edge Effect™ Kit for leaders sets out to meet this goal by bridging boundary-related principles from nature with real-world organizational needs.

Edge Effect

I have never heard of the "edge effect" concept being explored in boardrooms. Concepts in the Edge Effect™ Kit are drawn from nature; the edge effect is one such concept. It was first defined by Odum in the 1950s when research revealed how important and distinctive edges in nature can be. Many ecological boundaries (think of a forest meeting a meadow or a an estuary where a river meets the sea) have more life forms—both in terms of types and numbers—than the communities on either side. They can be tremendously productive environments. We see—but rarely acknowledge and support—this sort of richness where different work units, departments, organizations or even sectors make space for healthy interaction.

The goals of Edge Effect™ Kit include:

a. Increased familiarity with complex system concepts in general, and boundaries in particular, with an emphasis on re-framed thinking for current trends, issues and opportunities;
b. Increased familiarity with boundary critique, with an emphasis on re-framed thinking for current trends, issues and opportunities;]
c. Use of boundary concepts and critique for specific issues and opportunities in organizations and communities; and
d. Fresh, new ways of tackling long term challenges where traditional tools have been inappropriate and ineffective.

Products under development are

1. A book written for leaders and change facilitators;
2. Images for reflective work;
3. Concept-specific art cards with
   a. Questions;
   b. Quotes, and;
   c. Short stories/cases;
4. Process maps to illustrate ways in which materials can be used; and
5. The "Ecosystem" Café as a modified World Café process. The World Café process emerged at the intersection of intellectual capital and social justice, and is documented on http://www.theworldcafe.com/. The ecosystem café—developed with the knowledge of the world café core team—follows the same principles and makes explicit reference to principles from nature behind the process, and makes explicit use of principles from nature in the conversation processes.

Complexity science emerged through conversations with scientists from many disciplines, including scientists working with the natural world. Images from natural systems are used to engage people on levels not possible with text. The evolving suite of images is from artists and photographers who have considerable experience with the natural world.

The series of cards is designed for use in Ecosystem Cafés and related workshop settings. There are three types of cards. The *boundary critique cards* present the normative theory to guide conversations and reframing of issues. They are fundamental building blocks for other work. For example, leaders can become more conscious of the boundaries they are choosing to construct, strengthen or reify, and the implications of doing so. In the context of this kit, the boundary critique cards may be particularly valuable for enabling knowledge flow amongst community members and the formal organizations in which they work. Figure 2 shows a sample boundary critique card.

The *boundary concept series* presents different ways of understanding boundaries (for example "the bridge," "the eddy," and "intellectual estuaries," which are rich meeting places between communities). Figure 3 shows a sample card from the boundary concept series. The front of the card is standard for each concept; the backs of each card set have varied questions, quotes and real world examples about that concept. In books for practitioners, boundary concepts are usually treated in quite focused ways, such and boundaries one should erect to protect oneself (from unfair treatment, or for work-life balance for example) or boundaries that are barriers to innovation (as in the boundaryless organization). The concept series helps leaders recognize there are many other ways of understanding boundaries and related concepts, and different ways of understanding will be of value in different contexts. These cards expand the leader's conceptual frameworks and vocabularies.

The *boundary strategy cards* present specific ways of working with boundaries. This can be illustrated through one initial priority in the development of this kit: to work with the challenging barriers between communities or networks and member organizations. Important knowledge work can take place in the rich edges of trans-organizational communities of practice, but this does not automatically benefit the formal organizations in which community members work. Community members may meet weekly for coffee, at a pub after work, at conferences or interact regularly through social media. For example, there might be an open source platform for ongoing conversations, and monthly teleconferences for more intense and focused interaction. Such communities can be great incubators for knowledge generation, sharing and innovation. At present, the benefits of such learning are too frequently unrealized in the "home" organizations. Special effort is needed for this kind of knowledge flow and application of learning. These strategy cards build on the boundary critique and boundary concept cards. They share brief stories, quotes and questions grounded in boundary strategies used by respected leaders.

Sample Card: Boundary Critique
To Stimulate Small Group Dialogue
(front side)

Sample Text for Boundary
Critique Series (back side)

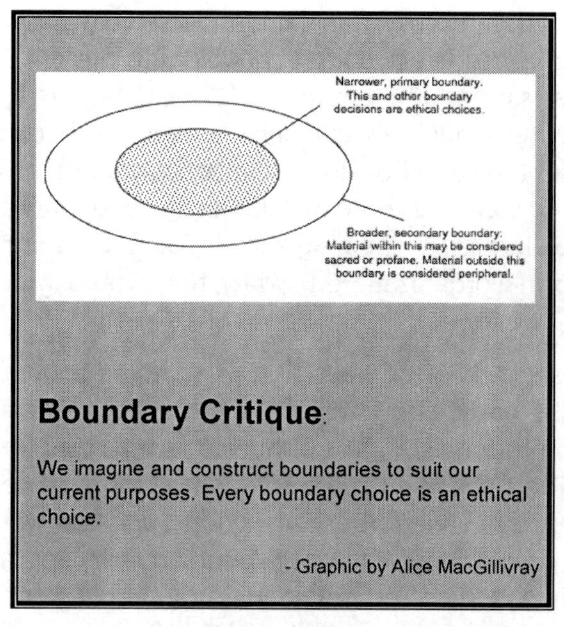

Question:
Which three groups in your organization are least respected or least listened to?

Figure 2 *Sample Boundary Critique card*

Sample Card: The Bridge
To Stimulate Small Group Dialogue
(front side)

Sample Text for Bridge Series
(back side)

Bridge:

The bridge spans a divide, enabling interaction, learning, and change.

"The Japanese Gardens at Butchart Gardens"
-with permission of the artist: Alice Saltiel

Question:
Who can speak more than one "language" within your organization (e.g., planning and operations, or science and policy, or jargon and plain public language)?

Figure 3 *Sample Boundary Concept Card (The Bridge)*

Sample Card: Habitat Crossing
To Stimulate Small Group Dialogue
(front side)

Habitat Crossing:

Deliberate movement back and forth between ecosystems.

"Shorebird Rhythms"
-Alice MacGillivray

Sample Text for Habitat Crossing
Series
(back side)

Question:
How do you model the flow of knowledge back and forth between your organization and your communities of practice and networks?

Figure 4 *Sample Application Card (Habitat Crossing)*

In pilot work with Edge Effect™ tools, I introduced boundary critique and boundary concepts, and facilitated use of the cards in workshops to address organizational challenges. Some comments have been very positive. Several participants have told me they needed more preparatory work and context than initially provided. They wanted a gradual introduction with time for exposure to new concepts and research, dialogue, action and reflection. This additional context is being realized through the book, process maps and other support materials.

Shift from mechanical language, metaphors, and assumptions to appropriate use of principles from nature may not happen easily. Concepts such as "replacing an employee" are supported by refined, long-term human resources practices that give little or no attention to the tacit knowledge and social networks of the departing employee. The vision statement, five-year plan, and cascading measurable objectives are so steeped in the traditions of some organizations that few question the work involved in regularly revising goals and justifying missed targets. Moreover, in complex environments one cannot predict exactly what tools introduced in what ways will enable optimal learning. So the Edge Effect™ Kit is deliberately loosely designed with many options to be explored by thoughtful leaders and attentive facilitators.

Figure 5 shows how the Edge Effect™ Kit translates learning from systems and complexity science into practical tools for leaders who want to build their strengths in complex systems.

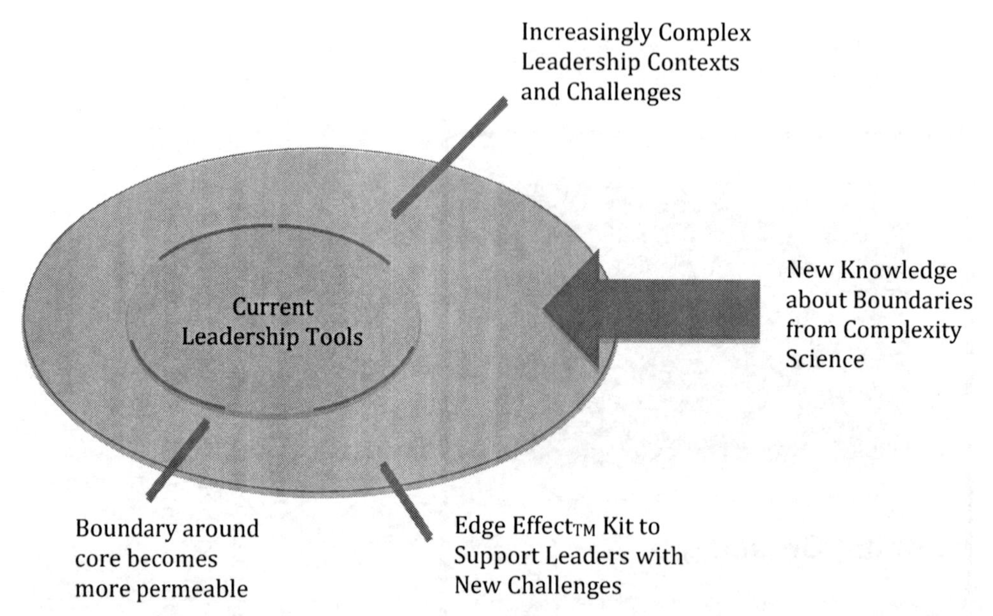

Figure 5 *Edge Effect™ Kit Draws on Complexity for Complex Leadership Work*

References

Bentz, V.M., and Shapiro, J.J. (1998). *Mindful Inquiry in Social Research*, ISBN 9780761904090.

International Leadership Association, http://www.ila-net.org/.

Kotter, J. P. (1996). *Leading Change*, ISBN 9780875847474.

Kotter, J. P. (2001). "What leaders really do," *Harvard Business Review*, ISSN 0017-8012, (December): 85-96.

MacGillivray, A. (2009). *Perceptions and Uses of Boundaries by Respected Leaders: A Transdisciplinary Inquiry*, PhD Dissertation: Fielding Graduate University, AAT 3399314.

McKelvey, B. (2002). "Complexity and leadership," in M. Lissack (ed.), *The Interaction of Complexity and Management*, ISBN 9781567204278, pp. 85-90.

Midgley, G. (2000). *Systemic Intervention: Philosophy, Methodology and Practice*, ISBN 9780306464881.

Richardson, K. and Cilliers, P. (2001). "Special editors' introduction: What is complexity science? A view from different directions," *Emergence*, ISSN 1532-7000, 3(1), 5-23.

Richardson, K.A. and Lissack, M. (2001). "On the status of boundaries, both natural and organizational: A complex systems perspective," *Emergence*, ISSN 1532-7000, 3(4): 6-18.

Tormala, Z. (January 27, 2011). HBR Blog IdeaCast, http://blogs.hbr.org/ideacast/2011/01/the-persuasive-power-of-uncert.html.

Ulrich, W. (2000). "Reflective practice in the civil society: The contribution of critically systemic thinking," *Reflective Practice*, ISSN 1470-1103, 1(2): 247-256.

Wittenberg-Cox, A. (2010). "BP lost its women before it lost its brand (and its stock price)," http://www.20-firstblog.com/?p=374&utm_source=Boomerang&utm_medium=Email&utm_term=More&utm_content=sk@soraya-kandan.com&utm_campaign=BP+Lost+its+Women+before+it+Lost+its+Brand.

Alice MacGillivray is a scholar-practitioner with a consulting proprietorship based in western Canada. Her early work with natural system conservation shaped her work as a systems thinker. Alice works with challenging problems and opportunities, leaving organizations with greater capability, capacity and resilience. Her work involves the strengthening of leadership throughout organizations or communities, and improving the generation and flow of knowledge across boundaries. It is frequently inspired by principles and examples from natural systems. Alice directed Knowledge Management Graduate programs in the Science, Technology and Environment Division at Royal Roads University, and now works with Royal Roads as an Associate faculty member and with Capella as a Visiting Scholar. Her research interests centre on the intersections of leadership, knowledge management and complexity thinking. She has published in peer reviewed books and journals including E:CO (Emergence: Complexity and Organization) and The Learning Organization. Alice has presented at several international conferences hosted by organizations including The Society for the Social Studies of Science, The International Society for Systems Sciences, The Association for the Advancement of Computing in Education and McMaster University's World Congress on Intellectual Capital. Some of her recent empirical work builds on Midgley's theory of boundary critique. Alice's degrees include Masters of Arts in Leadership and Human Development, and a PhD in Human and Organizational Systems from Fielding Graduate University.

9. Value Crafting: A Tool To Develop Sustainable Work Based On Organizational Values

Sjaña S. Holloway, Frans M. van Eijnatten & Marijn van Loon
Eindhoven University of Technology, NLD

This chapter describes a method to craft work using organizational values in a business context. The method, which is based in complexity, uses the integral-theory framework to arrive at a cycle that consists of four steps, i.e., picking an organizational value, better understanding it, developing intentions what to do with it in a project context, and applying it in the daily work.

The method was further developed into a practical activity tool, which was empirically pilot-tested in two small innovation teams in a Dutch manufacturing company. Results indicate that the R&D engineers in the pilot test were successfully able to craft an organizational value by using the tool. However, initially they experienced some problems with understanding the practical relevance of organizational values.

Therefore, future research should increase the basic understanding and practical usability of the tool.

Introduction

Whenever companies achieve success in the context of their industry, the need to regenerate the company is not a new phenomenon in business. As market demands grow more diverse, corporations must stay competitive, and retain market share. This also begins to trigger the needs for broader, more integrated employee skill sets, deeper exploration of novel products, and more dedicated attraction of new customers (Cartwright & Schoenberg, 2006).

Organizational teams are confronted with a challenging mix of complex requirements and opportunities. For instance, changes in work activities or modes of work (purpose and participation) require both employees and management to come to terms with old and newly formed work practices. Keeping in mind these new practices should be of the same standard and quality in order to help and not hinder the already dynamic changes in the work place. This is not a linear process. In order to maintain and regenerate the human and organizational resources the question arises: How can people be sustained, strengthened, and enriched in their work? How can the meaning of work be flexible, while the actual task patterns keep changing all the time?

We will propose a method to craft work by using organizational values called Value Crafting. This tool, which uses values, may create sustainable work by strengthening employees' abilities to (re)generate resources through their work values.

Complexity Concepts

We would like to use a complexity perspective as point of departure for our value-crafting approach. In this chapter we define complexity in part as "the simultaneous uniqueness and integration of system elements", Kira & Van Eijnatten (2008b: 750). In doing so, we look to capture the unique adaptations of organizational values as the elements in a dynamic work design process. While there seems to be a contradiction in terms of uniqueness and integration, the complexity perspective makes apparent that the grouping of interrelated parts with intricate internal and external structure is a necessary feature of all elements. We see these contradictions in most everything around us (e.g., life and death; summer and winter). Within an organizational context, we seek to create and observe novelty. As such, we use individual unpredictability as a way to create change within a work group. For this reason, we specifically take the holon as our basic lens or principal vantage point. The tool is based in complexity and the holon unit serves as a basic unit of measurement. A holon is defined as an entity that is simultaneously whole and part of a bigger whole. It is both autonomous or preservative with respect to one aspect, and dependent or adaptive with respect to another aspect (cf., Wilber, 1996; Edwards, 2003, 2007, 2009). This means that there are three different holons being addressed. The individual holon, represents the lowest aggregate level within the organization and most easily an agent capable of continuous transformative emergent patterns of behavior. The group holon is comprised of several individual holons that can also lead to transformative emergent patterns of behavior. Lastly, the organizational holon has a greater aggregate level comprised of several individual holons so it, too, is capable of emergent transformative change although most likely at a slower pace. Reality, seen as a complex system, essentially is thought of as being composed of holons organized within 'natural hierarchies of holons' or holarchies (Wilber, 1996; Edwards, 2007). This implies that connectivity is considered a basic characteristic (Fitzgerald & Van Eijnatten, 1998; Fitzgerald, 2002; Van Eijnatten, 2004). The result of this connectivity may trigger behavior from the interactions between individual, group, or organizational holons within holarchies.

We have adopted the holon as our basic unit of analysis, which means that we intend to model individuals, groups, and organizations in an integral way, i.e., a) by combining both *exterior* or objective observables and interiors or subjective introspections; and b) by bringing together both *agentic* or preservative and communion or relational, adaptive aspects, see Figure 1A.

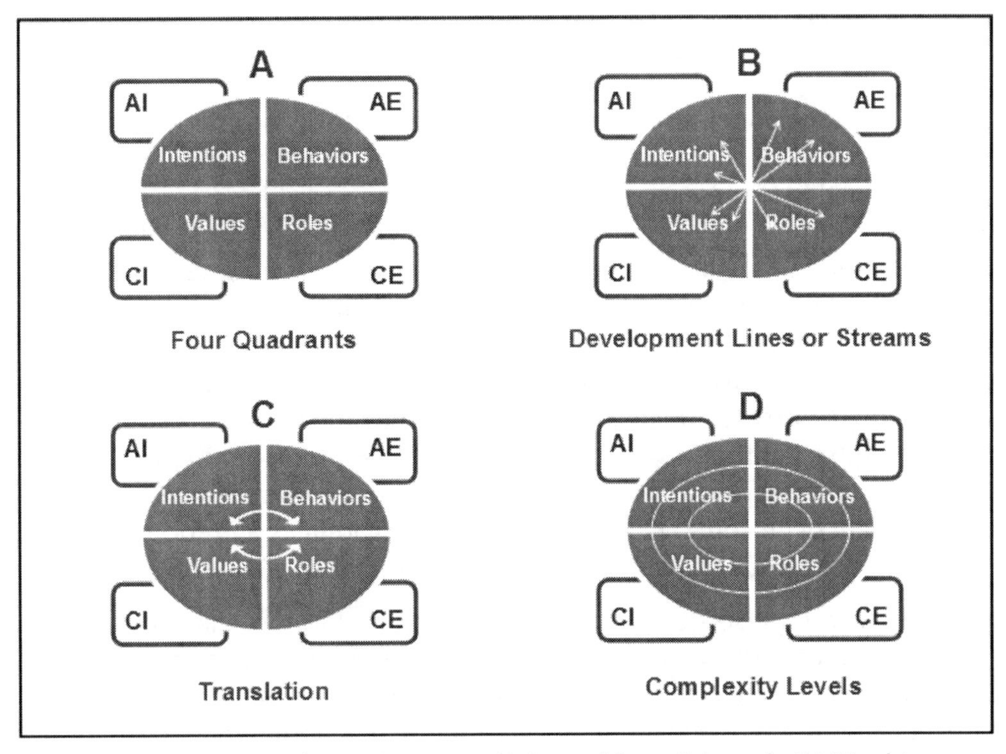

Figure 1 *Complexity Concepts (Adapted from Edwards, 2005a, b)*

Legend: AI = Agentic Interior — *Streams or Lines*
AE = Agentic Exterior — *Complexity Levels*
CI = Communion Interior — *Translations between Quadrants*
CE = Communion Exterior

Wilber (1996) called this the 'All Quadrants All Levels' (AQAL) framework; Edwards (2003) coined it the Integral Theory perspective (Cacioppe & Edwards, 2005a, b). In this chapter, we treat both conceptual frames as equivalents, although we are primarily focused on Edwards' interpretation of the holon domains.

We define sustainable human development as a balanced growth of human resources in all four quadrants of the AQAL framework / Integral Theory perspective. Sustainable human development seen from this integral perspective can be analyzed in a single matrix with the following four quadrants, see Figure 1 (Edwards, 2005a, b, 2007, 2009):

- *Agentic Interior (AI) resources* refer to intentions, reflexes, and thoughts (the consciousness self).
- *Agentic Exterior (AE) resources* apply to behaviors, physiology, and directed action (the behavioral self).

- *Communion Interior (CI) resources* look at worldview, myths, and values (the meaning-making self).
- *Communion Exterior (CE) resources* are associated with social presence, social role, and activity (the social self).

Both job and personal resources flow and weave paths of interior introspections, exterior observable behaviors, adaptive aspects of self preservation, and the perpetuation of meaning for the common good. Individuals "create, protect, foster, and nurture their resources...to sustain well-being", Hobfoll (2009: 95). The motivation to maintain balance in all four quadrants may be facilitated by this process, but it also brings to light the idea that the individual is "the smallest level of analysis", Hobfoll (2009: 96). Individuals are nested in relationships and networks with other individuals, groups, and organizations, as well as their sense of their ability to impact and control their environment successfully (Hobfoll *et al.*, 2003; Xanthopoulou *et al.*, 2009) is built on the premise that an individual's sustainable work ability is a multidimensional construct, founded on the development of personal resources relating to individuality and sociality, and to mental models, emotions, and behavior patterns (Kira *et al.*, 2010).

Translation and transcendence have been suggested as mechanisms by which human resources are created, developed, and at times discontinued (Kira et al., 2010). Transcendence is the process of discontinuous growth into different levels of complexity (Van Eijnatten, 2004). It is the transformational change that comes from double- and/or triple-loop learning (Argyris & Schön, 1978). Transcendence levels in each of the four quadrants allow for in-depth examination of human growth and maturation (Edwards, 2007), see Figures 1B and 1C. A *Transcendence Level* is a general measure of higher and lower levels and any specific level can have structure. Transcendence levels tend to unfold in a sequence and thus progress through stages at each level of complexity (Van Eijnatten, 2004). Complexity levels are not rigidly separated from each other but overlap and are fluid. Put in another way, levels are abstract measures that are fluid, yet are qualitatively distinct classes of recurrent patterns.

Translation is the process of flow (Csikszentmihalyi, 1990) within or into other quadrants, where capacities proceed to ripple through a similar level of development in all quadrants. The translation of resources provides a means for internalization and externalization of resources (Nonaka, 1994; Nonaka *et al.*, 2000).

Recent Applications Of Work Design

Job *crafting* is an approach to work design that seeks to capture what employees do to redesign their own work in ways that can foster their engagement, satisfaction, resilience, and ability to thrive (Spreitzer *et al.*, 2005). It is a means to utilize opportunities to customize work by changing tasks, interactions, and how the work is viewed. Wrzesniewski & Dutton (2001) defined job crafting as:

> *The physical and cognitive changes individuals make in the task or relational boundaries of their work. Changing task boundaries means altering the form or number of activities one engages in while doing the job, whereas changing cognitive task boundaries refers to altering how one sees the job [...], and changing relational boundaries means exercising discretion over with whom one interacts while doing the job* (Wrzesniewski & Dutton, 2001: 179-180).

According to Berg *et al.* (2008), job-crafting dimensions include a) *task crafting* by changing the scope of your job; b) *relational crafting* by altering the extent or nature of the interactions; and c) *cognitive crafting* by changing the way one thinks about one's job. In terms of the integral theory perspective described by Edwards (2005a, b), job crafting covers three out of four of the previously distinguished holon domains, see Figure 2A:

- Changes in the agency-exterior domain: concrete tasks and work activities.
- Changes in the agency-interior domain: comprehension and mental models of work (both purpose and priorities).
- Changes in the communion-exterior domain: social roles, relationships, connections, and collaborations.

Grant & Parker (2009) distinguished relational perspectives (socially embedded jobs, roles, and tasks) and proactive perspectives (workers' initiatives to change the way in which the work is done).

Kira & Van Eijnatten (2008a, b) further extended the development of job design to include both the contributions of socio-technical systems and the prospects of a chaordic systems approach by the use of complexity, in order to capture patterns of pro-activity and co-creation needed in modern work environments. Since work is increasingly becoming a responsibility instead of a fixed set of tasks, Kira *et al.* (2010) broadened the concept of *job* crafting into *work* crafting. Focusing on a whole career, work crafting includes not only the challenges employees encounter in terms of behavior, social roles, and intentionality in the short run, but also in the long run. In addition to that, their culturally shared meaning of the work context could be a focus, as well (Kira & Forslin, 2008). In terms of the Integral Theory perspective of Edwards (2005a, b), work crafting involves all four holon domains, including changes in the communion-interior (CI) domain: sense of belonging, cultural embeddedness, and shared meaning of values. In work crafting, collaboration is unstructured and focused more closely on the individual holon's ability to translate between quadrants and transcend into different levels of complexity, see Figure 2B. As can be seen in the figure, translation is possible between all quadrants.

Within a company, different people and departments intensely work together to keep the organization on track in a complex competitive environment. The organization's culture can play a role in the development of the organization's value development. Organizational culture describes many aspects of its mem-

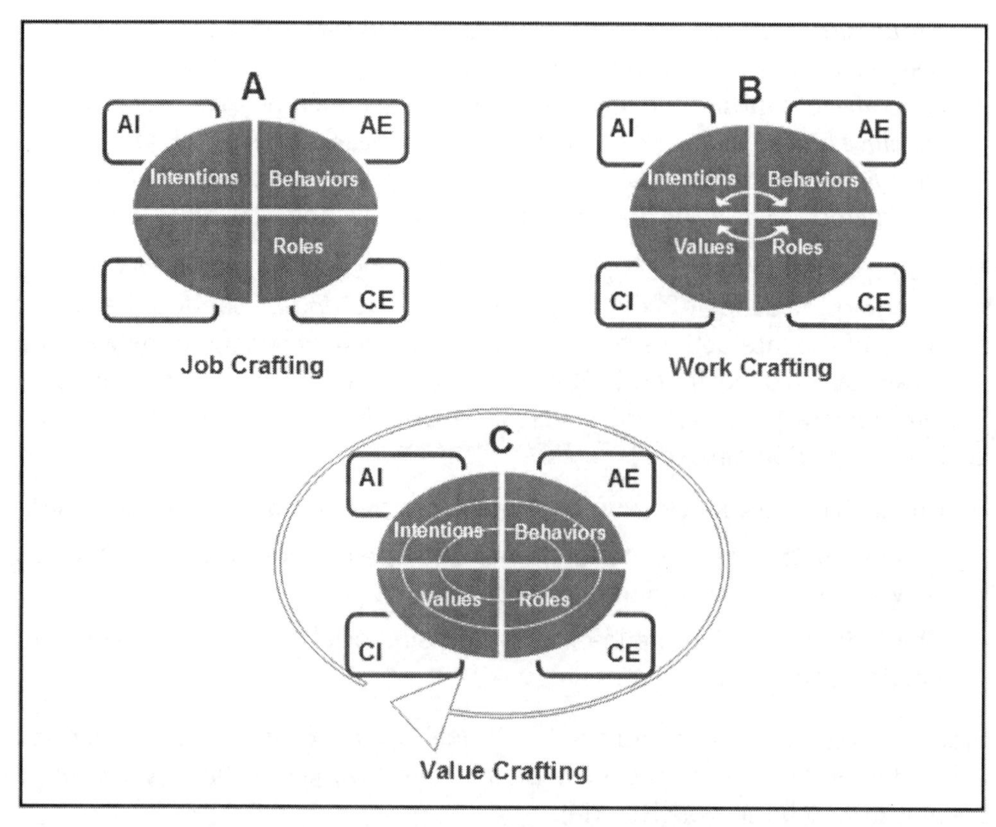

Figure 2 *Job Crafting (A), Work Crafting (B), and Value Crafting (C), seen through an Integral Lens*

Legend: AI = Agentic Interior
AE = Agentic Exterior
CI = Communion Interior
CE = Communion Exterior
Translations between Quadrants
Complexity Levels
Crafting Cycle

bers' attitudes, experiences, beliefs, and personal and cultural values within the organization. Organizational culture has been described in various ways: as a way to distinguish members from one another (Hofstede, 1984); a collection of fundamental beliefs and values within an organization (Wallace *et al.*, 1999); the implemented beliefs and values of an organization's members (Schneider *et al.*, 1996), and a result of internal and external factors from the environment beyond managerial control (Turnipseed, 1988). When one takes these descriptions into account, the development of values within the organizational setting is important and can affect peoples daily behavior. Values in this context can be seen as having the attributes of stability and instability as they can "change in accordance to the changing physical, social and spiritual environment of the individuals and groups that embrace them" (Seevers, 2000: 71).

The collective use of values within a corporate culture may facilitate better communication and shared knowledge. Using the method to craft work with organizational values, we include the regeneration of resources associated with world-view, myths and values (the meaning-making self). Value crafting is not crafting values but crafting work using values. In *Value crafting* it might be a means to create a new corporate culture by prompting the development and diffusion of organizational values within an organization, see Figure 2C. In value crafting, the individual-, collective-, or organizational-holon's abilities to translate into other quadrants, and transcend to different levels of complexity are proposed to work together. A value is picked, and translated into intentions/ambitions, behaviors and roles, and then transcends to a collective value meaning, see Figure 2C. The collective value meaning is then thought to transcend to a next level of complexity for the individuals and the group itself.

Value-crafting is defined as an activity tool that can be used to better solidify the formulation of work-group intentions/ambitions, the creation of action plans, reflect on those plans, and the examination of social roles taken by individuals and the group as a collective action. In this, we utilize organizational values as part of the cultural quadrant in all holons as a starting point to accomplish these tasks.

Motivation For Empirical Study

The presented value-crafting cycle comprises the desire to add to the development of participative sustainable work practices in the field of work design. In order for the model to gain any traction in the business world, the need to uncover its effect in an empirical study was the necessary next step. In this section, we describe how value crafting contributed to four-quadrant development with respect to innovation teams in a manufacturing organization.

In order to apply value crafting in a work setting, we have developed a practical tool that consists of a series of workshops.

Value-Crafting Cycle

We have combined the above-mentioned concepts in order to design a method for value crafting as a structured, cyclic intervention in an organization (Van Loon, 2010). We distinguish the following four steps:

1. *Select and interpret a value*: In the first step, a work group selects an organizational value, and develops an intention/ambition about what to do with the value.
2. *Develop intentions and actions*: In the next step, an intentional goal is set by the work group in order to change its behaviors, and to develop new work practices based on the value.
3. *Reflect on value and relate to own behavior*: In the third step, the work group reaches consensus about the reflections of the value, potential behavior re-

sults, and what may be possible by using the value. Consensus is used here to include different realizations and reflections about a particular value.

4. *Adopt and understand the value*: In the fourth step, the adaptation of the value is achieved by the work group. The collective experience and development help to give permanence not only to the value but also to the sustainable credence within the work group. A new value-selection process may start.

In the first step, the work group starts with the selection of a value from the set the organization is modeling. The individuals in the work group start with the formulation of their individual intentions/ambitions with regard to the selected value.

In the second step, the work group itself develops intentional goals in order to change its behavior. These intentional goals are developed by the work group. While the group is developing the intentional goals they have to consider any existing organizational constraints to the actions they may want to take such as procedures.

In the third step, the work group reflects on the possibility for the group to really change the behavior in order to improve performance. The work group itself becomes aware of all interactions that are influencing its behavior (i.e. continual discussion and communication), and thereby also its performance. These ongoing interactions are what make the group whole, and through its discussions and iterations the group tries to make sense of the value.

In the fourth step, the work group adopts the value. This means that by adopting the value the work group has learned to contribute to the organization in a way that fits with the organizational values. That may become the new standard, which may increase stability in the team, and may help the individuals to strengthen the organizational values in their operations.

The *interactions* between individuals, work groups and the organization allow for the translation and transcendence of values. Participation in the rapid changes is needed to keep a group process moving. In the third step, at the work-group level, reflection is the incremental movements toward the next level of comprehension needed for long-term adaptation and growth for all. Sharing is thought to help create a cultural artifact that sustains the continual development of values. This is a flexible process; no work group is the same, and consensus can change as the work group changes size or as cultural differences emerge. The cycle and process can be a flexible cultural artifact, and the meaning of the values can also change with the context. There is also no absolute starting point but the cycle needs to be completed.

We have examined the application of the value-crafting cycle both qualitatively and quantitatively. We first qualitatively analyzed observations carried out within the R&D teams interventions in the organization. Subsequently, two independent researchers did a quantitative analysis based on the integral theory lens by means of the Critical Incidents Technique (CIT).

Method used in the Value-Crafting Intervention

In a Dutch manufacturing organization, two innovation teams, each consisting of 6 R&D engineers, were recruited within individual subsidiaries of the same organization. Within those teams, an effort to bring the theoretical and conceptual models of value crafting into practice was carried out. Within each of the two teams, we tried to use the same methodology. This means that the group characteristics (such as group size) were the same, the value-crafting cycle was applied in the same way, and also the evaluation of the results in both teams was identical. The only difference in the approach between both teams was that the value-crafting cycle was adapted to the teams' projects goals, which were different in both teams.

In the workshops, the teams were stimulated to discuss and brainstorm about a particular value, and to assess the team work based on the four holon quadrants. The workshops were developed to combine both *exterior* or objective observables and interiors or subjective introspections and to bring together either agentic or preservative and communion or relational adaptive aspects as well. Each workshop was designed to encapsulate one of the quadrant domains. The first workshop started with a discussion about the group's interpretation of the selected value, and continued with the identification and formulation of the group's ambitions with regard to the selected value. The second workshop consisted of formulating the group's targets in the innovation project, and of improving these targets, inspired by the group's intentions/ambitions with regard to the value. The third workshop was designed to facilitate the group's iteration and reflection on the realization of the new targets in this step of the value-crafting cycle. The group could identify factors in the environment that hindered the realization of its targets. Then, plans were made to deal with these factors, such that the targets could be realized in the future. In the fourth workshop, the group was meant to reflect on its experiences during the value-crafting process, and to become aware of the potential benefits of usage of the organizational value for its work performance.

Results Of The Value-Crafting Intervention

The evaluation of the effects of value crafting was executed in three ways, by means of a questionnaire, observations, and through semi-structured focus-group interviews. A questionnaire was filled out by the individuals in the group at the start and at the end of the value-crafting process. This questionnaire evaluated which of the four quadrants had the highest priority for the individuals, in their practicing of a chosen value. On the individual level, the questionnaires indicated that for both teams' similar shifts in attitudes occurred.

At the initial measurement, before the value-crafting process was started, individuals in both teams had the highest scores on the cultural quadrant. After the value-crafting process was completed, the individuals in both teams answered that their intentions/ambitions and actions had a higher priority than before, at the expense of the cultural and social quadrants. This result suggests that the individuals became more aware of the importance of the value for them, and consequently had increased their ambitions and actions.

Observations were conducted in situ by the researcher, but also afterwards, by re-analyzing audio and video recordings that had been taken during the workshop sessions. To summarize the qualitative analysis of observations in Team A, the group develops an interpretation of the organizational value, and dialogues between the members lead to a unanimous definition of the group's intention/ambition with regard to the organizational value. The group goes on to adopt perspectives of the organization, looks closer at its project performance and makes plans to further execute its intentions and the feasibility to maintain these goals in the future. They also reflect member by member on the concrete contributions of the intentions and the benefits, if any, of using values in their daily work as a team.

In Team B, the group had difficulties developing clear interpretations of the value. It also used dialogue between individual members, and decided on a rather vague definition of what the groups' intentions/ambitions were with regard to the values it chose to work with. The team adopted a perspective of the value for the group itself and developed intentions on the premise of improving performance. However, Team B did not believe they adopted or developed anything in response to values or value crafting. In the end the team became aware of the possible benefits of its chosen value when working with others; this seemed to trigger a desire to try to use values to understand the other departments and employees.

Special attention was paid to reactions of engineers in the workshops, which might indicate shifts in the attitude of the teams towards the values. These shifts might be considered as signals of increased adoption of the value in the actual work. For Team A, the focus was toward the organization and the creation of best solutions, "our ambition...value...for the customer and the organization", while Team B focused on the team, "our ambition is to feel engaged... we do not...improve the commitment of the team by specific actions." Both teams felt pressured by time constraints and an increase in short-term pressures and goals that needed to be met. The consistent tone between the two teams was that Team A focused on the organization as a whole and where the team fit in it, and the focus of Team B was to understand how the team could relate to groups and employees outside of the immediate team.

The effects of value crafting at the work-group level was evaluated through semi-structured focus-group interviews. Both groups were asked a number of questions that revealed their experiences with regard to the value-crafting process. This happened in the fourth workshop, during which the groups were supposed to discover the benefits of using the value for improvements of their performance. On the group level, the focus groups revealed four main conclusions. First, that especially the social quadrant of the teams had revealed the highest growth during the value-crafting sessions. Both teams recognized that their highest potential for extra performance was in the relation with the environment. The teams' becoming aware of these benefits was a pre-requisite for the engineers to become willing to use the value-crafting methodology. One of the teams did not recognize these benefits up until the final workshop. These engineers had been reluctantly participating in the value-crafting sessions until they realized the added value in the fourth session. Both teams had difficulties with tacit knowledge. Talking about tacit knowledge was something that not all engineers were used to doing. Some of them talked about tacit concepts quite easily, but most engineers preferred discussions about explicit knowledge, only. The last conclusion is that organizational value crafting is especially helpful in increasing awareness of the important role organizational values play in work decisions. Thereby, value crafting is creating the opportunity to bring organizational values into practice. This is only successful if the right conditions apply. Too much focus on short-term goals is hindering the opportunity to use values.

The observations revealed that most steps occurred as they were formulated in the conceptual model. In contrast to the conceptual model, the teams have not been able to change their behaviors and social roles according to the value. However, this does not mean that the conceptual model is wrong, but it has been suggested that the timeframe was too short (in one team) and the willingness was too low (in the other team) to realize changes within the research period. Both teams have announced that they will change their behaviors in the nearby future, because they have realized that the value can bring benefits for their performance.

Secondary Analysis With Critical Incident Observations

During the time of the value-crafting interventions, audio and video recordings were used to capture these sessions for additional analysis. Two independent researchers were given the task to evaluate the videos of those sessions with Team A and Team B (Bolder *et al.*, 2010). The Critical Incident Technique (CIT) was employed to analyze the videos. First introduced to the social sciences by Flanagan (1954), CIT uses clear procedures to collect, content-analyzes and classify observations. This qualitative research approach makes it easier to identify and examine the consequences of perceived or significant incidents in observations.

Gremler (2004) suggested the development of the following five phases to analyze critical incidents. Phase 1, define the problem; in this case the problem definition was based on the preceding project by Van Loon (2010). Phase 2, the study design; the theoretical background was put into context of the video material. Phase 3, data collection; the data was collected through observation of audio and video material that was recorded during the intervention project by Van Loon (2010). Phase 4, data analysis and interpretation; the data was analyzed through observation and coding of the audio and video material. The video material was interpreted according to a list of potentially relevant critical incident occurrences that were developed according to the integral theory. Phase 5, results; the results are reported by using Critical Incident Technique as suggested by Gremler (2004).

Regarding the structure of the videos, there were videotapes of eight sessions. Each videotaped session had an intervention and observation video. The intervention video was an explanation of what the session's intentions were for the teams, and was not used for evaluation. The observation video contained the teams working with value crafting. The eight observation videos, one per team for each step of the four value-crafting cycle sessions were each approximately one hour in length. Independent observers, watched through the integral theory lens using the NVivo (2008) software program. NVivo allowed the researchers to chronological place their comments on a timeline. This was important as it made the possibilities of finding intersecting segments between the two independent observers easier as well as more concise, based not only on verbal but also on silent cues. The observers watched all videos independently before sharing their findings, and about eight to sixteen episodes were systematically observed.

In order to gather the data used for the observations in this research, a list of forty-one critical incidents were developed. To obtain the episodes that were coded, the observers watched the videos through their theoretical lens. Each observer individually selected parts that were interesting in terms of value crafting and the integral theoretical lens. Then, those selected parts were compared between the two observers. A total of 78 critical incidents were recorded between the two observers, 46 for Team A and 31 for Team B.

It was important that critical incidents captured the value-crafting cycle. When the critical incidents were developed for integral theory, it was important to notice when an individual was speaking and when the group was speaking. With the integral theory, critical incidents were developed for both individual aspects and group aspects to capture the translation effects between the four quadrants.

When the quadrants are activated by the use of values in the value-crafting cycle the translation effects are activated. These can be seen as arrows within the steps of the value-crafting cycle in figure 3. The movement through the cycle acts as a movement through the four quadrants of holon development. In other words, the translation effects of the cycle are representative of the holon aspects of intentions, behaviors, roles, and values.

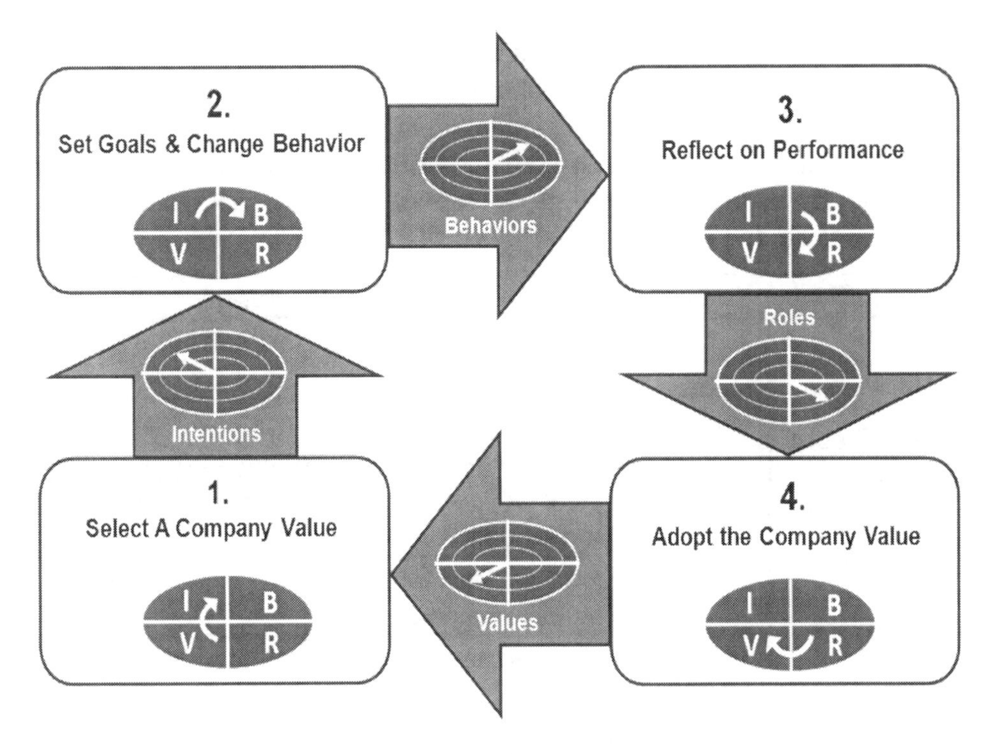

Figure 3 *The Value-Crafting Cycle seen through an Integral Lens*

Legend: V = Values
I = Intentions
B = Behaviors
R = Roles

Transcendence
Increase in Complexity Level

Translation
Transfer into other Quadrants

Results of the Secondary Analysis with Critical Incidents Observations

The evaluations of the value-crafting sessions were carried out over a three-month period by two independent observers. Table 1 gives a summary of the types of critical incident in the observations. Within the first few minutes of the video observations of Team A, episode 1.1, the critical incidents are primarily about intentions and an attempt to go from step 1 to step 2 in the value-crafting cycle. In episode A 1.2 translation seems to take place. The group starts with a group intention; (I.G) members then start expressing some thoughts about the proposal (B.G). This is then followed by a personal intention expressed by a member of the group (I.I.), Next, the whole group reflects on how things are proceeding and decides it needs to work in a different way (B.G.). In a very short period of time the group moves through steps 1-3 of the cycle and intentions and behaviors are expressed by the group. The group goes back to step 1 in the cycle and selects a new organizational value to work with, one individual (R.I.) takes some role within the group and group intentions (B.G) are reflected on in the conversation. At this point there are two levels of activity because there is an individual who takes a role while the group is in a dialogue about its behavior and goal setting. Subsequently the behavior seems changed within the group. In this process we see a transition of intentions, behaviors and a role taken by a member of the group representing the movement through steps 1, 2 and 3 in the cycle.

Type of Effect	Cat	Critical incident
Integral Theory effect on an individual level	I.I.	Here we look for critical incidents that show individual intentions.
	B.I	Here we look for critical incidents that specify a person's behavior or what actions are taken.
	R.I	Here we look for critical incidents that show individuals' roles and activities.
	V.I	Here we look for the use and expression of values and norms
Integral theory effect on a group level	I.G	Here we need to look for critical incidents that show group intentions. An important aspect is here if the conversation is consistent in its use of "we, us, our..."
	B.G	Here we look for group actions and behaviors. An important aspect is here if the conversation is consistent in its use of "we, us, our..."
	R.G	Here we look for critical incidents that show group roles and activities. An important aspect is here if the conversation is consistent in its use of "we, us, our..."
	V.G	Here we look for the group's use and expression of values and norms. An important aspect is here if the conversation is consistent in its use of "we, us, our..."

Table 1 *Description Of Critical Incidents In Terms Of Integral Theory (Adapted from Bolder et al., 2010: 35)*

There were distinctions between the two observed groups. Although we are only now giving a brief summary of Team A findings, we were able to make a comparison between the two groups that were observed. The critical incidents in Team A were more group oriented than in the observation videos of Team B. In Team A, there were many instances of iterations between the first three steps of the value-crafting cycle, hence instances of individual and group intentions, goal setting, and reflection were more prevalent in Team A's sessions. At the individual level, there were sixteen instances where the intentions of individuals were clearly expressed and an individual's decision to take a role within the group happened three times. At the group level, there were eleven instances where people as a group made an intention for the group and thirteen instances where this then led to some action or behavior; three instances where the group took a role about how it would see the value within the group or the work it would do; and one instance when a decision was made about the value within the group.

In Team B there were fewer iterations and conversation. At the individual level, there were five instances where intentions were expressed; one instance for a behavior and value use; and eleven instances where individuals took roles for the group. On the group level, there were five instances where intentions were expressed, seven instances where behavior to take action were expressed; and one instance where the group took some role as to how it would go forward.

Here, again, we do not discount the conceptual model, as we think that the timeframe for both groups was short and willingness of one of the groups may have been a hindrance. While we also take into consideration that from an integral point of view as we see the use of values as a starting point to create behavior, intentions and roles, it can make sense that steps one, two, and three in the cycle would be more active. To capture the adaptation process and see the use of the value in daily work will require more time. Further studies are planned to better capture a longitudinal view of value use in the organization that participated in this study.

Discussion

In this chapter we have suggested the concept of value crafting as a dynamic alternative to job design. Also, we have developed and pilot-tested a method for crafting organizational values in a business context. The method has been transformed into a practical tool that makes use of workshops as its main mode of intervention. Thus, this chapter serves two important ends: it presents a tool that is theoretically based in complexity, and combines it with a real-world application.

Value crafting might be used in work organizations for different purposes. It can be a helpful means for continuous development and change of existing organizational values in a work setting. The value-crafting cycle stimulates the translation of these values into all domains of human development. It also tries to account for the discontinuous phase changes that may occur in decision-making processes. The balanced development of behaviors, intentions, roles, and values may allow for shifts in complexity, and might propagate both the dissipation of old, and the emergence of new patterns of both thinking and working in a company. Value crafting can also be used for the introduction of new organizational values following a major organizational transformation, for instance a merger.

The value-crafting cycle is supposed to support the creation, exploration, and subsequent adoption of new organizational values.

The chapter's thesis is that value crafting might be helpful in creating sustainable work by strengthening employees' abilities to (re)generate resources through their work values. However, this thesis should be treated with great care. Basically, this chapter is about the development of a new idea, and the implementation of a first prototype. Also, the pilot research itself has strong methodological limitations. Therefore, any claims should be very modest. Future research should more generally apply, systematically analyze, and rigorously test value crafting in different business contexts. Future application work might focus on developing a manual for value crafting that explicitly informs engineers about the importance of organizational values as being 'soft' intangible aspects with 'hard' measurable impacts in contemporary working life. Future application work might also focus on further improving the look-and-feel of the value-crafting tool. For instance, it might be developed into sort of a card game, in which different sets of playing cards represent different organizational values, and in which each set represents all four quadrants, i.e., values, intentions, behaviors, and roles.

Acknowledgements

The authors wish to thank Stijn Verkuilen for his collaboration in the research that has eventually led to this article. The authors would also like to thank Eric Bolder, Jeroen Houter, Robin Koonen, Rob Lunenburg, Michiel Voorthuijsen, and Bart Winter for their help in developing procedures and analyzing data for the critical incident analysis study. This research is part of a PhD project that is funded by the company in which the research was carried out.

References

Argyris, C. and Schön, D. (1978). *Organizational Learning: A Theory of Action Perspective*, ISBN 9780201001747.

Berg, J.M., Wrezesniewski, A. and Dutton, J.E. (2010). "Perceiving and responding to challenges in job crafting at different ranks: When proactivity requires adaptivity," *Journal of Organizational Behavior*, ISSN 1099-1379, 31(2/3): 158-186.

Bolder, E.H., Houter, J. den, Koonen, R.J.M., Lunenburg, R.A.A., van Voorthuijsen, M. and Winter, B.W.J. (2010). "The effects of Value Crafting: Value crafting through the lens of panarchy systems thinking, integral theory, and Nonaka's leaning theory," Eindhoven University of Technology, The Netherlands, internal research report.

Cacioppe, R. and Edwards, M.G. (2005a). "Adjusting blurred visions: A typology of integral approaches to organizations," *Journal of Organizational Change Management*, ISSN 0953-4814, 18(3): 230-246.

Cacioppe, R. and Edwards, M.G. (2005b). "Seeking the holy grail of organizational development: A synthesis of integral theory, spiral dynamics, corporate transformation and action inquiry," *Leadership and Organization Development Journal*, ISSN 0143-7739, 26(2): 86-105.

Cartwright, S. and Schoenberg, R. (2006). "Thirty years of mergers and acquisitions research: Recent advances and future opportunities," *British Journal of Management*, ISSN 1467-8551, 17(1): S1-S6.

Csikszentmihaly, M. (1990). *Flow: The Psychology of Optimal Experience*, ISBN 9780060920432.

Edwards, M.G. (2003). "Through AQAL eyes: Part 3 - Applied integral holonics," http://www.integralworld.net/edwards7.html.

Edwards, M.G. (2005a), "The integral holon: A holonomic approach to organizational change and transformation," *Journal of Organizational Change Management*, ISSN 0953-4814, 18(3): 269-288.

Edwards, M.G. (2005b), "A future in balance: Integral theory and global development pathologies," in R. Slaughter, (ed.), *Knowledge Base of Futures Studies, Volume Three: Directions and Outlooks*, ISBN 9780958665407.

Edwards, M.G. (2007). "'It's just a phase I'm going through': Integral lenses and phase transitions in organizational transformation," paper presented at the *Seventh Annual Meeting of the European Chaos and Complexity in Organizations Network* (ECCON): Phase Transitions in Organizations, Bergen aan Zee, The Netherlands, October 19-21.

Edwards, M.G. (2009). *An Integral Metatheory for Organizational Transformation*, ISBN 9780415801737.

Eijnatten, F.M. van (2004). "Chaordic systems thinking: Some suggestions for a complexity framework to inform a learning organization," *The Learning Organization*, ISSN 0969-6474, 11(6): 430-449.

Fitzgerald, L.A. (2002). "Chaos: The lens that transcends," *Journal of Organizational Change Management*, ISSN 0953-4814, 15(4): 339-358.

Fitzgerald, L.A. and Eijnatten, F.M. van (1998). "Letting go for control: The art of managing in the chaordic enterprise," *International Journal of Business Transformation*, ISSN 1367-3262, 1(4): 261-270.

Flanagan, J.C. (1954). "The critical incident technique," *Psychology Bulletin*, ISSN 0033-2909, 51: 327-358.

Grant, A.M. and Parker, S.K. (2009). "Redesigning work design theories: The rise of relational and proactive perspectives," *Academy of Management Annals*, ISSN 1941-6520, 3: 317-375.

Gremler, D.D. (2004). "The critical incident technique in service research," *Journal of Service Research*, ISSN 1094-6705, 7(1): 65-89.

Hobfoll, S. (2009). "Social Support: The movie," *Journal of Social and Personal Relationships*, ISSN 0022-3514, 26 (1): 93-101.

Hobfoll, S.E., Johnson, R.J., Ennis, N., and Jackson, A.P. (2003). "Resource loss, resource gain, and emotional outcomes among inner city women," *Journal of Personality and Social Psychology*, ISSN 0022-3514, 84(3): 632-643.

Hofstede, G. (1984). *Culture's Consequences: International Differences in Work-Related Values*, ISBN 9780803914445.

Kira, M. and Eijnatten, F.M. van (2008a). "Sustained by work: Individual and social sustainability in work organizations," in P. Docherty, M. Kira and A.B. Shani (eds.), *Creating Sustainable Work Systems*, ISBN 9780415772723, pp. 233-246.

Kira, M. and Eijnatten, F.M. van (2008b). "Socially sustainable work organizations: A chaordic systems approach," *Systems Research and Behavioral Science*, ISSN 1092-7026, 25(6): 743-756.

Kira, M. and Forslin, J. (2008). "Seeking regenerative work in the post- bureaucratic transition," *Journal of Organizational Change Management*, ISSN 0953-4814, 21(1): 76-91

Kira, M., Eijnatten, F.M. van and Balkin, D.B. (2010). "Crafting sustainable work: Development of personal resources," *Journal of Organizational Change Management*, ISSN 0953-4814, 23(5), 616-632.

Loon, M. van (2010). *An Exploratory Research about the Short-Term Effects of Value Crafting in the Manufacturing Firm*, Eindhoven University of Technology, The Netherlands, master thesis report.

Nonaka, I. (1994). "A dynamic theory of organizational knowledge creation," *Organization Science*, ISSN 1047-7039, 5(1): 14-37.

Nonaka, I., Toyama, R. and Konno, N. (2000). "SECI, ba and leadership: A unified model of dynamic knowledge creation," *Long Range Planning*, ISSN 0024-6301, 33(1): 5-34.

NVivo 8 (2008). http://www.qsrinternational.com/products_nvivo.aspx.

Schneider, B., Brief, A.P. and Guzzo, R.A. (1996). "Creating a climate and culture for sustainable organizational change," *Organizational Dynamics*, ISSN 0090-2616, 24(4): 7-19

Seevers, S.B. (2000). "Identifying and clarifying organizational values," *Journal of Agricultural Education*, ISSN 1042-0541, 41(3):71.

Spreitzer, G., Sutcliffe, K., Dutton, J.E., Sonenshein, S. and Grant, A.M. (2005). "A socially embedded model of thriving at work," *Organization Science*, ISSN 1047-7039, 16: 537-549.

Turnipseed, D. (1988), "An integrated, interactive model of organizational climate, culture, and effectiveness," *Leadership and Organization Development Journal*, ISSN 0143-7739, 9: 17-21.

Wallace, J., Hunt, J. and Richards, C. (1999). "The relationship between organizational culture, organizational climate and managerial values," *International Journal of Public Sector Management*, ISSN 0951-3558, 12(7): 548-564.

Wilber, K. (1996). *A Brief History of Everything*, ISBN 9781570621871.

Wrzesniewski, A. and Dutton, J.E. (2001). "Crafting a job: Revisioning employees as active crafters of their work," *Academy of Management Review*, ISSN 0001-4273, 26(2): 179-201.

Xanthopoulou, D., Bakker, A.B., Demerouti, E., Schaufeli, W.B. (2009). "Reciprocal relationships between job resources, personal resources, and work engagement," *Journal of Vocational Behavior*, ISSN 0001-8791, 74: 235-244.

Sjaña S. Holloway is a PhD student in the BETA Research School at Eindhoven University of Technology, The Netherlands. She has an M.Sc. degree in Health Psychology from Leiden University, The Netherlands. Her current research interest is in the area of complexity science focusing on the creation of sustainable work systems with the desire to merge the studies of innovation management with the regeneration of human capital for the benefit of both. She has worked as a mental health clinician, consultant, and researcher for the last ten years.

Frans M. van Eijnatten is an Associate Professor of Organizational Renewal and Complexity at Eindhoven University of Technology, The Netherlands. He has a Ph.D. degree in Organizational Behavior from Radboud University, Nijmegen, The Netherlands. His current research interests focus on sustainable work, seen through a complexity lens. He has published about these topics in the Journal of Organizational Change Management, Systems Research and Behavioral Science, and The Learning Organization. He is founder and coordinator of the European Chaos and Complexity in Organizations Network (ECCON).

Marijn van Loon graduated in the Master Program Innovation Management from Eindhoven University of Technology, The Netherlands. He conducted an exploratory study about the effects of value crafting in a Dutch company. His main interests are human resource management, innovation management, and organizational renewal.

10. The Middle Ground: Embracing Complexity In The Real World

Tim Dalmau[1] *& Jill Tideman*[2]
1 Dalmau Network Group, AUS
2 Jill Tideman & Associates, AUS

The interactions of people in organizations around certain types of problems and issues can be seen as analogous to a complex adaptive system. Traditional approaches to management of these problems, whereby it is attempted to establish order and control through the actions of a few at the top has been shown repeatedly to fail. Many persist with this pattern of behavior, ironically often moving problems closer to the edge of chaos than closer towards certainty.

The first step in changing this pattern is to recognize that the style of engagement needs to shift to that more like a set of complex responsive processes. The methodology outlined in this chapter is one such approach, a dialogue and discernment that invariably handles complexity well and leads to positive outcomes. It has been widely used in a variety of settings and sectors and it is highly adaptable to a range of purposes.

Introduction And Context

It is important, we believe, to set the context for what is to be described. We are consultants, not currently academics (although one of us was in a former life), and the people we work with tend to be Board members, senior executives and middle management in large companies in the resources sector, steelmaking, healthcare systems, the financial services sector, the environmental sector and government generally.

We do this work in many different countries and contexts. Through good luck and circumstance we have had the privilege of working closely with Dick Knowles[1] who introduced us to the Process Enneagram as a tool, some 15 years

1. It is hard to over-estimate the contribution that Dr. Richard Knowles has made to our thinking on this model that he created. His book, *The Leadership Dance* (2002) outlines the origins of the model and its application to the field of self-organizing leadership. The work outlined in this chapter builds squarely on his original work, but has also comes from the active involvement of literally thousands of people in exploring, using and refining it in many different settings and countries. Pre-eminent among these is Steven Zuieback to whom we owe a great debt in helping us find new, different and practical ways to use it. Moreover, Ralph Stacey has been extremely formative of our thinking for the last 15 years or so. We are

ago. This model intrigued us and we have made contributions to its development along the way. We soon began to realize that the model seemed particularly applicable to settings where there was a disconnect between the formal organizational system and the hidden informal network of relationships, where there was disagreement about the nature of the problem at hand and uncertainty as to the best solution set.

The premise on which this chapter is based is that most executives with whom we work experience a mismatch (at some level) between the mental models they use to direct activity or discourse within the groups for which they are responsible and the way the system of which they are a part actually works. They lack language often to articulate this difference and it is expressed sometimes as a search for a tool or approach that will somehow reconcile these differences.

Whilst there are many conceptual traps in responding to this need, we believe that over the last 15 years or so, the approach outlined in this chapter provides a pathway that will engender cohesion, allow for emergence, embrace paradox, permit surprise and foster outcomes that simultaneously foster rational achievement, emotional commitment and social cohesion. This chapter relates the approach to current themes from complexity and complex responsive processes and then describes its characteristics, use and deployment.

Perspectives From Complexity

It is not our intention to canvas in detail the field of systems theory, complex adaptive systems theory or the emerging field known as complex responsive processes. That work has been done by others far more qualified and experienced than us to do so, except to say that we are polytheists around the models which arise from these endeavors, choosing whatever framework will both help us best understand the work we do and will lead to practical application.

We happen to hold to a view of organizations as constantly emerging spaces where real individuals converse with one another as they struggle to adapt to challenges and dilemmas, becoming all that they are innately capable of becoming as individuals, as cohorts and as organizations (Neville & Dalmau, 2008). We expect that such spaces allow for both positive emergent solutions as well as fragmentation; for alignment as well as for alienation, but especially for surprise and temporary coherence as the actors involved build new meanings and actions therefrom[2]. What is important to note, however, is that we often encoun-

---

particularly grateful to Bob Dick for critical comment.

2. In reviewing the writing of Ralph Stacey from about 1995 to the present, we find conceptual comfort in using insights from complex adaptive systems theory as *analogues* for understanding the issues and challenges we face. His concept of organizations as spaces in which complex responsive processes occur, describes (often quite accurately) the *nature* of the emergent conversations in which we find ourselves.

ter a tension between the drive to solve problems at the altar of the 'quick fix' on the one hand and the realization that the problem is far more complicated than first thought. Indeed there are often cohorts with varying interests around the problem, quite different interpretations as to its nature and a growing awareness that there is no easy way of resolving these tensions.

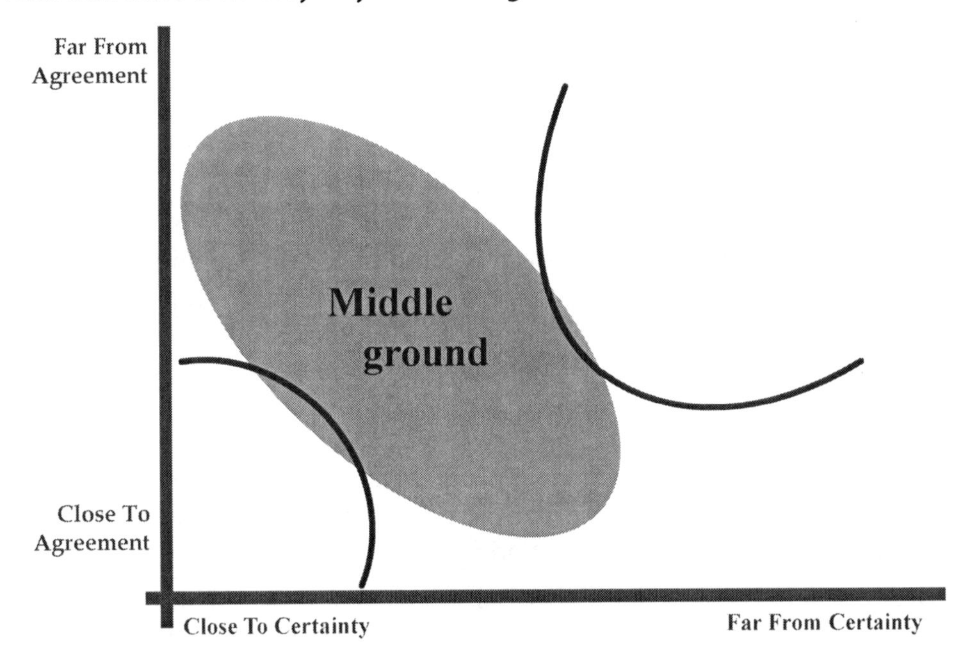

Figure 1 *Problem Types*

Stacey (1996) posits a way of distinguishing alternatives regarding both the context in which decisions are made and the degree of certainty surrounding them, as illustrated in Figure 1[3]. We have used this as a simple means[4] for distinguishing between those problems around which the dominant discourse of linear, top-down, leader directed decision-making is appropriate (what Hazy et al, 2007, describe as administrative or convergent leadership) and those problems where a more enabling or emergent disposition is called for. We have come to call this the "middle ground". In the top right area of Figure 1 are those issues around which there is little agreement and little certainty. Stacey suggests this is an area for avoidance, anarchy and randomness.

It is our contention that the social environments and problem types that many executives and managers face are not in the bottom left hand corner of convergent thinking and cohesive social networks at all, but rather in the middle ground. They are problems and issues around which reasonable levels of uncertainty exist as to the likely efficacy of any known strategy and reasonable levels of disagreement arise from the hidden informal social networks of the organiza-

3. Adapted from Stacey (1996: 47).

4. Easily accessible as a distinction to all in organizations from board members to operators.

tion as to the nature of the problem. It is also our contention that the pervasive and dominant model of change management that drives so much modification to structure, strategy and process in the corporate world[5] largely fails to achieve its desired outcomes because:

1. It is often conceived and executed within the frames of thinking that sit in the bottom left hand corner rather than the middle ground, and;
2. It lacks the appropriate style of leadership to support what emerges.

It is not that such thinking about change is inappropriate. It is appropriate for those problems and issues that sit in this bottom left hand corner and are susceptible to resolution by linear, mechanistic modes—the dominant discourse as Stacey calls it. Rather, this approach is often based on a failure to comprehend that so much change is large, complex and engenders a range of reactions among stakeholders that squarely define it as a middle ground phenomenon.

A few recent examples illustrate the type of problems we often find ourselves considering,

- The total re-working of the mining program at a bauxite mine near the end of its life, hampered by a low grade ore body and close to a community that has low tolerance for noise pollution from heavy machinery;
- The engagement of over 150 medical, nursing and allied health professionals to influence over 3000 health professionals regarding patients who require early and radical secondary intervention for early stage renal failure;
- The integration of 17 different country practices into one firm in the professional services sector;
- The reformulation of a global marketing strategy for a global financial services firm;
- The re-development of a close knit team of health professionals working with the medical staff and families of organ donors as they struggle to avoid introjecting into their own relationships the existential angst, confusion, anger, loss and dread that is naturally part of this type of work;
- The establishment of guiding principles and actions that generate a step change improvement in safety performance and an unplanned increase in the robustness of relationships among the leadership cohort of a steel manufacturing company;
- The re-constitution of a Board and the leadership of a multi-national company after about half of the Board members had been indicted for criminal offences;

5. The dominant model of change management in many of the organizations where we work is usually a derivative of John Kotter's work

- The re-patterning of engagement with households in a neighborhood of a large Australian city to stimulate behavioral change and transport choices that reduce greenhouse emissions.

Organizations and their leaders can find themselves facing such problems arising from a variety of situations, including neglect, a crisis, persistent conflict[6], cycles of failure, persistent disconnects between what is espoused and what is practiced (Dick & Dalmau, 1990), or something that is outside the current know-how of the group (Heifetz, 2009).

It is further proposed that such problems or issues and the semantic space in which they are created can be informed by the models of both complex adaptive systems theory *and* an understanding of complex responsive processes Stacey (2001). By system we mean a conceptual framework or entity that sees organizations as collections of individuals, that at times seem analogous to living entities and where complicated interactions among human beings in the "ordinariness" of their day to day communication seem to produce a level of self-organization at a larger level.

It is our experience that such problems and the social spaces in which they arise often have the following characteristics:

- The issue at hand is complex: there are many enmeshed and interacting parts some of which have either amplifying (positive feedback) or dampening (negative feedback) effects;
- The future is under construction in the minds of the players involved; often it is either wholly or partially unknowable in real and pragmatic ways;
- The drive to convergent thinking is balanced by an awareness of the very different (and often quite emotionally invested) interests that stakeholder groups have around the issue;
- It is beginning to dawn on the executives and managers involved that they actually can't control for the future and that there are real limits to their drive for efficiency;
- Establishing the boundary of the system is often fraught because of the various hidden and informal interests around the issue;
- Social strange attractors (Stacey, 2010) are at play in the form of established cultures, i.e., sets of assumptions, beliefs and perspectives that, in turn, guide action.

Snowden and Boone (2007) describe a complex system as one that:

- Involves interacting elements;
- That are non-linear;

---

6. Ronald Heifetz: http://www.youtube.com/watch?v=QfLLDvn0pl8,

- With emergent properties;
- Arising from a history of internal engagement of the parts and with the environment, and;
- Where what happens next cannot be predicted.

We contend that the situations we often face and that are the subjects of this chapter tend to look like complex systems and we take from Stacey (2010) the thought that complex adaptive systems theory, in particular, provides us with some useful analogues.

The forces described above are being played out in the very local and personal interactions between the players involved over time. The following characteristics are typical of the situations susceptible to the approach we describe:

- More often than not problems get to be resolved by ordinary people in organizations sitting down together and communicating in real time face to face[7].
- Their interaction (arising from their diversity) produces new forms of knowledge and a coherence to their shared understanding of the problem
- The focus of conversation is the current actual activity in which the players are engaged and the emerging future they desire to resolve tensions between reality and desire.
- It is in the conversation about these tensions that the players come to see new forms and patterns emerge, surprise often occurs, and the result is a shared will to act in some new manner. Mostly this produces new and more effective forms and processes, but it can also occasionally lead to less effective ones. In other forums, we have suggested this can be stated as creating rational, emotional and social outcomes for those involved.
- They are conversations in which their success depends on two critical variables, having the right people in the room and having a real and strong diversity of views among those voices.
- Among these right people in the room are a broad spectrum of both overt and covert power potential.
- Although the conversation is loosely directed, nevertheless self-organization and the emergence of both stable and novel patterns occur at the boundary of order and disorder in the interaction (Suchman, 2002). Making the diversity public is a key feature of the conversation and one which thereby fosters the emergence of new meaning and surprise.

---

7. We have consistently found that successful resolution of these complaints invariably involves people sitting down together and interacting with one another to create shared meaning, i.e., engaging with one another. Often this is all that is required to get a first approximation of a good answer which can than be further built upon.

- The interaction often starts with expression of affect and exploration of relationships. In other words feelings and emotions, meanings and personal constructs, beliefs and personal perspectives become part of the interaction: both the rational and the non-rational are grist for the mill.
- The end result is that the participants move more "towards" the issue, the problem or dilemma than away from it—they tend to personalize it more, attach some personal investment of their own to it and are far less likely to objectify it, i.e., remove themselves from it.

It is our contention that the previous lists seem to suggest the following: firstly, that the situations and issues we often find ourselves addressing with client personnel seem analogous in their characteristics to complex adaptive systems, and secondly that the nature of the process which emerges when we apply this process seems to bear many of the hallmarks of complex responsive processes. We do not say this is what occurs every time, but it does so more often than not. If not approached with both alertness and sensitivity it can quickly "degenerate" into either another variation of mediocre critical systems thinking. Worse still, if applied "mechanistically" then it is little more than another form of convergent thinking and management.

In the context of the distinctions outlined so far, we now describe a practical approach to embracing complexity in the real world.

The Process Enneagram

The Process Enneagram is, in our experience, a powerful conversation model, a planning framework and a diagnostic tool that has been used extensively in a range of organizational settings internationally over the last 10-15 years. It is particularly well suited to engage with those types of problems and situations that are analogous to the complex problem spaces described above.

It has been used in diagnosis, planning, facilitation or engagement, conversation mapping, coaching, integrating and clarifying issues, and galvanizing and coalescing shared will. It has been used in hostile union-management conflict resolution, safety assessments, client diagnosis, project planning and review, facilitating mergers and acquisitions, roll outs of enterprise resource planning systems (e.g., SAP, JDE), re-aligning world-wide marketing strategies, team building, and change processes including the re-configuration of whole industries within and across countries. People from accountants to coal miners, health workers, steel makers, corporate executives, marketing professionals, IT professionals, social workers, administrators, politicians and community workers have used the Process Enneagram with success and satisfaction. In almost all cases of its use, it produces results in the social space of the conversation similar to those described in the previous section of this chapter.

A Whole Of System View

At the most fundamental level the Process Enneagram enables either those engaged in the dialogue or those using it to generate a picture of the whole system that is their focus or interest, whether it be a bauxite mine, a metropolitan health service, a work team or a neighborhood. The significance of this should not be under-estimated as many organizational and management practices (what Stacey calls the dominant discourse) start out by breaking problems down into their constituent elements, believing that if they are re-constructed in some new manner then the "problem" will be solved—the focus is on the parts not the whole.

There is a place for this mindset, but mostly where there is both high social cohesion and stability of power relationships combined with high predictability as to the efficacy of a solution set—what we might call the bottom left hand corner of Figure 1. This approach however, often fails to acknowledge or understand the powerful and hidden human, cultural and social forces at work in organizations. It is these forces where there is less agreement and cohesion that tend to undo well-intentioned mechanistic approaches.

The use of Process Enneagram model as an approach to middle ground problems helps those involved to see "the system" with all its rational and non-rational elements as one, in a simple and visual manner. And when it is supported with appropriate leadership from power figures in the organization, it tends to produce coherent, surprising and engaging solutions.

A Way Of Engaging With Complex 'Middle Ground' Problems

As already outlined we contest that the Process Enneagram is a practical process framework and methodology for addressing complex or "middle ground" problems, especially those where there is reasonable uncertainty and reasonable to high disagreement among those involved as to the nature of the problem. This approach is not to be confused with the more conceptual distinctions found in some of the literature in the field, e.g., Snowden and Boone (2007), Olson and Eoyang (2001), or even more prescriptive approaches such as those described by Macdonald, Burke and Stewart (2008). It is operational and hands-on in its engagement of the players. Most of the people with whom we work seek for simplicity in any distinctions they are given, and are usually time poor. Consequently, we try to lead or guide their thinking using devices that make distinctions simply and quickly for them[8].

In real life, we suggest to personnel in client organizations that problems, issues or dilemmas that satisfy one or more of the following simplified criteria can be thought of as "middle ground" problems. We suggest that whenever:

8. A perilous activity on some occasions.

- The situation is complicated and enmeshed, and/or;
- The change sought is complicated, and/or;
- The outcomes are vague or unclear, and/or;
- There are unknown or unpredictable forces at work that can influence or interfere, and/or;
- People's feelings or reactions are likely to be triggered significantly, and/or;
- There is a need to equip, educate or train others to implement and sustain change, and/or;
- There are politics involved or likely, and/or;
- Individuals or groups have the potential to feel disenfranchised as a result of change.

it is highly likely that a different mindset and methodology is required to engage for resolution—a mindset informed analogously from complex adaptive systems theory and a methodology consistent with the nature of human interaction in such circumstances, i.e., complex responsive processes of interaction.

With the Process Enneagram forming our approach, we seek to ensure the conversation or thought process covers nine different domains of inquiry. These nine domains and the diagram on which they rest were developed by Dick Knowles as early as 1992[9] when he started to connect the work of Margaret Wheatley (2006) with that of the English philosopher J.G Bennett[10]. He also formed his views as to the practicalities of the model in the crucible of running a manufacturing plant for the DuPont company where he led an extraordinary and sustainable improvement in safety and performance across many dimensions.

When Dick Knowles first presented the enneagram to us in the mid-90s in Australia it contained the elements of identity, vision, tensions, principles and standards, work, information, learning and structures. The term process was added before the term enneagram in the late 90s to distinguish it and separate it entirely from the work done by others using the enneagram related to typing personality. There is no connection between the two. Tim Dalmau suggested that the word vision was too limiting and that intention was a better term that could encompass not only vision, but in the corporate world also ideas such as high level values, value add, strategic intent and customer value—concepts and ideas that tend to often drive human energy and action, as much as or sometimes more than vision. For a variety of linguistic and national-cultural reasons Tim Dalmau expanded the terminology on the nine points to cover:

---

9. Personal conversation with Dick Knowles.

10. This chapter, our use of the model outlined, and its efficacy do not depend in any way on its connections to the work of J. G. Bennett. Rather, as is described later, it is the nine points of focus combined with the order of engagement that makes the approach so useful.

- Identity and current state;
- Intention;
- Tensions and issues;
- Relationships and connections[11];
- Principles, ground rules and standards;
- The work;
- Information and will;
- Learning and sustainability, and;
- New context, structures and approach.

Now we have asserted that there are a class of problems often found in organizations in which the dominant reductionist mindset is not only ill-suited but will tend to drive them towards entropy. We have suggested that the nature of these problems can be informed analogously from complex adaptive systems theory. Moreover, we suggest the methodology or approach outlined in this chapter is, of its nature, both complex and responsive in the manner described by Stacey (2001).

Olson and Eoyang (2001) suggest that a complex adaptive system does not admit to any order or sequence of steps in a process or predictable, staged outcomes. Goldstein (2005) suggests otherwise when he states:

> *Self-organization (is) a term that suggests spontaneity and the inner-driven onset of new order. [A] careful inspection of research in complexity theory reveals the emergence of new order is more appropriately constructed than self-organized as such.*

We tend to agree with this point of view, and contend that conversations by real people in real settings, sensitively stimulated around these nine dimensions, allow for emergence, paradox and surprise to come forth. Paradox and surprise are important elements of deep learning—the contradictions, inconsistencies and absurdities to which they are attendant foster new perceptions, different perspectives and a level of disassociation that allow a group to move forward.

The nine dimensions tend to bring clarity to those in the conversation, stimulate commitment or a "drawing to" new aligned action, and enhance the functionality of the relationships of those involved. This result seems to result from, at a process level, the order of the conversation as represented in Figure 2. The order tends to produce these three types of outcomes: rational, emotional and

11. For some unknown reason, common usage of the term "relationships" in Australia does not allow for encompassing connections between non-human elements in a system and for this reason the original term was expanded to relationships and connections, to include such things as company to company, department to department, function to function interactions.

social and as such, the Process Enneagram shares this space with very few other processes, e.g., the process of dialogue fostered by Bill Isaacs (1999) and his colleagues and the early work in the 1950s of the Institute of Cultural Affairs[12]. Put another way, the order of a conversation or engagement seems to profoundly influence the outcomes of the engagement in terms of the emergent clarity of direction or purpose, the sense of commitment or attachment that a group feels towards the issue or problem at hand, and the willingness of the individuals to be part of the action going forward. A note of caution: using this model and approach to address problems in the bottom left hand corner of Figure 1 tends to over-complicate and confuse what should be routine rational problem solving.

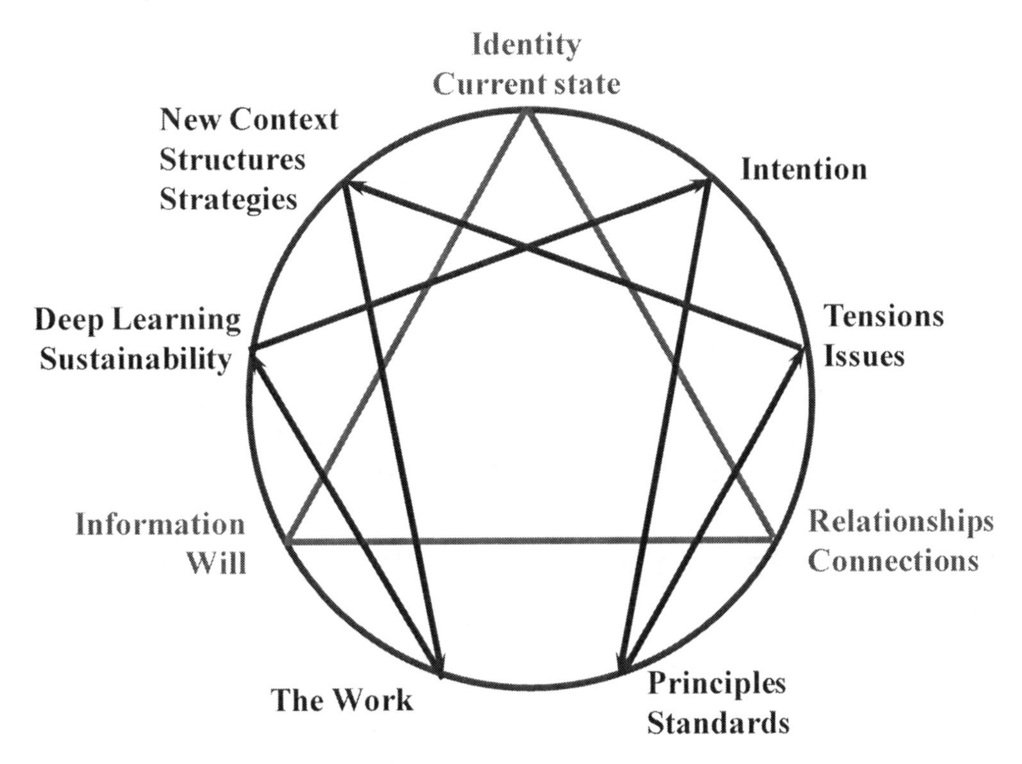

Figure 2 *A Process Enneagram*

Nine Points Of Inquiry

In its simplest representation the Process Enneagram can be viewed as nine areas of inquiry and conversation, which are as follows:

Identity and Current State: Sense of purpose and meaning—the area of focus that unleashes energy and commitment. The description of the present state in which the players find themselves and the historical forces and factors that have shaped them.

12. www.ica-usa.org.

Relationships and Connections: Description of the nature of relationships that the players have with other individuals, groups in and across organizations as a whole, both among people but also among units, functions, groups, and processes. This area includes both ideal and actual relationships.

Information and Will: Understandings about the importance of information, the relationship of information to individual and organizational effectiveness, and its impact in coalescing shared meaning to marshal concerted action.

Intention: What is it that the players want to create and achieve with people, within their areas of responsibility, around the problem or issue under consideration, or the organization as a whole?

Principles, Ground Rules and Standards: The priorities they hold, the principles they work to, the ground rules they seek to abide by and the standards that model their beliefs and aspirations within the organization, i.e., the underlying ground rules and priorities that guide (or should guide) behavior, strategies and tactics—both espoused and in-use (Dick & Dalmau, 1990).

Tensions and Issues: The existing dilemmas, constraints, contradictions and behavioral patterns that are currently keeping individuals, teams and the organization from realizing the desired outcomes.

New Contexts, Structures and Approaches: The creative perspectives, strategies, structures and approaches that model the espoused principles, ground rules and standards, resolve the tensions, and thus move the "system" toward the intended outcomes.

Work: The actions associated with the strategies that need to be implemented.

Deep Learning and Sustainability: This describes a process of ongoing reflection whereby individuals, groups and organizations can learn what is working relative to their outcomes and make course corrections based on the learning, combined with mechanisms for adjustment and regeneration to ensure sustainability.

The Green Triangle

The three points of the green triangle that underpin this model are identity/current state, relationships/connections and information/will. First developed by Meg Wheatley (1992), she originally called them relationships, information and vision.

Within organizations, there is a continuous shaping and re-shaping of the identity of the system itself, an evolving of connections and relationships and incessant flows of information. These intangible aspects of any social system have a profound influence on all other aspects of the system. Unfortunately in many organizations where the dominant discourse is that of a reductionist mentality (bottom left hand corner Figure 1) the leaders regularly seek to improve performance without paying attention to these three key variables. The first recourse seems to be to re-structure the hierarchy, re-engineer the procedures or re-vitalize the strategic plan, and often all three. This is done without attention to the impact of how relationships, identity and information interact together to create a dominant mindset that will often undo the best efforts and good intentions at organizational improvement.

These three factors need to be functional (at a minimum) in the group of players brought together to address a middle ground issue or problem. If not, they need attention first. In reality when we have players come together to address a middle ground issue, the degree of coherence among these three can vary enormously and it is usual to devote significant energy to these three elements first. Indeed, the green triangle in the diagram underpins the other six elements and, we suggest, should be foci of constant attention and energy for leaders. It seems common sense to us that if a group of people have re-asserted some current identity, opened up the information flow, and enhanced their relationships with one another they will be in a far better place to discover a shared intention, priorities, address issues and resolve how to move forward together.

The Conversation

In reality, a group comes together to address an issue, resolve a problem or create some new outcome that exists in the "middle ground". Depending on the issue the conversation could take anywhere from two hours to a few days. The conversation works through these nine points of inquiry more or less in the order indicated by the arrows—the ordering of the conversation is not prescriptive: common sense, surprise, paradox and emergent contradictions may dictate varying either the order or the amount of time spent on any given point. That said, the most powerful clarity and/or attachment seems to emerge when the order indicated in the diagram is followed.

In using the Process Enneagram to guide or map a conversation the normal starting point produces a shared and explicit understanding of the identity, history and current state of the group, the relationships and connections in the system that exist or are desired, and the access to and flow of information throughout the system along with the extent to which this information flow does or does not promote shared will to act. The conversation then moves on to the intention (or vision, objectives, aim or purpose) or possible intent that can emerge or is required about the issue. Kellner-Rogers[13] stated that too often we ask the question *"What is the problem and how do we fix it?"* when we should be asking the question *"What's possible here and who cares enough to make it happen?"* This simple shift in questioning at this point seems to move the conversation, from the bottom left hand corner to the middle ground, and starts to build an emer-

---

13. Myron Kellner-Rogers in presentation at Berkana Institute Dialogue, Sundance Center, Provo, Utah. October 2, 1997.

gent coherence. In other words, it tends to pre-empt the drive to reductionist, mechanistic, linear models of thinking.

The conversation then moves on to two highly related but distinct foci: the principles, ground rules and standards the players espouse and those they actually use (Dick & Dalmau, 1990). Next tensions, issues and dilemmas are explored. It is important to note that by the time these come to the fore the players have already considered the following:

1. The current state and how it came to be;
2. The nature and functionality of their relationships and connections;
3. The availability or otherwise of all the information that flows around the issue;
4. What they want to achieve, and;
5. The ground rules/principles they use and espouse.

In our experience the fact that they start examining "the problem" only after these previous five foci have been examined plays a very large role in both fostering a willingness to act in concert to create something different and in preventing deficiency and reductionist models of thought from taking hold in the group.

It seems to set the stage for emergence and novelty to come forth as the players consider the next part of their dialogue, viz: what new context is being called for, what approaches are needed and what strategy shall follow? The quality and efficacy of this part of the conversation depends on all previous six elements being covered thoroughly. It seems normal that groups then move to the phase of who is going to do what and this part of the conversation may look and familiar to those who take comfort in the bottom left hand corner of Figure 1, i.e., action planning. It is a natural result of all the steps that have preceded it. The conversation then moves to a deeper reflective phase. The players are invited to reflect on their own experience in the conversation, to assign meaning to it, to signify it in terms of other experiences that come to mind, and to examine how they will sustain the work that has started.

These conversations involve ordinary people sitting down together in real time. Providing that the right people are in the room—representing those who can affect outcomes or are affected by the outcomes and a diversity of mind set and interest groups. Then focussing on these nine domains of inquiry combined with the order in which they are addressed seems to produce new forms of knowledge, surfaces paradox and contradiction, and allows coherence to emerge at all three levels of rational, emotional and social outcomes. In our view this seems to be at a local level not unlike what Stacey (2001) describes as a complex responsive process. In saying this we are mindful that Stacey presents such processes as occurring naturally in the daily intercourse of the hidden informal system in organizations. We agree, and believe it is also possible to create social spaces that trigger and foster such processes.

It also needs to be noted that there are often issues and problems, which of their nature, cannot be addressed in one sitting, so to speak, no matter how long it goes for. This model allows for groups to iterate their conversation on a number of occasions, "spiralling" if you will, coming back to it with new insights, fresh understandings, and more divergent thought that enriches the conversation next time[14]. In our experience these become extremely significant and powerful conversations, and we are reminded of one such conversation of which we were part with a US-based steel company senior executive cohort that continued over 11 years!

The Process Enneagram In Use

The Process Enneagram is, at one level, a guide to a type of conversation that, we believe, has some key similarities to what Stacey (2001) describes as a complex responsive process. As such it can be as simple or complicated as is needed by the group and the issue/problem. Simple generic questions for the nine points of the Process Enneagram that can guide conversations is shown below in Figure 3. These questions can be tailored for the specific situation and setting.

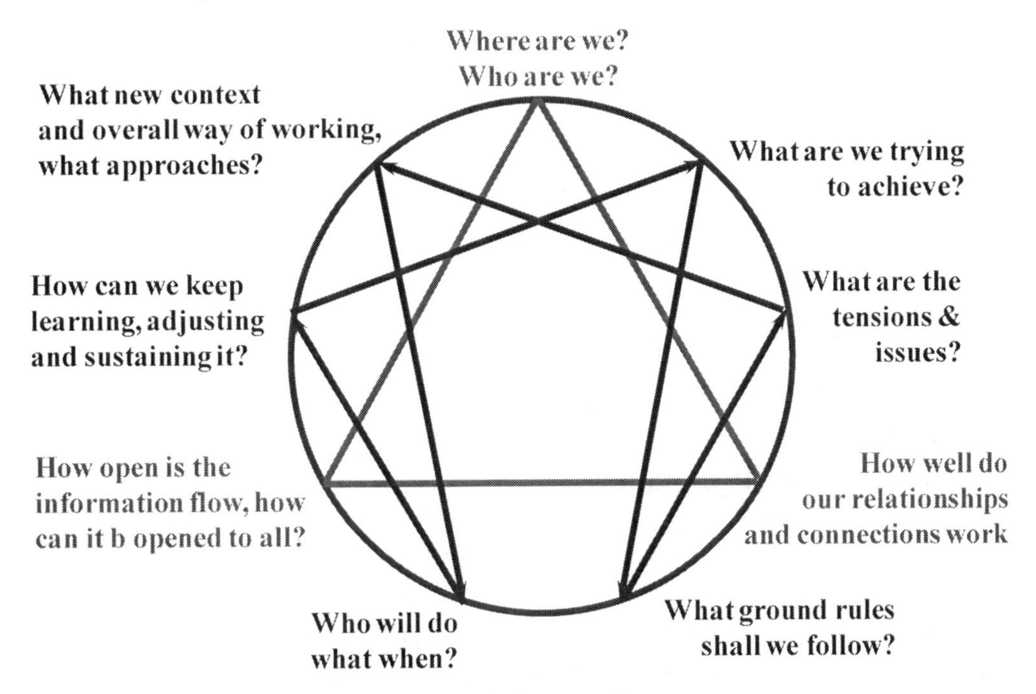

Figure 3 *A Simple Process Enneagram.*

14. Dick Knowles first mentioned this notion of a spiral, or corkscrew model of development to us in the mid 1990s. He said Charles Krone gave it to him as an image.

In addition the Process Enneagram can also be used as a way to record a conversation.

In our experience the starting point is often placing a small version of the diagram at the centre of a large canvas of flip chart paper (4 sheets taped together), and then over the course of the conversation, recording key points at the appropriate location on the Process Enneagram. Figure 4 shows a Process Enneagram that was created to guide and stimulate a conversation with 3 different coal mine leadership teams from a company in Queensland, Australia. Figure 5 is the summary created following a day-long conversation with each of the 3 coal mine leadership teams. It was collated from 3 similar one-page records to share with the corporate office and help them gain some insight into the challenges facing these teams and the next steps required to improve their performance.

A collection of templates, like those in Figure 4 has been collated into what we refer to as an *Enclave of Enneagrams*. These have been developed over the years by a range of people as a resource for practitioners or others in organizations to use. They cover the following types of situations, inter alia:

- Diagnosing and clarifying the problem;
- After action reviews;
- Company interviews;
- Strategic planning;
- Business planning;
- Organizational design;
- Facilitation design;
- Positioning an organization for sustainability;
- Reviewing change;
- Analysis of self as a leader;
- Project development;
- Coaching;
- Community development;
- Team development, and;
- Inter-group conflict[15].

In many instances using the Process Enneagram approach may require us to first have a one-on-one conversation with a range of people in an organization, such as members of a leadership team. These inquiries seek to establish the context of the interaction to come and who needs to be part of it, seeking always to involve key stakeholders and ensuring there will be dissonant and diverse voices present. They also seek to evaluate the type and style of leadership thinking in

15. The *Enclave of Enneagrams* can be viewed at www.dalmau.com/resources.

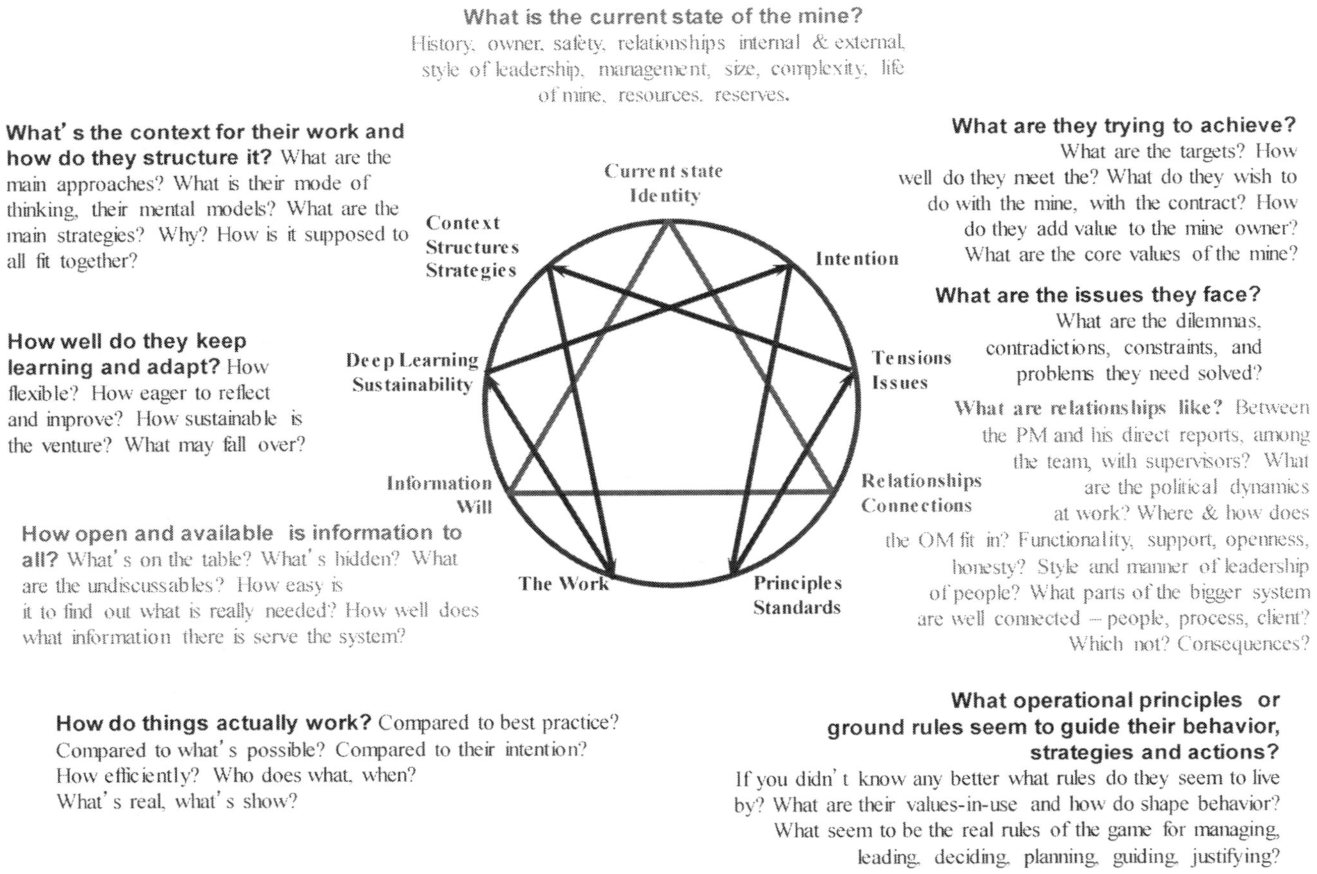

What is the current state of the mine?
History, owner, safety, relationships internal & external, style of leadership, management, size, complexity, life of mine, resources, reserves.

What's the context for their work and how do they structure it? What are the main approaches? What is their mode of thinking, their mental models? What are the main strategies? Why? How is it supposed to all fit together?

What are they trying to achieve?
What are the targets? How well do they meet the? What do they wish to do with the mine, with the contract? How do they add value to the mine owner? What are the core values of the mine?

How well do they keep learning and adapt? How flexible? How eager to reflect and improve? How sustainable is the venture? What may fall over?

What are the issues they face?
What are the dilemmas, contradictions, constraints, and problems they need solved?

What are relationships like? Between the PM and his direct reports, among the team, with supervisors? What are the political dynamics at work? Where & how does the OM fit in? Functionality, support, openness, honesty? Style and manner of leadership of people? What parts of the bigger system are well connected – people, process, client? Which not? Consequences?

How open and available is information to all? What's on the table? What's hidden? What are the undiscussables? How easy is it to find out what is really needed? How well does what information there is serve the system?

Current state
Identity

Intention

Tensions
Issues

Relationships
Connections

Principles
Standards

The Work

Information
Will

Deep Learning
Sustainability

Context
Structures
Strategies

How do things actually work? Compared to best practice? Compared to what's possible? Compared to their intention? How efficiently? Who does what, when? What's real, what's show?

What operational principles or ground rules seem to guide their behavior, strategies and actions?
If you didn't know any better what rules do they seem to live by? What are their values-in-use and how do shape behavior? What seem to be the real rules of the game for managing, leading, deciding, planning, guiding, justifying?

PM = Project Manager. **OM** = Operations Manager.

Figure 4 *Process Enneagram To Guide Conversations With Coal Mine Leadership Teams.*

LEADERSHIP & MANAGEMENT - 3 COAL PROJECTS, QLD

Has been little stability in leadership teams : PMs relatively new / inexperienced/unskilled (team members too). Until this work started no whole leadership team meetings, and no leadership team identity. Clients are challenging and demanding, and uncertain contract situation is a real challenge for some sites. Safety – in last 6 months safety record trending wrong way; leadership is challenged to engage staff effectively. Facing problems with productivity due to rain. All challenged by gaps and lack of experience in workforce, exacerbated by turnover. ABC, DE starting to make good progress now with team ; FG, less progress and some different issues. Working excessive hours, and still 'drowning. Demands from offsite management always urgent. Interpret this as inappropriate interference / lack of appropriate support

Continue to build l/ship team ; get our teams working better together. Improve our ability to work at the right level. Develop more collaborative and consistent processes for interacting with Client. Think, work and plan with a more long term view Develop / implement innovative recruitment strategies

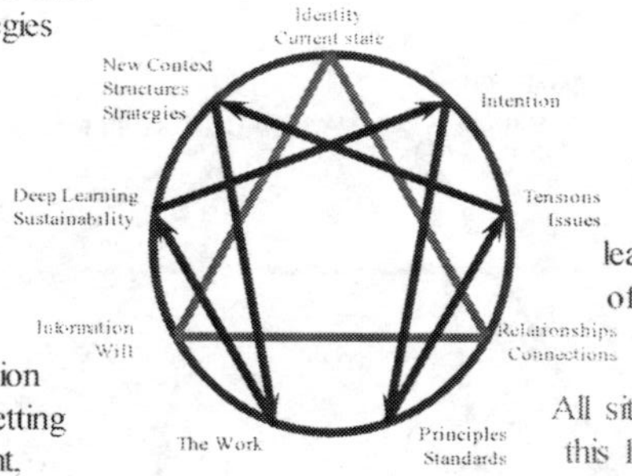

Meet production , HSE targets. Keep Client happy/Make a profit for Company. Make site a good place to work. Step up to changes resulting from restructure . Limited focus on positioning the project for the future

Identity
Current state

New Context
Structures
Strategies

Intention

How to step up to more complex role of PM, arising from re-structure (flow on). Client relations & contract management. Workload – lack of work/life balance (rosters). Recruitment, skills and capabilities. Management not leadership. Lack of accountability / understanding of role and expectations. Poor working conditions at ABC. Worker morale, loyalty

All creating discipline of regular l/ship meetings and work together. Socializing more as team and with their people

Deep Learning
Sustainability

Tensions
Issues

Information
Will

Relationships
Connections

Lack of whole team sharing of information / decision making – awareness now, and being addressed. Getting good consistent information out to crews a constant, widespread challenge (harder with constant turn-over). Information / requests flow both ways between site and Brisbane, but is anyone 'listening'?

The Work

Principles
Standards

All sites need to improve connections in l/ship team – this has started. Relationship with Client varies (good CV, poor and improving at .ABC, DE). Relationships with offsite management could be improved. Little connection between Site PMs. Connection with workforce need improvement everywhere – struggle to know how to do best

Have established regular l/ship meetings, and forward agendas ABC , DE moving forward with role clarification and shared team values discussion. 'Greenskins' recruitment strategy DE, Developing improved site communication strategies CV – need more support, other issues being resolved may assist

Principles in use v. desired principles
In use – work harder , will solve problems, actions rather than thinking /planning, management not l/ship, no follow through, train for compliance not development etc
Desired – plan, lead, follow through, use our peoples potential, develop our people etc

Figure 5 *Process Enneagram Summarizing The Record Of The Conversations With The Coal Mine Leadership Teams.*

the system. We have learned in the university of experience that some mental models around leadership are inimical to this type of work, others are neutral and some are quite supportive.

The Process Enneagram is used either explicitly, but usually implicitly, to guide an interaction. The questions put to the players are simply triggers to stimulate the conversation at deeper levels. The next step is to reflect back to the players a view of the system in focus that they have created from the various conversations. A diagram such as the one in Figure 5 makes it easy to tell the story of what has been uncovered. Often the reaction when the players see the whole picture is one of instant recognition, but also of surprise, shock, relief and sometimes denial.

Typically this whole of system view has not been visible to them before, and elements that were part of the hidden left hand conversation (Argyris & Schon, 1974) suddenly appear arising from, what players experience as, an authentic dialogue with little defensive routines evident. As a visual representation, this map on the wall allows for further group engagement and discussion.

In many circumstances the group decides that, although this paints an accurate picture of the current state of their system or problem or issue, it is not what they desire. A second dialogue is facilitated with the whole group to create a picture of a desired state and the strategies and actions that they need to put in place to achieve their intention and address the concerns. Some of these actions will, rightly, address aspects of the problem that lie within the bottom left hand corner of Figure 1. What is distinctive and different, is that the visual representation of this interaction for the players inevitably triggers a deeper conversation about the "system" as a whole and fosters shared meaning attribution, thus promoting rational, emotional and social outcomes.

The guidance or facilitation of this interaction is a sensitive task, balancing open-ended questioning with more pointed exploration, combined at times with unstructured dialogue. It has been our experience that this is best left to run its way to completion before the visual representation of the interaction is put to the group. These conversations can be tough at times, but if the facilitator can maintain a disassociated state then the diversity of voices in the room will tend to ensure that

- Paradox and confusion are embraced;
- Surprise and new mental models emerge;
- A detached whole of system viewpoint is painted and;
- Actions around which the players can align are produced.

At a deeper level, however, we also tend to see more functional relationships and a higher willingness to act (at least around the issue concerned) with greater shared alignment. Dee Hock (1999) asserts that real change is not about reengineering or re-organizing but about re-conceiving and it is our consistent experience that this approach inevitably leads to a re-conception of the issue or problem at hand, and thereby releases new possibilities and ways for the players involved to be and work together.

Accessing This Approach And Methodology

In response to demand by clients and our own desire to equip people to solve their own problems, we have found we can teach people to use the Process Enneagram without them having to understand in detail complex adaptive systems theory or the underlying science and mathematics on which it is based.

It is true that we thereby run the risk that the theoretical integrity and/or approach is simplified in such a way as to compromise the quality and success of the outcomes. Insurance against this is hard to find with any model once it leaves its creator's hands. Stacey (2001, 2010) identifies a similar concern when he describes the use of complexity theory in various guises as simply the historical dominant discourse of the bottom left hand corner dressed in new linguistic clothes. There is no doubt that having a working grasp of complex adaptive systems frameworks is required for the use of this framework—what Stacey (1996) describes as "extraordinary management". Along with this goes the style of guidance/facilitation described above and a willingness to embrace emergence, surprise and paradox at a very basic level as it arises from fierce conversation and dialogue.

Prior to teaching this model and its process to potential practitioners it is useful for them to have had the prior experience of being part of a dialogue that has been guided by the Process Enneagram. More than one experience under different circumstances is even more beneficial. Debriefing the process and providing some context in terms of systems thinking and Stacey's model should follow. A critical point is to help them to identify how they know if they have a complex 'middle ground' problem. Hence we recommend the use of a checklist similar to that referred to earlier in this chapter.

There are four types of situations in an organization that a person may routinely find that the application of Process Enneagram would be useful. These include

- Leading a dialogue (such as a strategic conversation in a workgroup);
- Leading a short working session around a specific complex issue;
- When tackling a complex issue alone or with a few others other, or when;
- Using the Process Enneagram with people who have fewer years schooling (e.g., engaging shop-floor operators around safety).

In addition to teaching the associated process facilitation skills we have created a guide around these four scenarios[16]. The guide is a mixture of facilitation and process tips (for example, how to decide who needs to be involved, reminders to set explicit rational, social and emotional outcomes for the dialogue, what needs to be done in preparation, language to use) and the Process Enneagram questions (plus alternatives) for all nine points, in the correct order.

Within this guide a workbook approach is provided for when there are a very few people who wish to "think/talk" together through a complex middle ground issue, such as one that lies at the heart of an improvement they wish to make. The workbook is set out in such a way that it steps the user through the Process Enneagram. This approach has been particularly useful when equipping a range of people in an organization to sit down with a few others in a dialogue around a major organizational change that fits the criteria of being a middle ground issue. Finally some suggestions are made for working with less well-educated people. It seems that the number of years a person spends at school statistically improves two things: number of words in their vocabulary and their capacity to make finer internal visual distinctions as a form of thought. It has little to do with intelligence. The guide again provides tips for adapting the facilitation and wording for groups such as this.

Beyond The Conversation

This chapter focuses largely on the use of a methodology and approach to foster interaction and emergent responses. It asserts consistency with complex adaptive systems theory and has some of the characteristics of complex responsive processes. It is essentially about a conversational tool for a form of dialogue.

However, this is not the only use of the underlying framework. Whilst each of these other uses could require another paper similar to this, it is worth noting that the Process Enneagram can also be used as a basis for:

1. A framework of thought consistent with complex adaptive systems theory;
2. An approach to planning large and complicated changes;
3. A way of diagnosing a complicated and extensive problem;
4. An approach to designing and guiding the process facilitation of many different workshop formats, and;
5. A framework for leadership.

It is this last use that is worthy of further brief comment in this chapter as the type and style of leadership in the client organization can radically affect the efficacy of this approach.

16. Available at www.dalmau.com/resources.

A review of leadership theory will quickly reveal a progression over time through models of transactional leadership (Berne, 1963), situational leadership (Blake & Mouton, 1972), transformational leadership (Burns, 2009), to servant leadership (Greenleaf, 1977). Derived mostly from the individualistic psychology of the thinkers in the north-eastern regions of the US, these models tend to presume that the individual leader lies at fulcrum of change and improvement, and is capable of making the required change.

Alongside this trend there appeared the neo-emergent theorists who propose a fundamental dynamic between the leader and the system for which s/he is responsible. These are founded on concepts of stewardship (Block, 2002), systems dynamics (Senge, 2006), and adaptive leadership (Heifetz *et al.*, 2009). The dynamic to which they point is far more than that addressed with the technical techniques of transactional and transformational leadership, but this group of thinkers seem to stop short of the largely European and UK tradition of seeing the system as worthy of as much, or more, consideration than the individual. In essence they are still caught in the "psychology of the individual" of North America.

When one steps across this metaphoric Atlantic gap a very different form of leadership starts to emerge and is represented in the work of Capra (2004), Stacey (1996, 2010), Wheatley (1998), Knowles (2002) and Snowden and Boone (2007). For these thinkers, leadership is something that resides in the "system". It is the role of the leader to foster, stimulate and entice this from the system. For example, Wheatley talks of leaderful organizations. It is not possible to work in this frame without a high level of disassociation that evokes in the leader a whole-of-system viewpoint, a tolerance of ambiguity, surprise and contradiction and a recognition that one is not in control.

It is our contention that leaders who operate in the first of these three paradigms will not usually be able to capture the potential that lies within the methodology and approach outlined in this chapter. Whilst it is possible for an outside facilitator or consultant to approach a middle ground problem with the methodology outlined herein it is likely the work will be either unsupported or unwittingly undermined by the leader involved.

Those who operate from the second group of mental models can quickly grasp the nature of a middle ground problem and are likely to at least be neutral in their impact on the resolution of the issues or problems involved. In our experience, however, it is those operate from the third group of mental models, i.e., who see leadership as residing in a living system and waiting to be evoked that better understand and actively embrace this approach and the theoretical frameworks on which it is based. They are therefore more likely to be able to capture the very best of working with a complex adaptive system.

Reprise

As we have been at pains to emphasize, the value of the Process Enneagram lies in its usefulness for addressing complex problems in organizations and groups. In our experience it reliably delivers rational, social and emotional outcomes. Its distinctiveness lies in its applicability to what we call middle ground problems—those about which there is limited agreement and limited predictability or certainty as to outcome. As such it is an approach that, in our experience, lies within the domain of complexity theory, and tends to foster and embrace complex responsive interactions and conversations in organizations. Its success relies on both the appropriate leadership style and mindset and on sensitive, flexible and disassociated facilitation.

References

Argyris, C and Schon, D. (1974). *Theory in Practice: Increasing Professional Effectiveness*, ISBN 9780875892306.

Berne, E. (1963). *The Structure and Dynamics of Organizations and Groups*, ISBN 9780345284730 (1979).

Blake, R. and Mouton, J. (1972). *The Managerial Grid*, ISBN 9780872014749.

Block P. (1993). *Stewardship: Choosing Service Over Self Interest*, ISBN 9781881052869.

Burns, J. (2009). *Leadership*, ISBN 9780061965579 (2010).

Capra, F. (2004). *The Hidden Connections: A Science for Sustainable Living*, ISBN 9780385494724.

Dick, B. and Dalmau, T. (1990). *Values In Action: Applying the Ideas of Argyris and Schon*, ISBN 9781875260126 (1999).

Goldstein, J.A. (2005). "Emergence, creativity and the logic of following and negating," *The Innovation Journal: The Public Sector Innovation Journal*, ISSN 1715-3816, 10(3): Article 31.

Greenleaf, R. (1977). *Servant Leadership: A Journey into the Nature of Legitimate Power and Greatness*, ISBN 9780809125272 (1983).

Hazy, J., Goldstein, J.A. and Lichtenstein, B. (2007). *Complex Leadership Theory: New Perspectives from Complexity Science on Social and Organizational Effectiveness*, ISBN 9780979168864.

Heifetz, R. Linsky, M and Grashow A. (2009). *The Practice of Adaptive Leadership: Tools and Tactics for Changing Your Organization and the World*, ISBN 9781422105764.

Hock, D. (1999). The Chaordic Organization, ISBN 9781583760956 (2000).

Isaacs, W. (1999). *Dialogue: The Art of Thinking Together*, ISBN 9780385479998.

Knowles, R. (2002). *The Leadership Dance: Pathways to Extraordinary Leadership Effectiveness*, ISBN 9780972120401.

Kotter, J. (1996). *Leading Change*, ISBN 9780875847474.

Macdonald, I., Burke, C. and Stewart, K (2008). *Systems Leadership: Creating Positive Organizations*, ISBN 9780566087004.

Neville, B. and Dalmau, T. (2008). *Olympus Inc.: Intervening for Cultural Change in Organizations*, ISBN 9781921142963.

Olson, E.E. and Eoyang, G. (2001). *Facilitating Organizational Change: Lessons from Complexity Science*, ISBN 9780787953300.

Senge, P. (2006). *The Fifth Discipline*, ISBN 9780385517256.

Snowden, D. and Boone, M. (2007). "A leader's framework for decision making," *Harvard Business Review*, ISSN 0017-8012, November.

Stacey, R.D. (1996). *Strategic Management and Organizational Dynamics*, ISBN 9780273708117 (2007).

Stacey, R.D. (2001). *Complex Responsive Processes in Organizations*, ISBN 9780415249195.

Stacey, R.D. (2010). Complexity and Organizational Reality: Uncertainty and the Need to Rethink Management After the Collapse of Investment Capital, ISBN 9780415556477.

Suchman, A. (2002). "An introduction to complex responsive process: Theory and implications for organizational change initiatives," Relationship Centered Health Care, 42 Audubon Street Rochester, NY 14610 USA.

Wheatley, M. (1992). *A Simpler Way*, ISBN 9781576750506 (1998).

Wheatley, M.J. (1992). *Leadership and the New Science: Discovering Order in a Chaotic World*, ISBN 9781576753446 (2006).

Tim Dalmau has worked in a variety of positions in private enterprise and higher education. In the last 30 years he has worked primarily as a consultant to public and private sectors in many different countries. He is regarded as one of Australia's foremost management consultants. He is the author or coauthor of an extensive selection of books, handbooks, and other publications widely used by senior executives and change practitioners. His practice covers the world, and he spends much of his time advising leaders of global corporations, working with Boards and senior executive teams of these organizations and designing systems for change and improvement. He holds a B. Sc., B.A., Grad. Dip Ed. and MBA.

Jill Tideman has 25 years experience in facilitating sustainable change in human systems in the both private and public sectors. She has worked in the engineering, infrastructure and resource sector, and consulting in sustainability, safety and transport policy in Australia and the U.K. She has been a senior executive in the SA Department of Transport Energy and Infrastructure for 20 years in areas of strategic change, environmental policy and strategy, and community programs. Jill has also worked in the Cabinet Office of the Department of Premier and Cabinet. Her consulting career has seen her contribute to the life and success of companies and individuals in the resources sector, steel industry, sugar production, environmental and fauna/flora conservation, engineering, mining, education and road transport in Australia, the USA and the UK. She holds a B.A. (Hons).

11. Envisaging Futures: An Analysis Of The Use Of Computational Models In Complex Public Decision Making Processes

Lasse Gerrits & Rebecca Moody
Department of Public Administration, Erasmus University Rotterdam, HOL

Public decision making takes place in an erratic, complex world. Increasingly, public decision makers deploy computational models in order to make sense of the possible consequences of decisions. Such models are rarely available 'off-the-shelf' and are often developed for specific cases. As such they are tailor-made and the process of building and using the models becomes integral part of the public decision making process. The research presented in this paper analyzes six cases of modeling and generating outcomes to determine the influence of setting boundary conditions in the designing of models on the use and outcomes of the models in public policy processes.

Introduction

Public decision makers are under pressure to make decisions on issues characterized by a) complex causation in terms of possible causes and consequences; b) consequently uncertain futures; and c) controversies. Mapping possible outcomes of decisions prior to decision making would be a welcome solution for consequences of the decision could be predicted. Reality however is full of unseen and unobserved relationships and it can render responses to policy decisions as a surprise to the decision makers. Making decisions about the future can be likened to 'jumping into the unknown' (cf. Flood, 1999).

One way to deal with such uncertainties is by developing and deploying advanced computational models, which are becoming increasingly popular in public decision making (Orlikowski, 1992; Winner, 1977; Markus & Robey, 1983). As modeling in terms of knowledge, accessibility and common practice is increasing it is understood generic models lack the specific contingent variables needed to explain cases (Richardson, 2008). Consequently, computational models are often tailor-made to specific conditions to understand the development of that specific case. The characteristics of a model are (re-) considered in the acquisition and construction of the models; therefore they become negotiable.

We summarize the setting of the scope, functionality and deployment of the computational model under the header of 'setting boundary conditions'. This is a process of negotiation where it is determined what the model should cover and what not, what it should generate and what not, and to what extent its outcomes will be used. In other words: the limitations of the model and the possible outcomes it can generate are predetermined. This includes the variables incorporated into the model and the used data, such as choosing between existing datasets, whose datasets or the generating of new data. It is a phase during which the modelers design the model to suit the needs of the public officials asking for the model. Consequently, the model and its characteristics become part of the public decision making process and therefore subject to the values, pressures norms and beliefs inherent to the policy process. From this perspective the models are not mere suppliers of neutral data informing policy but instead one of the components of the political decision making. The aim of this research is to determine the influence of setting boundary conditions on final policy decisions. The main question of this paper is: How does the setting of boundary conditions in the designing and using models influence the process of public decision making?

Six European cases are analyzed to answer this question. The paper is structured as follows. We will first discuss the theoretical background by blending notions from complexity theory with notions from Public Administration ("Modeling Complexity for the Real World "). The methodological approach is discussed in "Methodology", followed by "Case Studies", "Analysis" and ending with "Conclusions".

Modeling Complexity For The Real World

To understand the process of setting boundary conditions it is necessary to understand the nature of complexity that public decision makers deal with. For this we modify an example from Rescher (1998). Assume that the world consisted of two elements that are unchangeably related to each other. This world could be described using the following series: "abababab..." This hypothetical world makes any description easy because it requires very few descriptive components and because the series is endlessly repetitive, the remainder can be described without actually having real knowledge of it. In other words: the remainder of the series can be predicted using limited information. Of course, the real world does not consist of two elements so using "abababab..." to describe it does not justice to it. There are more elements in the world and each element has multiple subtypes. Therefore any reconstruction of reality in terms of modeling is representing a limited picture (Cilliers, 2001). What constitutes, limits, and explains a complex development is relative to the observer's point of view, it is the human observer who draws boundaries in complexity in order to make sense of it (Luhmann, 1990, 1995; Pel, 2009). The issue is that the remainder of the series contains information that is of paramount importance to explain the outcomes (Byrne, 2002). Reality often deals with a number of different variables explaining outcomes and situations. Therefore predicting situations which include a large number of variables becomes very difficult. Not only the question of what the different variables are but also the causality between these variables and the chronological order of this causality make predicting outcomes of decision making difficult (Rescher, 1998; Simon, 1976; March & Simon, 1993; Dror, 1968).

In the real world, the number and nature of the elements defining an emerging structure or process are not fixed but changeable instead. Reality is open, the constituent elements of the real world are connected ad infinitum with other elements and the elements that are being observed do not define its real borders (Byrne, 2005). However, human cognition is too limited to process the enormous information that comes with complexity of reality so humans make implicit and explicit choices in what is taken into account. Therefore they are bound in the degree of rationality they can display. Among others Simon, 1976; 1957; Dror, 1968; Lindblom, 1959; and March & Simon, 1993 recognized that public decision makers are prone to simplification in the options they consider because of this cognitive limitation. Others (cf. Sharkansky, 2002) have even suggested an evolutionary explanation for this simplification. While with computer models it is assumed by many that this bound in rationality can be lessened it can also be argued that this is not necessarily the case (Moody, 2010).

Any representation that could describe a complex system accurately would have to be as large as the system itself (Rescher, 1998). This is where the reality of computational models clashes with social reality. Byrne (2005) makes a distinction between simplistic and complex complexity that outlines the clash. Simplistic complexity is essentially complexity within closed systems, with the emergence of structures and processes depending entirely on the (fixed) variables defined within the system. In other words: it is complexity that stems from a limited series. Many of the archetypes of complexity theory that are often referred to, such as the computational simulations by Reynolds or Langton (Smith & Stevens, 1996), are examples of simplistic complexity. While simplistic complexity is functional in demonstrating the principles of complexity, it does not resemble social reality.

Complex complexity does resemble social reality because it departs from the perspective that any demarcation in the series describing complexity is arbitrary to a certain but indeterminable degree. Therefore, complexity arises not only from the constituent elements of a system but also from the fact that this constitution is dynamic in itself and causes fundamental uncertainty. The initial idea behind complexity theory—a limited system or set of rules can create complexity that cannot be explained by breaking this complexity down into separate elements—is therefore amended with the notion that the origins of complexity are discursive to the extent that it is not possible to discern the afore-mentioned simple elements. What is included and excluded for research is subjective but has a real impact on the outcomes generated.

A problem lies in the notion that while we might learn more, the new information will account for new questions. Each iteration requires more considerations for policy makers, leading to a proliferation of regulations and institutions to cover each new instance (Norgaard, 1984). Thus the complexity of the world and human's understanding of this complexity is carried over into the actual public decision making. We propose a distinction between complexity in data describing the real world (epistemic mode), complexity in the model trying to mimic the real world (ontological and functional mode), and complexity in the public decision making process trying to deal with the model and its workings (functional mode).

Complexity And Public Decision Making

The previous section discussed how models become necessary simplifications of the real world to aid people in making decisions. The complexity mentioned does not only encompass modeling and generating data but also the public decision making process itself. Earlier attempts to understand the complexity of public decision making as a systemic and bounded activity include Kiel (1989, 1994), Overman (1996), Morçöl (2002, 2008a, 2008b), Morçöl & Dennard (2000), Gerrits (2008), Teisman, Van Buuren & Gerrits (2009) and Meek (2010). Richardson, Dennard and Morçöl (2008) explicitly address the use of complexity-informed tools for public decision making. The main issue boils down to the fact that synoptic blue-print-like decision making does not suit the complex character of both the world that decision makers want to influence and the workings of the governing institutions themselves. Synoptic decision making assumes a model would deliver (non-biased) data, judged by decision makers to generate alternatives, of which the best alternative is chosen and consequently carried out. Having determined the impact, the decision makers change or consolidate accordingly. This rationalist approach is brought forward on several occasions while using computational models (March & Simon, 1993; Winner, 1977; Beniger, 1986; Goodhue *et al.*, 1992; Chen, 2005)

In reality, however, public decision making is often a messy affair with little room for neat, sequential steps. Public decision makers struggle with too little time to make well-informed decisions (cf. Morçöl, 2008b). Computational models bring the promise of shedding light on the object of decision making. The assumption is that a computational model is a value free tool that will provide a neutral oversight of all available alternatives with their consequences. Therefore it is believed by some that these models will decrease the bounds in rationality that decision makers face and that public policy making will become a more rational process in which all consequences are foreseen prior to decision making (Ware, 2000; Moody, 2010; Beniger, 1986; Goodhue *et al.*, 1992).

This line of reasoning assumes that there exist value free alternatives and consequences to be provided. While public decision making is also a political process in practice we see that computational models, next to not being able to include all variables needed for complete consequences also suffer from limits on the side of political values. It must be noted that the designer of the model is not a neutral object either and becomes able to influence the model (Winner, 1977; Chen, 2005; Ware, 2000; Wright, 2008). Known margins of error can be manipulated towards political values and the necessary choice which needs to be made on which variables to include in the model might be value driven as well. This demonstrates that next to the complexity in public decision making as well as the complexity of reality at large also political values need to be considered in order to understand the setting of boundary conditions in designing computational models. More specifically, next to the complexity in data and model as mentioned in the previous sections, there is also complexity in the actual decision making process because of the multiple actors with diverging norms, beliefs and interests. Following the discussion, we have identified a number of variables in each of the three categories of complexity. They are summarized in table 1.

Complexity...	Variable	Explanation
... in data	Culture	Dominant ideas about how complexity can be understood.
	Margin of error	The margin of error that is accepted and communicated to the client.
... in model	Ownership	The organization(s) that commission(s) and own(s) the model.
	Values and beliefs	The normative stances of the builders and commissioners of the model.
... in decision making	Authority	The organization(s) that make(s) the policy decision following the modeling.
	Multi-actors	The other actors that have an influence on the decision making process.

Table 1 *Core Variables*

Methodology

In order to answer our main question and to determine how the setting of boundary conditions in the design of models influence outcomes we have analyzed six case studies. The cases have been selected on the basis of three criteria. Firstly: a new model, tailor-made for this specific situation was deployed for the first time in each case. This allowed us to analyze the way the model has evolved. Secondly, the cases are selected to be internationally comparable. Finally, the cases have been selected on the basis of the policy field they entail in order to conclude that complexity issues and setting boundaries are not limited to one field of policy but can be generalized throughout different fields of policy.

The research strategy used is based on a comparative case study approach, since this approach recognizes the complex nature of social phenomenon, instead of limiting the study to a specified set of variables (Byrne, 2005; Byrne & Ragin, 2009) This also helps us to improve the analytical validity (as opposed to statistical validity) of findings. The comparison of relevant similarities and differences may contribute to the development of a line of reasoning in which the validity of our conclusions is based on plausibility based on the empirical evidence. The cases studied in this paper are therefore treated as unique cases, whose uniqueness can be made more explicit through comparison.

To improve the validity further we triangulated research techniques. First, we conducted 69 semi-structured in-depth interviews. Our sample included officials who ordered the computational models, experts who developed and deployed the model and stakeholders who had an influence on the model and its outcomes. Secondly, we evaluated relevant policy papers and expert reports. After a short description of the case studies in the next section, we present a table with per case sorted by the different variables.

Case Studies

This section discusses the case studies. Each case has a brief introduction (below) and the main characteristics are presented in Table 2. In this table we have used different variables for analysis of the case studies, these variables refer to the complexity in data, in the model and in the decision making process. In Table 2 it will become clear that the variables are interconnected and that computational models are often trusted even though margins of error are very high. Throughout the case descriptions it will be demonstrated that often an agreed upon situation exists on who makes the model, how and with which data. This seems to reduce complexity in decision making processes, but will decrease trust. However, when his agreement is absent, the process becomes very complicated.

Case 1: Morphological Predictions In The Westerschelde (Belgium And The Netherlands)

The Westerschelde estuary runs from the Belgian port of Antwerpen through the Dutch province of Zeeland before flowing into the North Sea. It provides maritime access to one of the largest ports of Europe. The estuary has a limited depth and the Belgian port authorities are constantly seeking ways to deepen the main channel in the estuary to facilitate larger ships. However, the estuary is Dutch territory and the Dutch authorities are reluctant to deepen. They regard the estuary as a fragile complex system that has a high ecological value and fear that the ecology could be destroyed by a deepening operation. Such an operation could possibly topple the multi-channel riverbed into a single-channel riverbed. The multi-channel riverbed is regarded as the main driver for the specific and highly valued ecology of the estuary. Negotiations starting in the early 2000's included the extensive use of computational models to assess the extent to which a deepening would harm the morphology of the riverbed and with that the ecological value. Research relied heavily on computational models developed by Dutch research institutes. The Belgians argued for an empirical in-situ test but that test played a marginal role in the research.

Case 2: Morphological Predictions In The Unterelbe (Germany)

The Unterelbe estuary runs from the port of Hamburg through the federal states Niedersachsen and Schleswig-Holstein before flowing into the North Sea. It is the main access to the port of Hamburg. Similar to the first case, the port authorities are seeking for a deepening of the main channel to facilitate larger ships. Such a deepening was carried out in the 1990's but had resulted in severe (partly) unforeseen and unwanted changes to the estuary. It seemed as if the desire to deepen had influenced the outcomes of the computational model. The ensuing societal and political protests (from both NGOs and the neighboring federal states) had led to a different approach when considering a new deepening early 2000. Data from federal research institute Bundesanstalt für Wasserbau was shared (upon request) with other actors such as NGO's. They built their own model using the same data and came to different conclusions than the port authorities. This set off a debate on a more sustainable future for the estuary and slowed down the original plan of the policy makers. A separate project to develop a long-term vision complemented the deepening project.

Case 3: Flood-Risk Prediction (Germany And The Netherlands)

In the last two years it was decided to run one application in the Netherlands and Germany with the goal to predict and manage floods from rivers. Before this applications and authorities were very dispersed on the subject. The application named FLIWAS was to integrate different applications and organizations to make sure water management and flood prediction could be done more efficiently. FLIWAS was developed and the application will predict on the basis of weather conditions, satellite data, past results and the height of the water whether a flood will occur and what the damage would be in terms of economics, damage to landscape and lives. Also the application is able to calculate proper evacuation routes. The implementation of the application has resulted in the water sector becoming more integrated and being able to communicate to policy makers what the result of certain actions are. It is now more the case than before that water management professionals are invited to the negotiation table in matters of urban planning, where they are able, on the basis of predictions and scenario sketching to convince governments that some plans might not be wise.

Case 4: Determining The Implementation Of Congestion Charging In London (United Kingdom)

The city of London has had a large problem with congestion. In order to find a solution to this congestion problem the local government has come up with a plan to reduce congestion by imposing a charge on all vehicles which enter the zone in which the congestion is worst. A computational model was used to determine where this zone should be so the location of the zone would be most effective in not only reducing congestion but also gaining the government enough money to reinvest in public transportation and cycling facilities. On the basis of traffic data, alternative routes and public transportation plans the organization Traffic for London had decided on a zone in which the measures are implemented. The application to do so finds its basis in scenario sketching so different alternatives of the location of the zone could be viewed with their effects.

Case 5: Predicting And Containing The Outbreak Of Live Stock Diseases (Germany)

Due to European regulations, and after the outbreak of mouth and foot disease in the 1990s, which caused significant financial damage, the German government decided to centralize all information on contagious live stock diseases into one application, TSN (TierSeuchenNachrichten). The application holds information on farms and animals. Further the application will make scenarios on how to contain and prevent outbreaks of contagious diseases. On the basis of the contagiousness of the disease, the estimated health of animals, natural borders, wind and weather conditions and the location of farms decision can be taken on which measures to take. These measures include the killing of the animals, vaccination of the animals, or installing a buffer zone in which no traffic is allowed. The German government appointed the Friedrich Loeffler Institute with the task to develop and manage the application.

Case 6: Predicting Particulate Matter Concentrations (The Netherlands)

Particulate matter in recent years has become an issue more and more prone to attention. Due to European regulations the countries in the European Union are to make sure the concentration of particulate matter in the air does not exceed a set norm. Therefore whether buildings and roads can be built becomes dependent on this norm, not only for the effect on air quality by the building process but also for the effect of the plans once in use. Applications have been made to predict the potential concentrations of particulate matter after implementation of building plans, the outcome of the prediction determines whether a building can be built. The problem in this case lies in the fact that the way to calculate particulate matter to begin with is unclear, scientist are not sure on the calculation as of yet, the health effects are not clear as well, just as the prediction itself. Furthermore other non-governmental organizations have made their own application to predict concentrations, in which mostly the outcome differs significantly from the applications local governments use. This causes each building process to be reevaluated for their legitimacy and this causes a lot of distrust.

Analysis

When analyzing the empirical evidence from the cases a number of issues stand out. First we will look at the complexity in data, secondly to the complexity in the model and finally to the complexity in the decision making process. It must be noted that these three complexities cannot be seen separately from each other, but for the sake of analysis they will be analyzed separately in order to come to a conclusion which encompasses all.

Complexity In Data

In complexity of data we have distinguished between *culture* and *margin of error*. There are some trends to be discovered. One of the first interesting features is based on trust, in cases, 1, 2, 3 and 4 a clear trust in the data and a clear reliance on modeling can be found. In the cases 5 and 6 however it can be seen that the reliance of modeling is not trusted or even distrusted. The reason for this difference can be found in the cases themselves. Where in the first four cases the actors dealing with the model share the same goal in terms of policy making, this is not the case in the last two cases. In these cases those with the application hold a different goal than those without the application or there are several groups of actors with applications with different goals. It may therefore be concluded that the trust in models increases when all actors involved hold the same political goal. When there are actors involved with different goals, or when there are different groups of actors with different goals who all own a different application, it is believed that the data as well as the model might be politically motivated.

When looking at the margins of error we see that the margins of error within these models and with the existing data are high in all cases. This can be explained by the large number of variables within the cases and their complex interrelation. In all cases those actors involved acknowledge these errors but also realize that politicians want to hear a nominal 'yes' or 'no' answer. Therefore these margins of errors disappear in the communication between the experts and policy makers as the experts simplify the presentation of their results. This shows that the reliance on the data and the model which is often high, is still subject to an acknowledged high margin of error. It is interesting however that in the cases in which the reliance on the model and the data is low, and where a culture of distrust exists these margins of error are emphasized. Not only are these margins emphasized but on the basis of these margins actors accuse each other of manipulation of the model and the data for their own political goal. Taking this into account it can be concluded that for complexity of the data we see that both the trust in the data as well as the margins of error and the group of actors and their goals are determinant for the outcome. When actors agree the trust in the data is high and margins of error are acknowledged but not communicated. When actors do not agree on the political goal, trust is low and the margins of error are emphasized and the manipulability of the data is communicated very frequently.

Complexity In The Model

When we look at the complexity in the model itself we have distinguished between *ownership* and *values and beliefs*. In terms of ownership it can be found that most models, except for case 6 are owned and developed by one organization. Therefore it is the case that they have a monopoly on the information generated by the model which grants them the power to use this monopoly in terms of decision making. Only in case 2 and 6 we see that other actors use the data and the information to build their own model. In both cases this has led to conflict. The situation in both cases has been conflictuous in nature, purely based on the political goals of actors (e.g., port authorities versus NGOs). In case 5 some issues of ownership have occurred as well, this can also be explained by the political goals of actors. In this case only those with the same political goal as the actors owning the application were granted access, those with other policy ideas were excluded. Ownership of the application therefore does influence the results of these applications but the main determinant is still the political agreement of actors involved. With agreement on the goal ownership does not seem to pose a problem or a threat to the application.

In terms of values and beliefs, we can find the same evidence as for trust of the data. Belief in the models is generally very high. Only in those cases in which actors hold different goals and the situation is conflictuous those opposed do not trust the model and accuse the owner of the model of distrustfulness, using the application for their own political motivation and manipulation of the model so their preferred outcome will prevail.

Complexity In The Decision Making Process

When we look at the complexity of the decision making process we have distinguished between *authority* and *multi-actor setting*. We can find that in terms of authority an interesting situation exist. In some cases, case 1, 4, and 5, there is a clear line of authority. Policy makers agree on the fact that this specific organization should provide for the data, the model and the results on which policy should be made. This is either institutionally arranged by legally granting these organizations this power, or arranged by an agreed way of conduct to which all involved agree.

In other cases the level of complexity is very high since there is no agreement, either intentionally or not, on where authority should lie. In the second case the work is carried by the main research institute that always does this work. How-

	Complexity In Data		Complexity In The Model		Complexity In The Decision Making Process	
	Culture	Margin of error	Ownership	Values and beliefs	Authority	Multi-actors
Case 1	Strong belief in computational models. Employees from the authorities and main research organization are often educated at the same university, hence sharing the same frames about models.	Margins of error are calculated nominally but are nonsensical to political decision makers who like to hear 'yes or no'. Aware of the uncertainties inherent to the outcomes of the calculations, analysts have a tendency to say 'no' just to make sure that no damage is done.	Research is automatically tendered to WL-Delft Hydraulics who owns and deploys the models Sobek and Delft-3D exclusively. There are almost no incentives to question this way of working.	Dominant belief that computational models are superior to other types of research because they can be manipulated without harming the Westerschelde. A tendency towards building increasingly complex models to mimic reality leads to longer development time and 'more research is needed'.	The Dutch government is the sole authority in deciding over the Westerschelde. Belgian authorities pressure the Dutch government to take actions favorable to the Belgians but demands are only met after long deliberations.	Economical and social interests exert considerable pressure to generate outcomes in favor of one of the sides. Ambiguous outcomes of the models are explained as indicating a 'no deepening because too much uncertainty' (Dutch) or 'a deepening is possible because there are no (negative) outcomes visible' (Belgian).
Case 2	Increasing reliance on modeling to generate answers. Data collection and modeling is done by the federal Bundesanstalt für Wasserbau (BAW) and the inter-state ARGE-Elbe. BAW thinks that data should be publicly available and shares it with anyone interested in using it, including NGO's.	Actual developments in the Unterelbe show that the margins of error were larger than initially communicated. Uncertainties were masked by the pressure to show that a deepening without negative consequences was possible. The actual developments lead to NGOs questioning the soundness of the decision making.	Research is mainly done by BAW, who develop their own models. Semi-openness about the data and the way the models work lead to other actors using the same material to build their own models.	Dominant belief in computational models but understanding that they are limited. Empirical in-situ testing is still out of the question because of practical constraints and societal opposition.	The government of Hamburg needs cooperation from the federal level and from neighboring federal states to carry out the deepening operations. All actors have diverging interests and it is difficult for Hamburg to let their demands prevail.	For a long time, Hamburg had enough authority to do whatever it wished to do. Nowadays there are many actors with diverging interests who have gained some influence over the decision making process. Main opponents (NGO's) use modeling to provide counter-research to the research Hamburg comes up with.
Case 3	Strong belief in computational data. Different organizations dealing with water management share data and therefore the application gives an illusion of completeness.	The margins of error in the data are high. A prediction of flooding is based on the formula: chance * damage. Therefore it is unclear what the actual prediction is. The margins of error are high in general since they are based on a large number of variables which hold a margin or error all individually.	The application is owned by the water management sector. This empowers them to have a monopoly on this information, thereby forcing local and national governments to treat them as equal partners in decision making.	Dominant belief in the application by the water management sector as well as for the government and the public. The application is trusted and seen as a legitimate way to assess threats.	Authority on urban planning formerly was with local governments. Since the application enables the water management sector to clearly communicate consequences of plans they are now part of the negotiation process and a force to be reckoned with.	The application resulted in the water management sector to form one block, often opposing building plans of local governments. Since both serve different interest decision making becomes conflictual.

Case 4	Strong belief in computational data. The scenarios were not sketched by civil engineers but by programmers. This accounted for a view on the application and data in which all data is seen as absolute truth.	The margins of error on the basis data of the application are low, traffic was monitored and the data therefore holds a small margin. The scenarios sketching is based on behavioral models: what would people do if… Therefore this data is highly questionable.	The application as well as the data are owned by Transport for London, they are invited to do so by the Labour party. Therefore only those in favor of the congestion charge are those with the data and the power to sketch scenarios.	Absolute belief in the model. Even those opposed to the charge are believers of the application. They just do not agree with the charge.	In the British system the only one in power to decide on the zone was the Labour party at the time. They delegated the task to Transport for London who held the same views. Authority therefore in this case was not challenged nor questioned	In this case the complications a large number of actors can cause is not present because of the institutional system giving only one party or actor power to decide.
Case 5	Within the government and the Friedrich Loeffler institute the data is heavily trusted. Other organizations who deal with animals and consumer quality as well as farmers believe the data to be politically motivated.	The margins of error are extremely high, for example: the estimated health of animals. Opponents of the implementation of the application claim the margins of error are consciously used for politically motivated outcomes.	The Friedrich Loeffler Institute owns the data and the applications and are unwilling to share with others. Therefore they have a monopoly on deciding on measures since others do not have any data. Furthermore they are accused of only granting those access to the application who hold the same political beliefs.	Dominant belief in the government that the data and the outcomes are neutral and correct. Opponents of the application claim they are politically motivated and will only serve the interest of the Friedrich Loeffler Institute and not those of other actors.	The Friedrich Loeffler Institute, since being the only one with the data, has the authority to propose measures to the government. The government will adopt the advice since it is the only way they can obtain information.	In the field of animal diseases there are many actors, the farmers, the consumer quality lobby, the meat industry, but also animal rights activists and the scientific community. All these actors are effectively shut out of the decision making process since they have to access to information or data.
Case 6	A culture of distrust exist. The data is not trusted by anybody, all parties accuse the other of manipulation of the data for their own political purposes.	The margins of error are extremely high. The question of what particulate matter is, is not answered by scientist yet, therefore the applications hold margins of errors which exponentially enforce each other. For this reason the application is highly manipulable.	The case of particulate matter becomes difficult because a large number of actors all hold their own application and their own data. There is no agreement who's data is trustworthy. Therefore there is no way for the government nor a judge to decide which party had the correct outcomes	The dominant belief is that any other party will cheat, manipulate, lie and trick all other parties with the application to make sure its political goal is achieved. There exists a lot of distrust.	Within the decision making process there is no clear line of who holds authority. The applications of local governments are trusted just as much as those from environmental activist. Negotiations therefore become complicated.	A large number of actors are in the field of particulate matter concentrations. They all distrust each other and accuse each other of manipulating the application and the data to serve their goal. Because the decision making process revolves around whether a building can be built, there are clear winners and losers in the process. A satisfactory solution has not been found and each project becomes a battle.

Table 2 *Overview of the Main Characteristics of the Case Studies*

ever, the research institute does not regard itself as an exclusive provider to the port authorities and the senate and has published its data to anybody who is interested. As a result, both the port authorities and NGOs were using the same data but arrived at different conclusions. This eroded the natural authority that an exclusive relationship between supplier and user normally provides. Furthermore, the Hamburg authorities had to deal with other federal states through with the river runs. Consequently, the diverging interests were exploited in the negotiation process. In the third case it is so that the water management sector is trusted and therefore is seen as legitimate by the national government to deliver results for policy making but because of their monopoly on the information they have overruled the local governments in several cases in terms of urban planning policy. Even though they are seen as legitimate, the existence and use of the model has given them the power to change an agreed upon arrangement, i.e. the dominance of municipalities to deal with urban planning. In the sixth case there is no clear line of who holds authority. This can be explained by the fact that there is no clear agreement on the data as well as the model and that there is no history of dealing with particulate matter. This makes the decision making process extremely complex since the results of the different models often conflict with one another. A decision seen as fair and legitimate by involved actors therefore, in this case cannot be made. A government official described the decisions made by the authorities as "rolling the dice".

A final factor is the number of actors and their relation with one another. Naturally a number of different interests can be found in each case and a clear trend on this variable is not to be found, it seems to be rather case specific. In case 1 we see that there is a high number of actors involved in the decision making process and that the diversity of the actors regarding their political goals and convictions is also high. This complicates the decision making process. Case 2 shows that the number of actors involved is somewhat lower in the first case but the main authorities share the same convictions, which ostensibly simplifies the decision making process. The fact that opponents have organized themselves efficiently and have had access to the same data but with different results means that in the end the decision making was as tiresome as in the first case. Case 3 and 5 provide us with insight on how a group of actors can become very powerful in the decision making process because they have the monopoly on the information. In case 4 it becomes clear that institutional arrangements can reduce complexity since only one organization has formal authority. Finally case 6 tells us that the lack of trust, the enormous difference in political opinion and the lack of one owner and authority make decision making so complex that a decision which is seen as legitimate by all actors becomes impossible.

Conclusions

We set to answer the question how the setting of boundary conditions in the designing and using models influences the process of public decision making. Analysis of the six cases shows that this influence is considerable. The key in this influence is the amount of diversity allowed in every dimension of the use of models, as measured by the variables defined in this research. Low diversity in the actors commissioning, developing and using the model ostensibly lower the perceived complexity of the whole process. In other words: if the actors involved share the same opinions, values and beliefs, they can agree on the boundary conditions relatively easily. Conversely, if the diversity of actors and their normative is high, the perceived complexity rises too.

We use the term 'perceived complexity' deliberately. We found that there is a real difference between what actors consider complex and what is complex in reality. In fact, every model used is a simplification of reality, no matter how sophisticated the model is. This simplification is inevitable because it would be impossible to mimic reality in exactly the same way as reality unfolds in time. But the fact that the model represents traces of real complexity rather than reality tends to disappear to the background once the diversity of actors decreases. This is because when actors think and believe the same things, they tend to think that their work encompasses all possible variety. In other words: being of the same mindset triggers unintentional selective blindness. Consequently, the models are not under close scrutiny and decisions made using a certain model reflect the biases that were unintentionally programmed into the model. For example, in the case of predicting the outbreak of livestock diseases, it appeared that the option 'clearing of animals' could never be a feasible outcome of the model whereas in reality it could be a possible answer.

A high diversity in actors raises the perceived complexity as multiple actors bring forward their own perspectives that are in many cases only partly convergent and downright contradictory in some cases. In other words: higher diversity leads to more obvious clashes of goals, beliefs and values. The models that are used and the results that the models generate are being questioned more explicitly and openly, consequently leading to a higher perceived complexity as it becomes much more difficult to reach a quick conclusion. The presence of diversity questions the setting of boundary conditions and opens the autopoietic nature of the actors commissioning, developing and using the model. Diversity or lack thereof is partly a design feature of the institutional dimension, partly an unintentional process between actors who trust and believe each other. As a design feature it emerges when the commissioning, developing and using models is clustered around one or a limited set of tightly coupled actors. Such a concentration of power, where both research and the final decision are strongly linked, causes the actors to develop a bias towards their own ideas. Whether this link is institutionally determined or not does not influence this. The models are consequently used as such. When such close links are contested or absent, the diversity raises because of the possibility of questioning current ideas and beliefs. As an unintentional process it emerges when actors develop relationships of trust and belief. Although actors are not aware of it, such relationships still promote convergence of thinking, thus decreasing contradictory ideas.

If anything, the current research shows that models and data never speak for themselves. On the contrary, they are heavily influenced by the social dynamics of the context they are developed in. This in turn influences the outcomes and subsequent policy decisions that are based on the outcomes. From our observations it follows that the complexity inherent to the subject, which includes the many different and conflicting interests and stakes, is not reflected in the policy decision making process. The final question is what could be done to improve the way boundary settings reflect some of the real complexity rather than the complexity as perceived by the actors who share the same normative stances.

1. It is important to recognize that computational models are normative because they can not mimic full reality and instead reflect the developers' and users' ideas. We observed often that belief in the model as the right descriptor and predictor of reality was almost absolute at the level of policy makers. "If it has a number it must be true". We argue that this number is as much a reflection of the developers' values and norms as it reflects real complexity. By accepting this, policy makers will learn that some variables are left out, even when they do not know which ones. Other actors with diverging interests and goals have a different perception of the problem and are therefore able to point at alternative variables that were originally left out. Since often it is institutionally arranged who is allowed to produce the model, policy makers are often unaware of these alternative problems, solutions and variables and might accept a given solution as the only or best solution while this may not be the case. It may seem contra intuitive to develop a certain model in cooperation with opposing actors but it will generate broader support and acceptance in the long run. As seen in the cases: if actors with diverging goals have been involved in the development of the model, they tend to support the outcomes regardless of the diverging goals.
2. It is important to recognize that the data and the relationships between variables are of complex nature. This means that it is possible models used by different actors generate entirely different outcomes. It is not as much a design flaw of the model but rather a consequence of the complexity of data and models. Again, it is necessary to involve actors with diverging goals in order to gain support for the outcomes, as mentioned in the previous point. Staging different models developed by different actors against each other is pointless in so far that none of individual the models is able to generate ubiquitous results that are supported by everyone involved.

3. It is important for experts to communicate the points mentioned above rather than trying to simplify their message to and trying to hide the normative biases. On top of that, it is important for policy makers to understand the real complexity of modeling (as shown in data, models and use in decision making) rather than trying to persuade modelers to deliver clear-cut answers. Sometimes reality is just complex, no matter how one wishes it to be simple. Therefore it is of importance that communication between modelers and policy makers improves and that they can share a common language. Where policy makers should be aware of the details and uncertainties of the model, modelers should communicate these more accurately. Ideally this will lead policy makers to be more specific in what they ask modelers to design and it will help them interpret results in a way policy making could improve. A model will envisage a future as can be reasonably estimated, but not the future as it will unfold in time.

References

Beniger, J.R. (1986). *The Control Revolution: Technological and Economic Origins of the Information Society*, ISBN 9780674169869.

Byrne, D. (2002). *Interpreting Quantitative Data*, ISBN 9780761962625.

Byrne, D. (2005). "Complexity, configurations and cases," *Theory, Culture & Society*, ISSN 0263-2764, 22(5): 95-111.

Byrne, D. and Ragin, C.C. (2009). *The Sage Handbook of Case-Based Methods*, ISBN 9781412930512.

Chen, C. (2005). "Top 10 unsolved information visualization problems," *IEEE Computer Graphics and Applications*, ISSN 0272-1716, (July/August): 12-16.

Cilliers, P. (2001). "Boundaries, hierarchies and networks in complex systems," *International Journal of Innovation Management*, ISSN 1363-9196, 5(2): 135-147.

Dennard, L., Richardson, K. A., Morçöl, G. (2008). *Complexity and Policy Analysis: Tools and Methods for Designing Robust Policies in a Complex World*, ISBN 9780981703220.

Dror, Y. (1968). *Public Policy Making Reexamined*, ISBN 9780878559282.

Flood, R.L. (1999). *Rethinking the Fifth Discipline: Learning Within the Unknowable*, ISBN 9780415185295.

Gerrits, L. (2008). *The Gentle Art of Coevolution*, ISBN 9789075289169.

Goodhue, D.L., Kirsch, L.J., Quillard, J.A., Wybo, M.D. (1992). "Strategic data planning: Lessons from the field," *MIS Quarterly*, ISSN 0276-7783, 16(1): 11-35.

Kiel, D.L. (1989). "Nonequilibrium theory and implications for public administration," *Public Administration Review*, ISSN 0033-3352, 49(6): 544-551.

Kiel, L.D. (1994). *Managing Chaos and Complexity in Government: A New Paradigm for Managing Change, Innovation and Organizational Renewal*, ISBN 9780787900236.

Lindblom, C.E. (1959). "The science of 'muddling through'," *Public Administration Review*, ISSN 0033-3352, 19(1): 79-88.

Luhmann, N. (1990). *Essays on Self-Reference*, ISBN 9780231063685.

Luhmann, N. (1995). *Social Systems*, ISBN 9780804726252.

March, J.G. and Simon, H.A. (1993). *Organizations*, ISBN 9780631186311.

Markus, M.L. and Robey, D. (1983). "The organizational validity of management information systems," *Human Relations: Studies towards the Integration of the Social Sciences*, ISSN 0018-7267, 36(3): 203-226.

Meek, J.W. (2010). "Complexity theory for public administration and policy," *Emergence, Complexity and Organization*, ISSN 1521-3250, 12(1): 1-4.

Moody, R. (2010). *Mapping Power: Geographical Information Systems, Agenda-Setting and Policy Design*, ISBN 9789085596127.

Morçöl, G. (2002). *A New Mind for Policy Analysis: Toward a Post-Newtonian and Postpositivist Epistemology and Methodology*, ISBN 9780275970123.

Morçöl, G. (2008a). "Complexity of public policy administration: Introduction to the special issue," *Public Administration Quarterly*, ISSN 0734-9149, 32(3): 305-313.

Morçöl, G. (2008b). "A complexity theory for policy analysis," in L. Dennard, K.A. Richardson and G. Morçöl (eds.), *Complexity and Policy Analysis: Tools and Methods for Designing Robust Policies in a Complex World*, ISBN 9780981703220.

Morçöl, G. and Dennard, L. (eds.) (2000). *New Sciences for Public Administration and Policy: Connections and Reflections*, ISBN 9781574200706.

Norgaard, R.B. (1984). "Coevolutionary development potential," *Land Economics*, ISSN 0023-7639, 60(2): 160-173.

Orlikowski, W.J. (1992). "The duality of technology: Rethinking the concept of technology in organizations," *Organization Science*, ISSN 1047-7039, 3(3): 398-427.

Overman, E.S. (1996). "The new science of management: Chaos and quantum theory and method," *Journal of Public Administration Research and Theory*, ISSN 1053-1858, 6(1): 75-89.

Pel, B. (2009). "The complexity of self-organization; boundary judgments in traffic management," in G.R. Teisman, A. van Buuren and L. Gerrits (eds.), *Managing Complex Governance Systems: Dynamics, Self-Organization and Coevolution in Public Investments*, ISBN 9780415459730, pp. 116-133

Rescher, N. (1998). *Complexity: A Philosophical Overview*, ISBN 9781560003779.

Richardson, K.A. (2008). "The role of 'waste' in complex systems," in L. Dennard, K.A. Richardson and G. Morçöl (eds.), *Complexity and Policy Analysis: Tools and Methods for Designing Robust Policies in a Complex World*, ISBN 9780981703220, pp. 55-68.

Sharkansky, I. (2002). *Politics and Policymaking: In Search of Simplicity*, ISBN 9781588260840.

Simon, H.A. (1957). *Models of Man: Social and Rational. Mathematical Essays on Rational Human Behavior in a Social Setting*, Wiley.

Simon, H.A. (1976). *Administrative Behavior. A Study of Decision-Making Processes in Administrative Organization*, ISBN 9780029290002.

Smith, T.S. and Stevens, G.T. (1996). "Emergence, self-organization, and social interaction: Arousal dependent structure in social systems," *Sociological Theory*, ISSN 0735-2751, 14(2): 131-153.

Teisman, G.R., van Buuren, A. and Gerrits, L. (2009). *Managing Complex Governance Systems: Dynamics, Self-Organization and Coevolution in Public Investments*, ISBN 9780415459730.

Ware, C. (2000). *Information Visualization: Perception for Design*, ISBN 9781558608191.

Winner, L. (1977). *Autonomous Technology: Technics-Out-Of-Control as a Theme in Political Thought*, ISBN 9780262730495.

Wright, R. (2008). "Data visualization," in M. Fuller (ed.), *Software Studies: A Lexicon*, ISBN 9780262062749, pp. 78-86.

Lasse Gerrits, Ph.D. M.Sc. BA (1979) works as an Assistant Professor in Public Administration at the Erasmus University Rotterdam, The Netherlands. He received his PhD degree in 2008 on an investigation into the way policy makers decide about port extensions in Europe. Currently, his main research focuses on the developing and applying complexity theory in the realm of public decision-making, among others in infrastructure projects, urban planning and sea port development. He was awarded an NWO-Veni innovational research grant for excellent young researchers in 2010. Next to research, he runs a Master's program called Governance & Management of Complex Systems. The main goal of this program is to educate future policy makers into complexity thinking and governance.

Rebecca Moody finished her Ph.D. project regarding the influence of Geographical Information Systems on policy design and agenda-setting at the department of Public Administration at the Erasmus University in Rotterdam in February 2010. She specifically focuses on how forms of ICT such as Geographical Information Systems can help or hinder issues from reaching the political agenda and how processes of policy design might be altered by the use of Geographical Information Systems. Furthermore, in 2007 she has cooperated in a research on the influence of Web 2.0 technologies on processes of micro mobilization. In this project it was researched how, individuals, hoping to gain political attention for their issue, used Web 2.0 technologies to attain their goal. In January 2010 Rebecca started her post doc position at the Erasmus Studio. In her research she will focus on visual culture. New technologies for distributing but also creating visual images can have a profound impact on the way public policy is made. This impact, whether it is present, and if so, in which way, will be researched. The relation between government and citizens will be looked at but also the relation between experts and scholars among each other, as well as in relationship to the government.

12. Leading Radical And Rapid Adaptability In A Turbulent Environment

Ramzi Fayed[1]*, Stephen Duns*[1,2] *& Gervase Pearce*[1]
1 Australian Graduate School of Leadership, AUS
2 Success Works, AUS

The level of turbulence (the pace of change of Complexity) that characterizes the global business environment has been steadily increasing over the past 50 years. Organizational leaders, confirmed in the recent IBM global survey of CEOs, now face the challenge of a pace of increase in complexity that exceeds their ability or preparedness to adapt within a feasible timeframe. In this emerging environment many leadership teams are likely to suffer from "Adaptive Stress". The Australian Graduate School of Leadership has sought to assist organizations prepare to cope with adaptive stress. The authors of this paper, as a consequence of their field experiences and research findings, support the need for a fundamental re-think of strategic leadership's adaptive activities. It is proposed leadership driven adaptation is:

- **The core role of enterprise leadership;**
- **Not a context dependent process;**
- **Reflected similarly at different organizational levels, and;**
- **To be synchronised with continuous improvement activities.**

We will argue that these propositions are more appropriate to as a basis for leadership action in the high turbulent environment of the 2010s. Indeed, as the level of turbulence increases, different leadership approaches to adaptation in different contexts including different organizational levels, and under reducing response lead times, we argue that there is no alternative other than to converge toward a duplicated inter-related hierarchy of an adaptive leadership framework. In this framework, similar adaptive activities are undertaken at different organizational levels in sync with incremental activities. The Adaptive Fractal Leadership (AFL) framework draws on ideas from complexity, social and quantum theories. Critical to adaptive success are the shared values, leadership theory and leadership practice guidelines that foster self-organization, innovation and commitment.

Introduction

The increasing pace of change of complexity, that is the turbulence that has characterized organizational environments over the past 50 years, has created the potential for increasing levels of organizational adaptive stress. Adaptive stress occurs when the level of turbulence exceeds leadership's current ability or preparedness to make necessary adaptive adjustments. Successful leadership in a turbulent environment secures sustainability by radical and rapid adaptation. Doing more of the same more efficiently or leveraging current relative market power are not prescriptions for longer term adaptive sustainability.

The challenge is particularly acute for those organizations that still undertake a significant proportion of their activity within a traditional hierarchical command structures and related culture, this is still typical of major banks, government departments and large mature not for profits all which have been shielded one way or another from change pressures.

Adaptive stress can be ignored for an extended period if the organization has sufficient market power that can be leveraged; usually the consequence of favorable legislation, monopolies, oligopolies or, when organizations have the capacity to externalize a significant proportion of their total real costs without community imposed penalties.

The problem with ignoring adaptive stress by utilizing current market power is that market power positions and legislative support for such positions can change rapidly with dramatic impact and typically with insufficient time to design and implement the radical adaptive adjustments necessary.

Organizational adaptation in a turbulent environment requires collaborative, timely and radical adjustments to all organizational value realization and related support workflow processes and dependent stakeholder relational networks. These adaptations must be technically, financially, competitively and ethically feasible, and require appropriate lead time that can only be gained through an understanding of change patterns.

This paper initially overviews the evolution of organizational adaptive response strategies in an increasingly turbulent environment over the past 50 years, setting the scene for the introduction of the "Adaptive Fractal Leadership" framework and its related leadership methods and practices that are designed to achieve adaptive sustainability in turbulent environments.

Evolution of Adaptive Organizational Leadership Strategies

After World War II, the linkage between organizational strategy and structure was confirmed through the pioneering work of Chandler (1962). His conclusion that structure follows strategy was based on historical analysis of a number of large-scale American organizations. Put another way, for an organization to successfully adapt to an evolving environment, it will need an appropriate competitive strategy implemented via a structure that supports the proposed strategy shift. It is interesting to note that several decades later, the work of Prahalad & Hamel (1990), Stalk *et al.* (1992) and Eisenhardt & Martin (2000), amongst others, established that capability was also an important explicit intervening factor and that strategy, structure and capability were all interrelated.

The 1960s were dominated by a linear growth mindset; the amount of energy supplied, the number of telephones connected and the staffing requirements in 5 years were all assessed by simply multiplying current one year growth by five. Strategic leadership in the guise of long range strategic planning fostered a simple linear perception of future adaptive requirements that appear justified given the linear nature of growth in the 1960s. By the end of the 1960s the adaptive responsiveness of many large scale enterprises had been desensitized and they were ill prepared to deal with the adaptive demands of the early 1970s.

In the 1970s, organizational adaptive pressure was initially triggered by the global oil crisis and the gap between the pace of developments in organization environments and the organizational ability to respond increased. Confidence in the ability to manage for the future declined sharply, as evidenced by the decline in resources committed to long range planning departments; indeed, some were dismantled during this period (Fayed, 1986) as organizations became embroiled in short term survival in an environment characterized by intensifying competition, increasing complexity of stakeholder expectations and increasingly complex interrelationships. Attempts to adapt by increasing the review frequency of strategic plans, making cosmetic adjustments to functional structures, "pretend" decentralizations including "temporary" decentralised empowerments often accompanied by inadequate investment in the development of the decision making capabilities and system re-design necessary to support decentralization, typically increasing adaptive stress rather than reduce it. When the "Grand" decentralization experiment was abandoned in favour of what was perceived to be lower risk re-centralisation frontline and middle management inevitably became disengaged and increasingly mobile.

In the 1980s, partly to reduce costs and partly to enhance competitiveness, many organizations reintroduced the divisional structures that had been utilized earlier in the 20th century by General Motors, General Electric and others. This time the re-structuring package came in the guise of Strategic Business Units (SBU); structures promoted as increasing market responsiveness and reducing fixed costs of centralised bureaucracies. The typical outcome, with few exceptions, was a significant increase in short term financial focus, as future value was absorbed to support current bonuses, coupled with increased difficulty of implementing organizational initiatives involving more than one SBU. In Australia during this phase one large scale construction organization with approximately 60 business units needed to coordinate the efforts of 16 of these units to deal with one major contract. Strategic investment in the development of core capabilities and the servicing of major clients suffered. The SBU structure and associated reward systems tended to discouraged investment in the development of future core capability and key relationships, as a consequence did little to enhance medium to longer term adaptive stress.

Responding to prevailing economic conditions in the 1990s and early 2000s, strategy was increasingly driven by short term cost management based on outsourcing and headcount reductions to increase process efficiency (Lean processes). These strategies had the effect of flattening organizations through "right sizing" and "de-layering" that paid little regard to the longer term cost consequences of increased workforce mobility—a development that significantly increased the cost of securing and retaining certain categories of staff. However, this development was an important contributor to the emergence of the "networked organization" and collaborative leadership.

The evidence drawn from the attempts to adjust to an increasingly turbulent environment over the past 50 years supports the conclusion that maladaptive organizational behavior is primarily the consequence of poor leadership practices therefore, what is required is a fundamental rethink of leadership practices.

Leadership practices are driven by values and beliefs. The increasing relative value of human capital, and other developments, are driving organizations towards the adoption of distributed leadership. However, there is also increasing recognition for organizational leaders to integrate incremental improvement actions more effectively with radical adaptation. For a period it was assumed we could resolve this issue by requiring management to deal with incremental improvement and leadership to deal with radical adaptation. In hindsight and it is always easier to re-assess such issues in hindsight, it is evident that this was not the answer. All successful leaders need to be at times effective managers and all successful managers at times need to be effective leaders, leadership and management are converging.

The increasingly turbulent environment requires a more appropriate organizational structure capable of fostering the convergence of leadership and management. The network organization has emerged to satisfy this requirement.

Over the past several decades, the authors have sought to reflect on and learn from their joint experiences and the experiences of others in how to deal with challenges that are central to effective leadership. We have concluded that the challenge of organizations adapting to their evolving environments under conditions of high environmental turbulence is now the core leadership challenge. To determine how best to address this challenge we need to ask:

- How is a locked-in mindset unfrozen?

	Functional Hierarchies	Networked Organisations
How environment is perceived	Environment relatively placid and predictability. Developments requiring a radical strategic response occur very infrequently.	Environment is relatively turbulent and as a consequence highly inter-related; Developments beyond the short term are largely unpredictable. However, higher level social and economic change cycles and patterns provide anticipatory potential.
Adaptive Response	Slow and progressive and successful when a sustainable competitive advantage based on unique or difficult to imitate capabilities has been secured. To achieve this adaptive response- Create a vision, specify a mission, establish objectives, propose strategies, develop support infrastructure, specify the roles required to implement the strategies and realise the vision. Structure an action plan (including the identification of agreed performance measures), monitor and adjust performance over time.	Strategic intent needs to be adaptive and reflected throughout the network. Because strategic intent evolves in cycles through fuzziness to clarity. Organisation wide tolerance of initial fuzziness is necessary. Continually seek to evolve the basis on which competitive advantages are secured by learning from experiences faster than competitors. Adopt an appropriate divergent scope of possibilities consideration as the pace of change accelerates. Foster emergence by adopting an open systems approach to innovation. Mitigate the potential risk consequences of all key interrelationships. Where possible use collaboration to limit risks.

Table 1 *Environmental Perception And Adaptive Response*
(Source: working paper, R. Fayed, 2010).

- How should consideration of a broader set of inter-related consequences be fostered?
- How can inherent self-organizing and self-integrating capabilities be rapidly mobilized?
- How can a meaningful, clearly specified and feasible shared intent be rapidly established?
- How can an effective collaborative effort in pursuit of the shared intent be ignited and sustained?
- How do we know when a new adaptive cycle is required?

The conclusions we have reached regarding the answers to these adaptive leadership challenges are embodied in the Adaptive Fractal Leadership framework presented below.

The Adaptive Fractal Leadership Framework

The Adaptive Fractal Leadership (AFL) framework evolved out of the individual experiences of the authors and represents their current view of what needs to be done to rapidly and radically respond to changing stakeholder expectations and market developments. The framework is further supported by several recent doctoral research programs.

The initial activity in the framework is triggered by a fuzzy statement of intent (vision and purpose) that is clarified and agreed through subsequent component framework activities. The end result is a clear, shared purpose and vision that is time bound and supported by quantified objectives that drive what needs to be done, learned and adjusted.

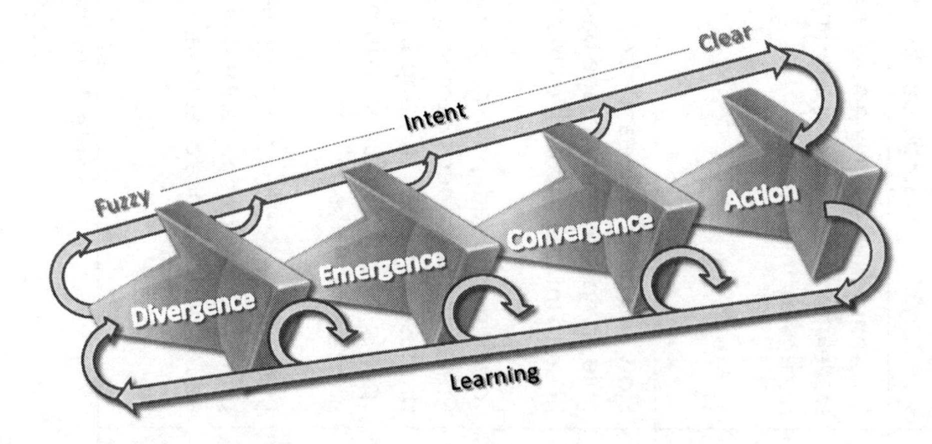

Figure 1 *The Adaptive Fractal Leadership Framework.*

The AFL framework involves four core activities and two support activities.

Core Activities

- **Appropriate Divergence**: This activity challenges current assumptions, expands current scope of considerations, develops stakeholder perception empathy and explores potential future competitive space, given a fuzzy initial intent. Seeks to make sense of weak signals and relevant emerging patterns. A key concept is to mobilise "collective intelligence", the ability to leverage a constructivist approach and integrate various perspectives of the "truth" into a shared and expanded world view.
- **Relevant Emergence**: Achieves a new level of understanding, or being, that allows a radical new approach (a quantum leap) in the ability to deal with the issues being confronted. Culminates in the identification of completely new competitive stakeholder related strategies.
- **Timely Convergence**: Convergent thinking means summarizing key points, sorting ideas into categories and arriving at general conclusions. The convergent phase is goal-oriented, focused and usually subject to time constraints. It is focused on securing results by developing a detailed action plan and budget that includes clarified purpose and vision, specific objectives, identified risks and proposed mitigation action, and a process for measuring outcomes against expectations.
- **Action and Adjustments:** This activity involves implementing the Action Plan previously developed whilst making ongoing adjustments driven by learning-in-action, resolving conflicts and adjusting to unanticipated developments.

Support Activities

- **Intent Clarification:** Shared purpose and vision are progressively clarified and driven by a deepening understanding of possibilities and behavioral support for values that are incorporated into a shared mindset.
- **Ongoing Learning**: The AFL framework allows for rapid refinement and learning by reflection–in-action as well as reflection-on-action. Learning takes place on two levels: learning loops need to be incorporated into each element of the framework in an iterative and formative pattern; and meta-learning needs to be undertaken regularly to determine the effectiveness of sub-process learning loops in fostering overall intent.

The Fractal Nature of the Leadership Framework

A fractal is a [geometric] pattern that is repeated at ever smaller scales. Rather like holograms that store the entire image in each part of the image, any part of a fractal can be repeatedly magnified, with each magnification resembling all or part of the original fractal. This phenomenon can be seen in objects like snowflakes. While fractals are a mathematical construct, they are also found in nature, such as snowflakes, river systems and the morphic fields of apple seeds and acorns. The AFL framework is proposed as an adaptive decision making framework inherent to all levels of sustainable human adaptive activity systems therefore, every leadership adaptive decision and consequential action at any level must involve an AFL framework structure if it is to contribute the survivability of the entity.

These AFL framework activities undertaken at different organizational level and in different organizational contexts are typically coordinated by and aligned to the specified overall organizational Strategic intent.

An overall organizational AFL cycle is initiated whenever a "strategic" development occurs, or is anticipated, that calls into question previous strategic assumptions relating for example to; the level of expected relevant economic activity, unexpected competitive behavior, the advent of a new transformational technology and so on.

Most large scale organizations have a significant proportion of their total activity committed to long standing value realization in which continuous incremental improvement plays an important role and reward systems are typically designed to foster short term results. Experience in organizations such as IBM has indicated that integrating an AFL type framework into this organizational unit, essential to the sustainability of this organizational unit, requires a separate centrally managed budgeting system due to the mismatch in the performance profiles as between a long standing business activity and an emergent business activity. However, irrespective of the stage of life cycle of a business if new or further radical adaptation is required an AFL type framework will be required.

Intent Clarification

The Fractal Framework proposes beginning with a fuzzy intent that is progressively clarified by learning that enhances self awareness and shared understanding of values as it moves through the framework of divergence, emergence and convergence.

The development of a shared purpose and vision involves the joint clarification of intent. Purpose is concerned with clarifying, in broad terms, what value is added, whilst vision is concerned with what the end result will look like.

A clearly specified and shared intent assists in determining which opportunities should be pursued and which should be ignored. All participants are at their best when their own values are in sync with a shared core purpose and vision that is meaningful and self-defining.

Clarity of intent and consistent perceived adherence to that intent is also crucial to authenticity. To achieve authenticity, clarity about personal values and high levels of self-awareness are critical.

Self-awareness is considered a necessary means to discovering a meaningful purpose (see Note 1: Self-Awareness). To acquire this self-knowledge, a deep self-understanding needs to be developed. Having clarity about identity needs and, most importantly, motivations empowers conscious and active decisions and necessary action for successful adaptation.

It is in the ongoing clarification of intent and learning that the model is dynamic, indeed adaptive in its own right. Rather than starting with the answer and seeking to justify it, the process framework refines a fuzzy intent with powerful questions and learning throughout the process. A key assumption is that the answers will emerge as a result of collective intelligence and be refined into a small number of possible futures, which can be tested, or prototyped, and lead to ongoing action as a result of learning and new understanding.

Ongoing Learning

In a knowledge economy, organizational learning that generates appropriate knowledge is the primary basis upon which sustainable adaptive competitiveness can be built. Definitions of knowledge suggest that knowledge is that which is acquired through the process of learning, however Argyris and Schon (1978) argue that learning takes place only when new knowledge is translated into new behavior that is becomes replicable (see Note 9: Argyris & Schon, 1978).

Incorporating learning loops is essential to ensure effective adaptation and genuine emergence of collective intelligence. The leader does not have the answer, but supports a process of co-creation of a possible answer that can be tested and refined.

Learning To Generate New Knowledge

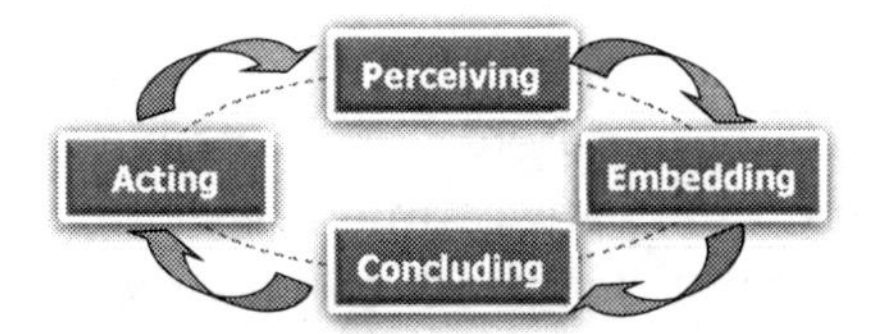

Figure 2 *de Geus's View on How Organizational Learning Occurs de Geus (1997).*

- Perceiving—feedback from customers on service provided; input from staff regarding products; benchmarking against competitors.
- Embedding—aligning individual mental models; aligning perspectives; new action determined.
- Concluding—understanding what other organizations do and how you can be different and distinctive in the area of interest; looking for patterns from which to draw conclusions.

- Acting—encouraging trial and error (experimentation); doing things differently.

NB: de Geus's view of learning, in which perception, sharing, trust and action coordination are important components, highlights the relationship between organizational culture, shared mindsets and organizational learning.

Generating Knowledge Through Relational Networks

Much of an organization's capacity for adaptation emerges from the learning that occurs through its social network relationships, which is also an important aspect of the organization's social capital. Certain relational networks self-organise around the production and institutionalization of knowledge. These networks of independent learners (communities of practice) collaborate with one another in self-organizing ways to develop and institutionalize shared knowledge. They challenge deep rooted assumptions and norms of the organization that were previously inaccessible either because they were unknown or unnameable. Where an individual's learning affects the individual's mental models, which in turn affects future learning, the process "*Individual double-loop learning*" occurs according to Argyris and Schon (1978). However, *Organizational double-loop learning* occurs when individual mental models are incorporated into the organization through shared mental models that become capable of affecting future organizational action:

- Problems are defined and resolved through the interacting contributions of the network members, mobilizing the 'social capital' of key 'problem owners'
- The capabilities of the network to deal with and perhaps prevent that class of problem situation are enhanced, thus growing the stock of social innovation capital (adapted from McElroy, 2002).

As with any other self-organizing system, social capital develops best when it is allowed to grow and is not hampered by excessive direction and constraint. Like plants, it is cultivated by its environment and inputs and perhaps shaped by judicious "pruning," but not designed, engineered or built. The appropriate analogies are organic rather than mechanical (adapted from McElroy, 2002)

These relational networks exist as social fields within groups that effective leadership can identify and influence. More work is required to understand how to identify and influence these social fields.

Learning And Shared Mental Models

Kim (1993) argues that organizational learning is dependent on individuals improving their mental models, how they make those mental models explicit and how they develop a shared mental model for the organization. This process, according to Kim, allows for organizational learning to be independent of any specific individual.

Kim (1993) poses an interesting question: Which of these organizations will be easier to rebuild to its former status? An organization in which all physical records disintegrate overnight—no reports, no manuals, and no records. The only things that remain are the physical assets and the people. In another organization people quit showing up for work and are replaced by new people.

According to Kim, it will be easier to rebuild the first organization because only its static memory has been destroyed, not the shared mental models. In the second organization, the new people will have their own mental models which will have no connection with the remaining organizational memory.

Kim concludes "The shared mental models are what make the rest of the organizational memory usable. Without these (shared) mental models...an organization will be incapacitated in both learning and action" (p. 45).

But how does an organization develop such shared mental models?

Mental models are difficult to articulate and share with others because they are a mixture of what is learned explicitly and what is absorbed implicitly (Kim, 1993). Leadership success in making mental models explicit depends on how well they are able to analyze complex dynamic systems, which requires the ability to understand the complexity that resides in the nature of the interrelationships whose causes and effects are distant in time and space (Senge, 1990).

Appropriate Divergence

In the divergent phase, powerful questions are needed to challenge thinking and encourage people to be open to new ideas. If the divergent phase is closed too soon, the degree of perspective expansion and empathy will be insufficient to trigger the emergence of creative insight with innovative potential. Ideally, a group will absorb the stress of divergence and stay in a divergent inquiry mode until agreed potential action has emerged. It should be noted that the elements of the Fractal Framework overlap and can involve either forward or backward iterations to secure the desired intent.

Divergent thinking typically generates alternatives, has free-for-all open discussion, gathers diverse points of view and unpacks the problem. Divergence is non-linear and needs "chaos time".

Engaging Stakeholders at Rational and Deeper Emotional Levels:

Having identified relevant stakeholders (see Note 2: Identifying Relevant Stakeholders), how to engage stakeholders emotionally, as well as just logically, now needs to be addressed and is clearly articulated in Theory U (Scharmer, 2007). Indeed, it is argued that engaging stakeholders at deeper emotional and spiritual levels is the only way to achieve profound and sustainable change. In addition to reason, emotional engagement is also required to successfully secure adaptation. As a consequence, the ability to logically and emotionally engage relevant stakeholders is a key practice requirement within the Fractal Framework for leadership. (See also: Note 3: Engaging Stakeholders at Rational and Deeper Emotional Levels and Note 7: The Art of Hosting and Harvesting Conversations that Matter (AoH))

Mobilizing "Collective Intelligence"

Accepting that there is no one correct or final version of the truth is a key idea of quantum theory and an accepted part of the constructivist paradigm. Leaders are often expected to have the answers, however we are all limited in our perspective by our individual values, experiences and knowledge. By entering a wider field of consciousness, we are more likely to overcome the limitations of our individual perspectives. Leaders need the skills to tap into the combined or collective intelligence of the people around them. This is made possible using the practices outlined, especially Open Space (Owen 1997), World Cafe (Brown, 2005) and Appreciative Inquiry (Cooperrider *et al.*, 2000), all of which are core methodologies of the Art of Hosting.

Another quantum theory concept is that our observations are distorted by what we are looking for and our version of the truth is not the only right answer. By gaining the perceptions of a broader range of people it is possible to encourage a higher order of wisdom to emerge. (See also Note 4: Mobilizing "Collective Intelligence")

Take a Holistic Systems Perspective

Every system has a purpose and living organic systems have evolving purposes (driven by adaptation). Feedback is essential to provide ongoing guidance for transformation processes. Transformation processes realize the value necessary to sustain the engagement of those stakeholders that, combined, made the delivered value possible.

Mechanistic systems have pre-specified transformation processes designed to achieve pre-specified outputs, e.g. when an organization is nationalized, its purpose, transformation activities, outputs and sometimes its inputs are specified in legislation. This limits the capacity of the organization to adapt and survive; therefore, every living system must have a degree of control over its capacity to adapt functions. The nature and performance of this adaptive mechanism is an important determinant of the system's survivability.

Every system stakeholder will have a characteristic perspective (that is, their own way of looking at and understanding the world, based on their language, culture, formative experiences and current expectations), which in turn shapes their views of what the purpose of the system is. It should be noted that outcomes can have unintended consequences; the greater the complexity, the greater will be the inter-connections and the harder it is to anticipate consequences. A greater capacity to respond rapidly to unintended consequences can reduce their harmful impacts. (See Note 5: Bank Queues: A Holistic Systems Thinking Example which provides a real world application as an example of the whole framework).

Map the Relevant Strategic Arena Network

Every strategic arena is based on a demand process. Once the demand process has been specified, a strategic arena can be constructed by identifying all supply processes serving the specified demand process. Supply processes include: the focal supply process (the supply process in which the organization is embedded), the complementor supply processes (supplying related demands), rival supply processes (that directly compete) and the substitute supply processes (that seek to substitute for the focal supply process).

All supply and demand processes have related stakeholders that comprise a strategic arena relational network that includes contributing, rival, substitute, complementary, regulatory and community stakeholders. Each strategic arena can be viewed from different levels of aggregation depending on how the demand process served by that specific arena is defined.

By mapping the strategic arena network the options across the relational network can be explored as part of the divergent exploration. It assists thinking within the current supply process into a wider domain of possible supply possibilities.

Relevant Emergence

Emergence is the process of deriving some new and coherent structures, patterns and properties in a complex system. For the purposes of organizational leadership, we define Emergence as the creation of a new level of understanding, capability or being. The idea of a quantum leap is a change that results in such emergence. A quantum leap is a quantum theory concept, incorporating non-linearity, strange attractors, connectedness and context. It is the influence of non-linearity and strange attractors that can cause emergence to be unpredictable.

The challenge for leadership is to create an environment that increases the probability of a quantum leap into an emergent state, a completely new level of understanding or being. It is this new emergent state that will, it is hoped, allow an organization to successfully adapt.

Quantum leaps involve moving to new states that are unknown, indeed unknowable, and therefore can be risky and dangerous. This reinforces the importance of collective clarity of intent, as it guides the direction of the emergence and allows the group (team or organization) to adapt to potential unpredicted emergent outcomes.

This adaptive challenge involves the ability to let go of control and allow self-organization to do what needs to be done, which takes courage. It goes against the comfort of Newtonian cause-and-effect relationships that have been integrated into the way we think.

Reinforce Collective Clarity Of Intent (Purpose And Vision)

Discovering purpose is to discover why something exists. Often we hurry to get into action before we properly understand why we need to take action. Gaining clarity of purpose, especially, gaining collective clarity, is critical for setting the right course for taking action. A purpose, therefore, becomes a navigational tool like a compass as it helps us to discover the direction of travel for our efforts so they can be of service.

Purpose can also been described as 'the glue' that brings people's contribution and efforts together. This is because it defines why we are working towards something and why it is worth working on this together. In fact, purpose becomes an invisible leader as it connects different actions taken and supports everyone to know why their contribution is valuable.

> *Collective clarity of purpose is the invisible leader*
>
> Mary Parker Follett

The greater good, as reflected in the vision statement, defines why we need to take action and provides a basis for clarifying the purpose to be pursued. Purpose becomes a powerful attractor that allows people to put their individual efforts to work on making a difference for all.

In an organization or a community many purposes coexist and often not enough effort is given to interconnect these purposes, so it can often feel that different and conflicting purposes are at play. It is therefore important to remember that different purposes are at play, for example:

- The purpose of the stakeholders that the organization serves;
- The purpose of the whole community / organization;
- The purpose of the core group, and;
- The purpose of each member of the core team.

Putting effort into gaining clarity and, specifically, collective clarity of purpose is critical to emergence. Seeking purpose is not something to be done once, either. As developments unfold, it may be necessary to rethink purpose.

Incorporate Processes To Support Innovation And Co-Creation

The perspective of organizations as organic, or living, systems has implications for innovation. Drawing on Kauffman's (1993) work, as living systems adapt within their environment, the elements alter "both the fitness and the fitness landscape" of each other. This reciprocal influence and change is a process of co-evolution and co-creation. Consequently, each element is partially dependent on the evolution of other related elements for their mutual coevolution.

Margaret Wheatley (1994) goes further and claims that a living system only accepts its own solutions. People will only accept and support those things they are part of creating. In addition, a living system "participates in the development of its neighbor" and an isolated system is doomed.

So in order for a new, or innovative, solution to be created and accepted it needs to be co-created. A participatory approach to innovation is necessary to ensure its success, both in terms of its development and acceptance.

The challenge for leadership is not to come up with the right answers, but to foster the right process to allow emergence to be co-created.

Timely Convergence

The convergent phase is goal-oriented, focused, linear, structured and usually subject to time constraints. It is focused on getting results and may require quick decisions. Convergent thinking means evaluating alternatives, summarizing key points, sorting ideas into categories and arriving at general conclusions. The convergence process is a narrowing of focus to a limited range of future scenarios, made possible by the new level of understanding generated through Emergence. Ultimately one, or perhaps a manageable few, of those scenarios will lead to action.

Exploring Possible Futures

A future scenario can be thought of as a fictional story about a plausible future and how it might develop—that is, the events that would lead to that future. Some stories will remain fiction, some may be clearly possible, others may become fact; we won't know what is true until it happens.

When this future is the desired future we call the scenario the organization's "vision" and how it might come about the "strategy" (van der Heijden 1996). Exploring a range of such scenarios can, in the face of uncertainty, help to expand the scope of thinking, identify potential strategies and facilitate the design of more flexible strategies that enhance risk mitigation.

Scenario thinking assumes an uncertain and unpredictable future—there are no "guaranteed solutions," only hypothesized possible futures. Scenario thinking is based on systems thinking and recognizes interactivity among multiple variables at multiple systemic levels. It increases comfort with uncertainty by providing several future outlooks to stimulate interest and exploration.

Because extrapolating from historical developments has limited value in predicting the future, scenario thinking does not rely solely on facts from or about environmental trends. It also draws heavily on the perceptions of those involved in the scenario development process. It emphasizes root causes and incorporates dynamic processes into analytical thinking. It is thus eminently suited to an "organic/systems" view (adapted from Morgan & Hunt, 2002).

Experience has repeatedly confirmed that those who are involved in the development of the scenarios are most committed and best able to grasp and use them. Much of the value comes from the incidental learning involved in the scenario development and application process. However, only some "end-users" are fully involved in the development process, which makes some issues more important than might at first be apparent, including:

- The titles chosen for the scenarios that are finally used as planning thought stimulation tools—these should be evocative and memorable but not misleading
- How the scenarios are expressed and conveyed to all those who might use them—they must be conveyed in a form that facilitates understanding.

(See Note 8: Simplified Approach to Scenario Development)

Specify Stakeholder Value Propositions

To achieve a desired level of performance requires securing the ongoing engagement, compliance and approval of the network of contributing and served stakeholders.

The requirements of different stakeholders can conflict. Organizational adaptation therefore demands that different stakeholder requirements be balanced and relevant leaders must ensure that each stakeholder gains enough to justify in their perception the effort, resources and risk they are expected to contribute in exchange for what they are offered.

Securing stakeholder engagement involves balancing as equitably as possible:

- Relative stakeholder power—with higher priority going to those with more power to influence the organization's own outcomes;
- Relative stakeholder legitimacy—with higher priority going to those with greater legitimacy; and
- Relative stakeholder urgency—with higher priority going to those with greater urgency.

Legitimacy depends on legislation, regulations and social norms (beliefs about what is right and wrong). One key basis for legitimacy is proportionality between a stakeholder's contribution to outcomes and the risks they incur against the benefits accorded them.

A stakeholder's risk is defined as his or her perception of the uncertainty and adverse consequences of engaging with an organization or project. The nature of this uncertainty and the potential adverse consequences are classified into a number of perceived risk dimensions by several authors (summarized by Ward & Sturrock, 1998). These risk dimensions have social, performance, financial, physical, psychological, temporal, economic and opportunistic elements.

Expectations within these dimensions vary over time and can cause changes in stakeholder requirements, possibly necessitating a rethink of the current business model. More specifically, changes in functional, emotional, payment, access, informational and regulatory requirements all need to be monitored over time.

Collectively, these jointly feasible value propositions are referred to as the organizational unit's "business model". To be successful these value propositions must be:

- Financially viable;
- Technically feasible;
- Strategically aligned;
- Continually adapting to changes in stakeholder requirements;
- Ethically considerate of other affected stakeholders, and;
- Compatible with regulatory intent and legalities.

Action and Adjustments

When scheduling major initiatives, sufficient time needs to be allowed for refining and preparing to communicate to the relevant leadership team the proposed action. It is also important to prepare to respond to questions from the decision-making leadership team or board.

One of the reasons why questions are not asked is because the leadership team or board members think they have understood what was said and have no reason to seek clarification. Leaders must therefore take the responsibility to ask questions too, to find out whether listeners have in fact understood what was intended and to create opportunities to correct misunderstandings. Leaders must take the trouble to find out what their followers think they have said, especially if they seek to minimise risks associated with major initiatives. Think of questions as an opportunity to address followers' concerns and as stepping stones to repeat and reinforce key points or deliver additional information. Importantly, make sure you understand the question being asked of you.

It is important for leaders to ensure there is a system in place to receive timely feedback and that there is time allocated to reflect on this feedback. Therefore, following the senior leadership team or board decision regarding the proposed initiative, the action group reviews; what went well and why; what could have been done better and how; what needs to be done to progress the proposed initiative and what will need to be done to further contribute to the organization's adaptive capabilities. This must be an on-going and iterative process.

Foster Self Organization

Many who write about leadership, organizations and complexity write about self-organization (Wheatley, 1997; Gharajedaghi, 2006; Pascale *et al.*, 2000; Uhl-Bien & Marion, 2008; Ackoff, 1997). It is an inevitable consequence of quantum and complexity theories. Complex adaptive systems thrive in a state of disequilibrium, as a consequence of strange attractors and quantum leaps. In such a state a fixed or rigid organization cannot work. The system needs to be free to adapt to the chaos, the changing life conditions.

Yet we persist in designing organizations based on a mechanistic model. We persist in the erroneous assumption that the laws of Newtonian physics will apply to organizations. Too many leaders persist in the related command and control leadership style.

Self organising behavior is induced by a combination of key role identification and specification, shared intent clarification and subjective team pressure imposed by an impending deadline within which a strategically significant value proposition must be designed and articulated.

A business, as an organization, and when considered as a complex adaptive system, needs to be given the freedom to self-organise in order to thrive. The analogy often used is of a flock of geese. A flock of geese in flight naturally form a 'V', the most efficient form to minimise wind resistance. There is no leader instructing geese to fly in a particular formation or which goose is to fly in which position. The goose at the front of the 'V' slips back when tired and another goose moves forward and so on and so on. It is a self-organizing system that works to a small number of simple rules. When those rules are replicated in computer simulations, the same formation appears. (See Note 6: Self Organization)

Communicate To Secure Required Action

Storytelling is a crucial leadership practice that supports a compelling case for decision-makers. Presentations, like reports, can be considered stories. The "story" needs to generate a sense of curiosity and unfold in a manner that generates interest and understanding. Logical inconsistencies can cause the audience to lose track of your argument or hinder them in receiving the message you are intending to send.

Why storytelling is considered important according to Denning (2001: xiii):

- *Nothing else worked;*
- *Charts left listeners bemused;*

- *Prose remained unread;*
- *Dialogue was just too laborious and slow;*
- *Time after time, when faced with the task of persuading a group of managers or front-line staff in a large organization to get enthusiastic about a major change, I found that storytelling was the only thing that worked* (Denning, 2001: xiii).

Denning (2001) has written extensively and compellingly about his experience of being relegated to the management of "information" in the World Bank and gaining traction with the use of stories to change the strategic intent of the organization and incorporate knowledge management into its mission statement. He develops a specific notion of organizational storytelling called the "Springboard" story (Denning, 2001). He outlines the elements for developing a Springboard story:

- The story should be brief and textureless;
- The story must be intelligible to the specific audience;
- The story should be inherently interesting;
- The story should spring the listener to a new level of understanding;
- The story should have a "happy ending";
- The story should embody the change message;
- The change message should be implicit;
- The listeners should be encouraged to identify with the protagonist;
- The story should deal with a specific individual or organization;
- The protagonist should be prototypical of the organization's main business;
- Other things being equal, true is better than invented, and;
- Test, test, test.

Denning (2005, 2007) takes storytelling even further to a model of leadership. He suggests that change is a constant and the best way for leaders to synthesise ideas and make sense is through effective business narrative.

The idea is also found in Flemming (2001), in which he suggests truly successful leaders of the future will progress beyond the recognition of the constancy of change to cultivating the qualities and skills that can maximise the potential hidden within change itself. "Yesterday's mantra of 'Change or Die' will evolve to a new mantra for the future: 'Live to Change'" (p. 34). Flemming asserts that one of the most powerful ways for leaders to make sense out of the complexity is to tap the power of one of humanity's oldest art-forms—storytelling. He goes on to say that few tools are as powerful and readily available to the leader as the use of personal and organizational narrative. He explains that leaders can communicate through narrative in two ways: first by listening and telling stories and second by effectively interpreting these stories to the organization.

Denning (2007) explains that storytelling can improve communication of complex messages. Traditionally, people present information in an abstract and analytical format of:

Define Problem >> Analyze Problem >> Recommend Solution

However research on this approach suggests that it elicits responses from people based on what they already believe. They will either accept it as true if it is aligned with their beliefs or reject some part of it if it does not. So if someone already disagrees then they will reject the definition of the problem, the analysis or the recommended solution.

Denning suggests the following approach is more effective:

Get Attention >> Stimulate Desire >> Reinforce with Reasons

> *When the language of leadership is deployed in this sequence, it can inspire enduring enthusiasm for a cause and spark action to start implementing it... leaders and their followers stay in communication and co-create the future by continuing the conversation* (Denning, 2007: 27).

Storytelling in organizations happens all the time. Good leadership directs the activity in a positive affirming way (Kaye, 1996). Organizations whose stories have characters and plots that enlighten and enrich their members through frequent communications are the healthiest storytelling systems. People in such organizations are likely to want to become an established part of the storytelling culture.

Conclusion

The AFL framework has been successfully applied in a number of environments and at a number of organizational levels. Experience of implementing the framework, and associated practices, has revealed that one of the keys to success is the ability of the facilitator to discern which specific practice to use in any given situation and at any given stage. It is also necessary to recognise that insights gained at a subsequent stage may require a re-think of conclusions drawn at an earlier stage. At this point, the framework is as much art as it is science. There is scope for further research in this regard to determine if leadership practice guidelines can be developed to more effectively support less experienced facilitators.

The need to take a holistic approach is also critical. Focusing on any one of the elements of the framework will provide ineffective results. For example, rushing to action before allowing time and space for divergence and emergence will stifle innovation and leave insufficient time to clarify, communicate and transfer ownership to those who will be involved in implementing the action that is required to secure the agreed intent. Sensitivity to the timing of the progression through the framework is essential.

We initially nominated 3 propositions in this paper that reflected a different way of thinking about adaptation in order to deal more effectively with increasing turbulence:

1. The core role of enterprise leadership is adaptation;
2. The appropriate approach to adaptation is neither context; nor
3. Level of organization dependent.

The AFL framework we presented supports each of these propositions whilst also acknowledging the importance of ensuring systemic linkages between adaptive actions at different levels through the linkage of intent that drives AFL activity at each level.

Given that the AFL framework specifies what needs to be addressed, we have incorporated below in Table 2 a summary of how it is suggested we could proceed. What questions should we be asking?

It is also recognized that sustainability is universally dependent on the continuing commitment and engagement of all involved stakeholders. The methods outlined above facilitate an on-going understanding of the different and evolving perceptions of stakeholders; this enhances the capacity to sustain stakeholder engagement.

The identified convergence that is occurring in the leadership of adaptation across all forms of organizations and at all organizational levels is manifested in the fractal nature of the proposed AFL processes framework.

Increasing turbulence creates increasingly complex interrelationships between chains of risks, best identified through the adoption of a systems view. Indeed, the global financial crisis was the consequence of the interrelationship between chains of related risks that were at least initially either not understood or ignored. The use of scenario analysis and systems thinking to identify related risk inter-relationships is clearly an important basis for formulating risk mitigation strategies associated with adaptive activity and an area that justifies further research.

Finally, as has been demonstrated learning and innovation are critical to adaptation. Learning and innovation within the AFL framework will only foster relevant adaptation when they jointly enhance the probability of achieving agreed meaningful value-driven intent with perceivable increased relative speed and effectiveness.

AFL FRAMEWORK STAGE	HOW DO WE PROCEED?
INTENT:	How do we evolve a succinct shared purpose and vision from an initial fuzzy intent? How do desired scenarios assist in clarifying the vision?
DIVERGENCE:	How do I explore the possibilities and make sense of the world around me? Am I viewing the organisation as a machine or as living Complex adaptive systems? How do I think systemically? Do we utilise the appropriate Harvesting tool? Which of these is appropriate? Strategic Arena Analysis ™? Theory U? Powerful Questions? Mental models?
EMERGENCE:	How do I tune into the patterns, metaphors and associations that help generate insights that radically change the way I and we see the world?
CONVERGENCE:	How do I do converge collaboratively on what needs to be done? Which method(s) should be used: Circle? World Café? Open Space Technology? Appreciative Inquiry? Collective mind-mapping? Pro Action Cafe? Action learning? Deep decision making?
ACTION and LEARNING:	How do **I** bring it all together and both foster and accelerate learning? Action Learning? Reflective Practice?

Table 2 *AFL Framework How To Proceed (Source: Fayed and Duns course notes)*

Notes

Note 1: Self-Awareness

The literature on self-awareness and leadership is extensive. It is widely agreed that self-awareness is an important part of leadership. (Avolio & Gardener, 2005; Block, 1996; Covey, 1992; George, 2003; Goleman, 1995; Hall, 2006; Heathfield, 2006; Heifetz & Linsky, 2002, Hock, 2005; Jaworski, 1998; Kerns, 2004; Owen, 2000; Seligman, 2002; Senge *et al.*, 2004; Simms, 2007). Goleman provides a strong link between self-awareness and leadership success in his work on emotional intelligence (Goleman, 1995, 1998). Hock introduces the concept of Chaordic Leadership, promotes self-awareness and suggests 50% of leadership effort should be on self, although this is largely based on his personal experience and anecdote (Hock, 2000).

Note 2: Identifying Relevant Stakeholders

Organizational stakeholders of interest include those that:

- Contribute directly to value realization (including those that provide various forms of funding);
- Compete directly and indirectly for the patronage of end-customers seeking similar or substitute value (rivals and substitutors);
- Deliver additional value either before or after the relevant purchase that complements the main value delivered (e.g., an alteration tailor provides services that complement the purchase of a garment that requires adjustment);
- Jointly utilise the same core resource to serve different market requirements (e.g., expertise in optics technology used by camera and photocopier suppliers);
- Are part of a demand process (e.g., purchasers, users and influencers); and
- Have an influence on the creation of regulations or are responsible for enforcing existing regulations.

Note 3: Engaging Stakeholders at Rational and Deeper Emotional Levels

Quantum theory argues that everything is connected through fields of energy. Complexity and systems theory also notes that relationships and context are as important as the elements or the parts. Synergy takes those ideas one step further: the interaction of two or more agents or forces is greater than the sum of their individual effects. The by-product is an evolving phenomenon that occurs when individuals work together in mutually enhancing ways to achieve success by inspiring one another to set and accomplish personal adaptive goals and a group vision. It is important to recognise and influence these connections to develop a collective view and gain the benefits of collective intelligence.

The Art of Hosting and Harvesting Conversations that Matter (AoH) is a set of methodologies that engage people at the deeper emotional levels and support the process of developing collective intelligence (see Note 7: The Art of Hosting and Harvesting Conversations that Matter (AoH)). Relevant assumptions that underpin the Art of Hosting and Harvesting include:

- **New solutions grow between chaos and order**: If we want to innovate, we have to be willing to let go of what we know and step into not knowing. In nature all innovation happens at the edge of chaos, or in the space between chaos and order (the chaordic path).
- **Conversations matter**: Conversation is the way we think and make meaning together. It is the way we build strong relationships that invite real collaboration.
- **Meaningful conversations lead to wise actions:** People who are involved and invited to work together only pay attention to that which is meaningful to them. Conversations that surface a shared clarity on issues of importance foster ownership and responsibility when ideas and solutions must be put into action. Actions that come out of collective clarity are sustainable.
- **Organisations are living systems**: When people join together in an enterprise or organisation they have more in common with a living system than with a machine.

Note 4: Mobilizing "Collective Intelligence"

Accepting that there is no one correct or final version of the truth is another key idea of quantum theory. Leaders are often expected to have the answers, however we are all limited in our perspective by our individual values, experiences and knowledge. By entering a wider field of consciousness, we are more likely to overcome the limitations of our individual perspectives. Leaders need the skills to tap into the combined or collective intelligence of the people around them.

Another quantum theory concept is that our observations are distorted by what we are looking for and our version of the truth is not the only right answer. By gaining the perceptions of a broader range of people it is possible to encourage a higher order of wisdom to emerge.

The psychologist Carl Jung envisaged a collective unconscious as the collective memory of humanity. He thought that people would be more tuned-in to members of their own family, race and social and cultural group, but there would also be a background resonance from all humanity. Morphic resonance theory argues that past forms and behaviours of organisms influence organisms in the present through direct connection across time and space. This proposition leads to a radical reaffirmation of Jung's concept of the collective unconscious, which moves us towards an explanation of the enduring aspects of culture that continue well after the people who created them have perished.

The notion of the mind as an open limbic system is also explored by Goleman (2006), where he develops a compelling argument demonstrating that people do not operate as separate entities.

The notion of collective consciousness is related to fields and connectedness. There is a temptation for leaders to remove the "bad apple" or the "dead wood" to improve performance. We are often surprised by how little impact that has. Culture is an enduring field in a group. Changing the way people react to their environment does not happen quickly and the effects of history continue to have an impact long after the event. This makes change difficult and complex. On the other hand, at some stage there will be an adaptive tipping point where a critical percentage of the total group suddenly cause a new "way of doing things around here" to emerge and become the norm.

Note 5: Bank Queues: A Holistic Systems Thinking Example

Consider how a bank might approach managing the length of queues in bank branches—a customer service quality issue. The simplest mechanistic approach recognizes that the average length of a queue depends on the average rate at which customers arrive at the end of the queue and the average time spent in the queue:

Average length = [average arrivals/hr] x [average time in queue]

Average time in the queue is a function of the average duration of transactions.

If this is all that is taken into account, given average arrival rates and a target for average queue length, a staffing number can be calculated to ensure that average queues stay around the target length. This fails in two obvious ways:

- It ignores variability (e.g., in demand by time of day or in duration of individual transactions);
- It creates a high level of overheads that cannot be reduced without changing branch processes.

Introducing Feedback and Handling Variability

It is possible to collect data on the rate of arrivals by time of day and try to roster more staff to cover the expected busy times, although this requires either part-time staff (remember when banks did not have any?) and/or multi-skilling.

To make multi-skilling effective, there has to be a simple feedback loop established, so that when the queue(s) exceed a specified level, staff move onto the counter; when queues fall below another level, staff can revert to other work. In theory (ignoring the variability of human behavior and other branch activities) a feedback loop can operate as strictly and mechanically as a thermostat.

More Radical Change

Introducing part-time staff and multi-skilling allows branches to stabilise maximum queue lengths in a fairly mechanistic way. However, to keep queues under the desired limits, enough staff members are needed to handle peak loads, which imposes wasteful overheads. Further consideration is needed.

A complex adaptive systems perspective suggests that strategic innovation is required in other areas: conditions of employment, training, duties, recruitment, pay structures, segmenting customers by type of transaction and providing alternative transaction channels.

Diverting customers to less staff-intensive ways of performing transactions is one strategy. ATMs, telephone and internet banking, improving information systems so that transactions take less time, splitting queues by function (account transactions versus enquiries) and increasing opening hours have significantly improved responsiveness to customers. In one way or another, however, all have short- and long-term unintended consequences that will need to be monitored and addressed if their negative consequences are to be minimized.

Detailed specification of the system will facilitate the identification of points of leverage where small investments can have large effects.

Note 6: Self Organization

Other examples from nature are offered too, such as the fire ant. (Pascale *et al.*, 2000). The fire ant arrived in North America before World War II. There has been a determined effort to eradicate, or even control, the species resulting in an astonishing array of survival strategies. The fire ant started in Miami and now has spread across 260 million acres in eleven states. The ant's sting is mild but they communicate to sting in groups, causing significant pain, and even death for babies, calves and other fauna. Their fertility levels fluctuate according to their life conditions. They can be extraordinarily prolific as necessary, and then when the danger is past the worker ants kill off the queen bees to limit reproduction.

A final example is that of the Spanish attempt to conquer the Navaho Indians. (Brafman & Beckstrom, 2006) Spanish conquerors were successful by identifying and executing the key leader(s) of their enemies. It was a successful strategy with the majority of people in the conquered nations or territory. They tried the same tactic on the Navaho Indians and it was a total failure. It took hundreds of years and virtual total genocide before they were finally defeated. One reason for their persistence was the fact that there is no identifiable leader. Instead a leader will emerge for a specific purpose, called a Nant'an.

This is a very different approach to leadership that is at odds with our command and control mechanistic view of organizations.

Note 7: The Art of Hosting and Harvesting Conversations that Matter (AoH)

AoH was developed by a group of people, although largely attributed to Toke Moeller and Monica Nissen of Interchange in Denmark. There is no published book that describes the package. There is a range of published material that supports the individual methodologies. (Baldwin, 1998; Brown, 2001; Cooperrider *et al.*, 2000; Hock, 2000; Owen, 1997, 2000; Srivastava & Cooperrider, 1999) In addition AoH is explicitly based on "Living Systems", which is a term used to include complex adaptive systems and complexity. The material presented is drawn largely from the training journal provided to participants of the training program. The core methodologies of AoH include Circle, World Café, Appreciative Inquiry and Open Space Technology and the practices of AoH are Reflective Practice and Presence.

Note 8: Simplified Approach to Scenario Development
(Adapted from Schoemaker 2002 and van der Heijden 1996)

Step 1. Select the most important demand side and supply side uncertainties

Rank separately the identified demand and supply side key uncertainties (typically 5 to 7 uncertainties). In order to determine rank order, within each of these two lists of uncertainties determine the degree of negative or positive impact on the achievement of desired outcomes and the likelihood of that impact occurring in the medium term (three to five years out). Highly ranked uncertainties will have a high probability of high medium term impact on desired outcomes.

Now select the top ranked demand and the top ranked supply uncertainty for use in the next step. This will require a discussion that seeks consensus by surfacing and understanding previously unarticulated concerns.

Step 2. Develop a Two-By-Two Scenario Matrix

For the two highest ranked uncertainties selected, specify two feasible extreme "outcomes"—one desired and one not desired—that should ideally define the range of feasible outcomes that might occur.

By creating a 2x2 matrix, four scenarios are then generated. The scenario defined by the most desired supply and demand side outcomes provides a basis for specifying a future desired vision, while the least desired supply and demand side outcomes scenario provides a basis for mitigating risks by formulating, for example, more flexible strategies. The "scenario matrix" below shows how contrasting outcomes for the two selected uncertainties are combined to generate four scenarios.

Uncertainties	Uncertainty 1 = Outcome 1 (e.g., Incremental)	Uncertainty 1 = Outcome 2 (e.g., Discontinuous)
Uncertainty 2 = Outcome 1 (e.g., Best)	Scenario A	Scenario B
Uncertainty 2 = Outcome 2 (e.g., Worst)	Scenario C	Scenario D

Diagram 3 *Scenario Matrix*
Source: IMIA

The example below comes from a scenario development exercise in the health-care industry. It is based on contrasting radical discontinuous changes with incremental changes:

- A medical technology "breakthrough" (discontinuous) is contrasted with incremental gains in medical technology
- A quantum change in patient information capture and use, along with parallel quantum changes in information capture and delivery in many arenas (discontinuous change), is contrasted with incremental improvements.

An Example: Healthcare	**Medical Technology = Incremental Gains**	**Medical Technology = Breakthrough**
Information Capture & Use = Incremental Improvement	**Scenario A:** Healthcare Held Hostage	**Scenario B:** Intel Inside
Information Capture & Use = Quantum Change	**Scenario C:** Consumers Take Control	**Scenario D:** The Last Frontier

Diagram 4 *Example Scenario Matrix*
Source: Schoemaker 2002.

Note 9: Argyris and Schon (1978)

Argyris and Schon (1978) present a theory in which organizational learning takes place through individual actors whose actions are shaped by their shared mental models. They argue that these shared models protect the status quo in organizations where prevailing conditions are rarely challenged and there is silent assent; in other words, these are places where very little learning takes place.

Senge (1990) describes mental models as "deeply held internal images of how the world works, which have a powerful influence on what we do because they also affect what we see." Mental models are more than memory as a passive storage device—these memories actually affect what we see and do.

Kim (1993) states that "mental models not only help us make sense of the world we see (but) they also restrict our understanding to that which makes sense within the mental model."

References

Ackoff, R.A. (1997). *From Mechanistic to Social Systemic Thinking*, ISBN 9781883823115.

Argyris, C and Schon, D.A. (1978). *Organizational Learning: A Theory of Action Perspective*, ISBN 9780201001747.

Avolio, B. and Gardner, W. (2005). "Authentic leadership: Getting to the root of positive forms of leadership," *The Leadership Quarterly*, ISSN 1048-9843, 16(3): 315-338.

Baldwin, C. (1998). *Calling the Circle: The First and Future Culture*, ISBN 9780553379006.

Block, P. (1996) *Stewardship: Choosing Service over Self-Interest*, ISBN 9781881052289.

Brafman O. and Beckstrom R. (2006). *The Starfish and the Spider: The Unstoppable Power of Leaderless Organizations*, ISBN 9781591841838.

Brown, J. (2005). *The World Café: Living Knowledge through Conversations That Matter*, ISBN 9781576752586.

Chandler, A., Jr. (1962). *Strategy and Structure: Chapters in the History of the American Industrial Enterprise*, ISBN 9780262530095.

Cooperrider, D. Sorensen, F., Jr., Whitney, D. and Yaeger, T. (2000). *Appreciative Inquiry: Rethinking Human Organization toward a Positive Theory of Change*, ISBN 9780875639314.

Covey, S. (1992). *Principle-Centered Leadership*, ISBN 9780671792800.

de Geus, A. (1997). *The Living Company*, ISBN 9781857881806.

Denning, S. (2001). *The Springboard Story: How Storytelling Ignites Action in Knowledge Era Organizations*, ISBN 9780750673556.

Denning, S. (2005). *The Leader's Guide to Storytelling: Mastering the Art and Discipline of Business Narrative*, ISBN 9780787976750.

Denning, S. (2007). *The Secret Language of Leadership: How Leaders Inspire Action through Narrative*, ISBN 9780787987893.

Eisenhardt, K.M. and Martin, J.A. (2000). "Dynamic capabilities: What are they?" *Strategic Management Journal*, ISSN 1097-0266, 21: 1105-1121.

Fayed R. (2010). "Transforming hierarchy into networks," Australian Graduate School of Leadership Working Paper.

Fayed, R. (1986). "When strategic planning comes to the crunch," Australian Institute of Management, Journal. Based on research amongst 15 major Australian organizations and sponsored by Telecom Australia, April 1986.

Flemming, D. (2001). "Narrative leadership: Using the power of stories," *Strategy and Leadership*, ISSN 1087-8572, 29(4): 34-36.

George, B. (2003). *Authentic Leadership: Rediscovering the Secrets to Creating Lasting Value*, ISBN 9780787975289.

Gharajedaghi, J. (2006). *Systems Thinking, Managing Chaos and Complexity: A Platform for Designing Business Architecture*, ISBN 9780750679732.

Goleman, D. (1995). *Emotional Intelligence*, ISBN 9780553095036.

Goleman, D. (1998). *Working with Emotional Intelligence*, ISBN 9780553104622.

Goleman, D. (2006). *Social Intelligence: The New Science of Human Relationships*, ISBN 9780553803525.

Hall, L.M. (2006). *Winning the Inner Game: Mastering the Inner Game for Peak Performance*, ISBN 9781890001315.

Heathfield, S.M. (2006). "Build an organization based on values," http://humanresources.about.com/od/strategicplanning1/a/organizvalues.htm.

Heifetz, R.A. and Linsky, M. (2002). *Leadership on the Line: Staying Alive through the Dangers of Leading*, ISBN 9781578514373.

Hock, D, (2000). "The art of chaordic leadership," *Leader to Leader*, ISSN 1087-8149, 15, http://www.leadertoleader.org/knowledgecenter/journal.aspx?ArticleID=62.

Jaworski, J. (1998). *Synchronicity: The Inner Path of Leadership*, ISBN 9781881052944.

Kauffman S. (1993). *The Origins of Order: Self-Organization and Selection in Evolution*, ISBN 9780195079517.

Kaye, M. (1996). The Story-telling Organization in Myth-makers and Story-tellers: How to Unleash the Power of Myths, Stories and Metaphors to Understand the Past, Envisage the Future and Create Lasting Change in Your Organization, ISBN 9781920698911.

Kerns, C.D., (2004). "Strengthening value-centered ethics (Part 2): Managerial leadership action roles: What, why, and how?" *Graziadio Business Report*, 7(2), http://gbr.pepperdine.edu/2010/08/strengthening-values-centered-leadership/.

Kim, D.H. (1993). "The link between individual and organizational learning," *Sloan Management Review*, ISSN 0019-848X, Fall(October 15):

McElroy, M.W. (2002). "Social innovation capital," *Journal of Intellectual Capital*, ISSN 1469-1930, 3(1): 30-39.

Morgan R.E. and Hunt S.D. (2002). "Determining marketing strategy: A cybernetic approach to scenario planning," *European Journal of Marketing*, ISSN 0309-0566, 36(4): 450-478.

Owen, H. (1997). *Open Space Technology*, ISBN 9781576750247.

Owen, H. (2000). *The Power of Spirit*, ISBN 9781576750902.

Pascale, R.T., Millemann, M. and Gioja, L. (2000). *Surfing the Edge of Chaos: The Laws of Nature and the New Laws of Business*, ISBN 9780609808832.

Prahalad, C.K. and Hamel, G. (1990). "The core competence of the corporation," *Harvard Business Review*, ISSN 0017-8012, 68(3): 79-91.

Scharmer, C.O. (2007). *Theory U: Leading from the Future as it Emerges, The Social Technology of Presencing*, ISBN 9780974239057.

Schoemaker P.H.J. (2002). *Profiting from Uncertainty: Strategies for Succeeding No Matter What the Future Brings*, ISBN 9780743223287.

Seligman, M. (2002). *Authentic Happiness*, ISBN 9781740511087.

Senge, P. (1990). *The Fifth Discipline: The Art and Practice of the Learning Organization*, ISBN 9780091827267.

Senge, P., Scharmer, C.O., Jaworski, J. and Flowers, B.S. (2004). *Presence: Human Purpose and the Field of the Future*, ISBN 9780974239019.

Simms, M. (2007). "Self-awareness important for capable leadership," *The Bellingham Business Journal*, June 1.

Srivastava, S. and Cooperrider, D. (1999). *Appreciative Management and Leadership*, ISBN 9781933403175.

Stalk, G., Evans, P. and Shulman, L.E. (1992). "Competing on capabilities: The new rules of corporate strategy," *Harvard Business Review*, ISSN 0017-8012, 90(2): 57-69.

Uhl-Bien, M. and Marion, R. (eds.) (2008). *Complexity Leadership Part 1: Conceptual Foundations*, ISBN 9781593117955.

van der Heijden K. (1996). *Scenarios: The Art of Strategic Conversation*, ISBN 9780471966395.

Ward P., and Sturrock F. (1998). "'She know what she wants...': Towards a female consumption risk-reducing strategy framework," *Marketing Intelligence and Planning*, ISSN 0263-4503, 16(5): 327-336.

Wheatley, M. (1997). *Leadership and the New Science*, ISBN 9781576750551.

Ramzi Fayed has led the Australian Graduate School of Leadership and its predecessor for the past several decades. He originally qualified as a physicist with a B.Sc. (Physics) from The Faculty of Technology at Manchester University in the UK and was also awarded an MSc and a PhD in Management Science by Manchester University. He has held a wide variety of honorary appointments, including journal editorial board positions, various Chairmanships, board positions on not for profits and global commercial enterprises. His senior academic appointments at a number of universities have included an Honorary Chair. He has been an invited Faculty Member on Advanced Management Programs in Australia, New Zealand, Asia, Europe, and North America. He has led or acted as principal advisor for over 90 major local and international commercial initiatives, presented papers to over 100 conferences and published booklets, cassette series, chapters in books and papers in peer reviewed journals. He has also pioneered a number of educational innovations in regard to postgraduate degree content and material delivery. He is a long standing Fellow of the Australian Institute of Management and a Foundation Fellow of the Australian Institute of Company Directors. He led the team in the mid 1960s that pioneered the design and implementation of, at that time, the world's largest computerised commercial application package to control the movement of motor vehicle spare parts from the five UK car manufacturers to 75 of their largest spare part warehouses across the UK.

Stephen Duns is a Partner, Social Innovation and Research, at Success Works and Director of the Australian Graduate School of Leadership. Stephen has a passion for helping individuals, organizations and communities to identify and achieve their potential. Stephen's career experience includes several Chief Executive Officer and general management roles in health and community services in Australia and the UK. He has also worked extensively as a consultant in the UK, Europe, USA and Australia. Stephen's recent voluntary interests include President of Joy 94.9, Director of Merri Community Health Services, teaching business governance in developing countries with Australian Business Volunteers, Chair of Austin Health's Human Research Ethics Committee, President and Chairman of the Board of Parkinson's Victoria and Parkinson's Australia and President of Cobaw Community Health Services. Stephen has a Doctorate in Business Leadership and has a Master of Business Administration, Bachelor of Arts (Philosophy) and Bachelor of Letters (Psychology) degrees. Stephen is a Fellow of the Australian Institute of Company Directors and Associate Fellow of the Australian College of Health Executives, a qualified SDi practitioner as well as an accredited Master Practitioner in Neurolinguistic Programming and Hypnotherapy.

Gervase Pearce is Partner of BCS Consulting and Director of the Australian Graduate School of Leadership. Gervase commenced his career as a commissioned officer in the Royal Australian Navy. In addition to operational and staff positions, he lead a number of major cultural change initiatives across the operational element of the RAN, including implementation of Women at Sea and Quality Management. Gervase has broad industry experience as an operational and senior manager with particular emphasis on dealing with complex and wicked problems related to strategy, strategic change and leadership. He has worked with assisting strategy development and implementation and organizational change initiatives in a range of industries including finance, heavy and light manufacturing, FMCG, emergency services and public sector. Gervase has a Master of Business Administration (MGSM), Bachelor of Financial Administration (UNE), is a member of the Australian Institute of Company Directors and was a former council member of the Futures Foundation (Australia).

13. Ethical Decision-Making And Metaphors: Enhancing Moral Consciousness Using Parables And Complexity Theory

Edwin E. Olson
Graduate School of Management & Technology, University of Maryland University College, USA

Metaphors from complexity theory and religious traditions are a useful tool in ethical decision-making. Within a framework of virtue ethics and moral agency proposed by Dyck and Wong (2010) this article demonstrates how parables from the Christian tradition can enhance organizational virtue. Parables challenge conventional wisdom and evoke alternative, responsible ways of viewing issues of injustice, diversity, inclusion of stakeholders, and transforming systems. Complexity metaphors discussed include open/closed systems, the uncertainty principle, adjacent possible, coupling, attractors, top-down causation, and fractals. The parables include the Rich Farmer, the Good Samaritan, the Feast, the Pearl of Great Price, and the Vineyard Laborers. The CDE model (Olson & Eoyang, 2001) and a case study illustrate the six steps involved in a "paraplexity" method to jointly apply complexity metaphors and parables. Further research and application is suggested using additional complexity concepts and metaphors and other parables, including those from other religious traditions.

Introduction

My focus in this article is on the nexus of complexity theory, spirituality, and organizational leadership in dealing with moral and ethical issues[1]. I will integrate the concepts from two sources—the parables from Christianity and the metaphors of complexity theory about the natural creativity of the universe. The intertwining of these two sources can unsettle and stir the listener with a reality of a spiritual presence in the world—where substantial change in moral consciousness is possible.

This article provides a tool to help us evolve as ethical[2] decision-makers, develop new levels of emergent understanding, a level of spirituality, and even a sense

---

1. I have written about the emergence of the sacred in science and religion (Olson, 2009) and the vital role of spirituality in social change (Olson, 2010).

2. Ethical decisions are guided by our notions of what we think the organization *should* be. Ethical decisions reflect our individual and collective values and preferences (Cilliers, 2004).

of the divine[3]. Narratives, stories, and parables ask us: How can this be? Why are people oppressed? How can we be liberated?

In times of uncertainty, the parables help us take adaptive action in the immediate and local demands of a moral situation. They reduce the complexity of the world around us to clarify appropriate responses. We can clearly observe what is happening, see the implications, and take action within our ability, scope of influence, and time and resource constraints.

Virtue Ethics

Anderson and Escher (2010), Harvard Business School graduates, have amplified the MBA oath, a pledge for graduating MBAs to do whatever it takes to lead ethically and with integrity. The authors provide three suggestions for overcoming moral blindness: first, make a commitment to do the small things right; second, use a more rigorous framework for decision-making including breaking down ethical decisions to their component parts of intention, means, and outcome; and third, become more self-aware. The third suggestion of self-awareness, including increased corporate awareness, is the focus of this article.

Dyck & Wong (2010) have provided a framework for the virtue ethics and moral agency embodied in the MBA oath. They are particularly interested in how the nexus of spirituality and organizational culture might promote specific behaviors that facilitate and enable virtue development. They discuss how four specific spiritual disciplines can be observed and practiced to help build an organization culture necessary to enable moral agency. Their analysis is derived from Aristotle for whom the goal of life was *eudemonia*—commonly translated as happiness, but a more contemporary translation given by Dyck and Wong might be "flourishing". Dyck & Wong argue that organizations flourish when its members practice and facilitate such character virtues as justice, self-control, practical wisdom, and courage.

They believe that their "virtuous cycle model", if implemented, over time will develop a virtuous organization. The model has four phases covering the four spiritual disciplines:

I Confession. Sensitively noticing shortcomings and injustices associated with current structures, systems, and practices. When we confess our own shortcomings we don't convey attitudes of superiority. We are aware of needs that are being overlooked and of unacceptable behavior that needs to be corrected.

---

3. Philip Clayton, the Ingraham professor at the Claremont School of Theology, calls this radically Emergentist theism. This theory posits "God" as the emergent property of spirituality in the universe—the universe becoming aware of itself.

II Worship. Being attuned to the inherent goodness in others and facilitating shared listening. Expecting others to be insightful, to consider what others are saying, listening to the voices of the Spirit, even if it might appear to be contrary to the facts.

III Guidance. Including others in discerning how to proceed. Seek the counsel and knowledge of stakeholders to understand existing structures and systems.

IV Celebration. Rejoicing after implementing ideas or procedures that might help overcome shortcoming in the current practices. This step of celebrating past successes leads back to the first step as we identify and confess new concerns for liberating social systems.

The authors believe that the spiritual disciplines can be put into practice in any organizational environment. Even in hostile environments people can confess shortcomings, find the good in others, seek guidance, and experiment with new structures or systems. They refer to these mini-cycles as moments of micro-emancipation that act as "yeast" to help organizations rise to become more virtuous. The model uses narratives and stories to form organizational memory which shapes the formation of organizational virtue. The authors call for work to operationalize the four spiritual disciplines and test whether and how their practice enhances organizational virtue. This article is a step in that direction through the use of parables.

Parables As Antenarratives

Ethical decisions are socially constructed, that is, knowledge and awareness of ethical neuro-pathways are created from the combined knowledge, experience, interactions, and observations of others (Weick, 1995; Galloway-Seiling, 2010; Gergen, 2009). The variations in the interaction with others become the source of new self-organized ethical decisions.

To promote variations, stories must be told that bring about new, diverse, contradicting, and alternative ways of seeing. Baskin (2010) in his discussion of organizational story-telling, which he calls "storied spaces", points out the value of antenarratives (alternative constructions of reality), if organizational leaders are able to create a truly open-ended, creative environment where participants feel safe to explore novel concepts that might disrupt the status quo. Storied spaces generate alternative viewpoints that reflect the interplay between dominant narratives and antenarratives. Baskin believes this offers a complexity of perception equal to the complexity of current environments.

As an antenarrative, parables can help us engage with uncertainty and emergence as it is happening and then to make difficult moral decisions. To confront the culture of fear in our time, we must envision new ethical metaphors for living, metaphors that evoke new, responsible ways of relating to one another (Hill, 2010). These metaphors would embrace and celebrate our evolutionary origins, question our parochial ideologies, question default ideas about race and class, and question our ideas about what strength and security means.

The parables attributed to Jesus were told in an oral environment where people were used to adapting stories, re-creating them, transforming them, utilizing them in ever new ways to communicate afresh and anew in every different context (Liebenberg, 2000). We can engage our own contexts as we consider the parables, since these structures invite polyvalency and nuanced understanding rather than promote fixed meanings. Rather than look for an ultimate, isolated, universal meaning of the parables, as if they are objects waiting to be discovered, I will use complexity concepts and parables attributed to Jesus[4] to frame alternative models for ethical decision-making.

Parables And Complexity Theory

At its simplest the parable is a metaphor or simile, drawn from nature or common life, arresting the hearer by its vividness and strangeness, and leaving the mind in sufficient doubt about its precise application to tease it into active thought (Dodd, 1961).

Historically, parables have been effective in addressing moral ignorance, weakness and hypocrisy and producing resilient moral character. However, a limitation in using the ancient wisdom for moral development today is that the metaphors in the parables have become so familiar that we no longer feel tension in the associations they draw upon.

We no longer hear or read them with a sense of first-timeness, unpreparedness, and surprise. Metaphors from complexity theory can unpack the underlying meaning in the parables and recast them in ways that are compelling to a modern audience. Complexity metaphors can stimulate intuitive understanding and insight and evoke a sense of wonder, beauty, and mystery. A complexity theory reframing of a parable creates variation in its meaning which helps us evolve and adapt our moral compass to fit the local situations we face.

The parables told by Jesus are about experiencing the paradoxical tension between his message which transcended cultural barriers and conventional wisdom of the day. Complexity theory also affirms that the creative tension between opposites is at the very heart of the self-organizing, bottom-up phenomenon that creates order and structure from the interaction between the system and its environment.

4. The parables selected are from the 21 parables identified by the Jesus Seminar (Funk, Scott & Butts, 1988) as those most likely told by the historical Jesus.

Using Complexity Concepts in Workshops

In workshops I have conducted I have seen how the parables, when presented with modern evocative and imaginative images, give participants an alternative perspective to guide their future decisions. For example, the parable of the small amount of yeast that leavened fifty pounds of flour suggests that evolutionary processes produce amazing and unexpected results from insignificant beginnings. Such a metaphor challenges decision-makers to look for the potential in faint signals.

I have chosen four parables to illustrate the four spiritual disciplines presented by Dyke & Wong. I then present a case using a fifth parable to illustrate the six steps in the paraplexity method I have developed.

I Confession—awareness of injustice. ***The Rich Farmer*** points out short-comings and injustices in societal patterns. It raises awareness by pointing out the lack of awareness of the farmer who is often called "the rich fool".

II Worship—looking for the good in others. ***The Good Samaritan*** focuses on the essential goodness in others.

III Guidance—including others in decisions. ***The Feast*** has a message of inclusion of those marginal to the system.

IV Celebration—transforming systems and structures. ***The Pearl of Great Price*** goes beyond the immediate "fixes" to ethical issues and holds out a vision of a higher consciousness - a paradigm shift - that can celebrate continuous improvement and evolution.

I. Confession—Awareness Of An Injustice: *The Rich Farmer*

> *There was a rich man who had much money. He said, "I shall put my money to use so that I may sow, reap, plant, and fill my storehouse with produce, with the result that I shall lack nothing." Such were his intentions, but that same night he died.* (Gospel of Thomas 63:1)

In many respects the farmer is to be admired. He plans, saves, and protects his belongings. He looks forward to a life of leisure, enjoying what he has acquired in a lifetime of work. Many would see him as prudent and a good steward of his resources—certainly he is not wasteful or reckless with his wealth.

The problem is—he is obsessed with his possessions. The whole meaning and value of his life is linked to his possessions. His sense of his self-worth and selfishness is expressed in a series of "I" and "my" statements. There is no separation between him and his possessions. He is hoarding everything for himself, not aware of or caring about the existence of anyone else. The parable, a monologue in which the rich farmer is talking only to himself, dramatizes his isolation. The farmer exhibits the ethical bias of over claiming credit (Banaji, Bazerman & Chugh, 2003) which leads to an overblown sense of entitlement.

In this parable Jesus is saying that, given the economic situation in Palestine, goods need distribution (Williams, 2008). The point of storing produce was to help the community, especially the poor, in times of famine. The notion that he controlled his resources is an illusion—ultimately they belong to society; it is only a question of how rapidly they will be distributed.

The parable is an example of a *closed system* with *impermeable boundaries* of presumed self-sufficiency. The farmer is undone by his lack of humility and empathy toward others. He is disconnected from community. True wealth and survival is found in functioning as an *open system* with *permeable boundaries* that allow give and take transactions with all of the stakeholders. This will build a pattern for a life that is sustainable.

As workshop participants discussed the parable, the concepts of open/closed systems and boundaries helped them think about retirement. Many have made a goal of making a lot of money early in life and retiring early, but to be truly rich they realized they must live productively, being mindful of their responsibilities to family, extended family, and communities—opening their boundaries. They also reflected on how they spend their money on consumer goods, a lavish lifestyle, and other inordinate expense.

The parable embodies the aspects of *Confession* in the Dyke & Wong model. Being aware of shortcomings, attitudes of superiority and unacceptable behavior and practices is needed in assessing the power and reward systems in organizations. As well as a private vice, greed is a public problem needing public solutions, such as the inheritance tax (McCullough, 2008).

II Worship—Looking For The Good In Others: The Good Samaritan (Peterson, 2002)

> *There was once a man traveling from Jerusalem to Jericho. On the way he was attacked by robbers. They took his clothes, beat him up, and went off leaving him half-dead. Luckily, a priest was on his way down the same road, but when he saw him he angled across to the other side. Then a Levite religious man showed up; he also avoided the injured man. A Samaritan traveling the road came on him. When he saw the man's condition, his heart went out to him. He gave him first aid, disinfecting and bandaging his wounds. Then he lifted him onto his donkey, led him to an inn, and made him comfortable. In the morning he took out two silver coins and gave them to the innkeeper, saying, 'Take good care of him. If it costs any more, put it on my bill—I'll pay you on my way back.'* (Luke 10:30b-35)

The parable is about the willingness and ability to see a person in need and respond. The phrase "when he saw him" shows up three times in the parable. They all "saw" the man's condition, so they are all on the same level playing field, but only the Samaritan responds. They all understood the man's situation, but only one acted to remedy it. So the first two are without excuse (Hultgren, 2000).

Holloway (2003) says that the heart of the story is a powerful Greek verb *esplanknise* which means at the sight of the naked and bleeding man the Samaritan's guts churned inside him with such ferocity that it simply obliterated the walls that ritually separated him from a supposed enemy. This gut churning is an example of the *Uncertainty* and intractability in complexity dynamics where outcomes are unpredictable.

As the Samaritan's guts churned, there were a number of possibilities. He could choose to walk away, call others for help, kill the man, or given assistance. He resolved his own uncertainty by choosing to take care of a person who was not only different from him but who was also a sworn enemy.

Jesus presents a new way of relating to reality in this parable. Grace, change, mercy, and aid come from unexpected sources at an unexpected time. The parable shows the spiritual and moral danger we are in when we allow our codes and traditions to dictate our behavior. Every code is provisional and discardable when faced with the uncertainty of new situations, but how much risk should we take? How much uncertainty can we handle?

In a workshop a participant who managed a large non-profit organization said applying *Uncertainty* to the parable made the parable larger. He realized the importance of letting go of the certainty of old ways of thinking that dictated his behavior and instead be more flexible, adaptive, and ready to improvise.

Kauffman (2008) describes our untapped potential as the *Adjacent Possible*. To survive in a changing environment, autonomous agents need to evolve to a higher level of complexity, to what is possible for them given their limitations and the environment surrounding them. The Samaritan moves into his Adjacent Possible. The workshop participant also saw how he could move into his Adjacent Possible of creating a new enterprise that would expand into new markets.

The parable illustrates the *Worship* phase of the Dyke & Wong model. Being attuned to the inherent goodness in others and acting on new adjacent possibilities is more likely if we let go of certainty and are not tied to our rules and worldview

III Guidance—Including Others In Decisions: *The Feast* (Peterson, 2002)

> *There was once a man who threw a great dinner party and invited many. When it was time for dinner, he sent out his servant to the invited guests, saying, Come on in; the food's on the table. Then they all began to beg off, one after another making excuses. The first said, I bought a piece of property and need to look it over. Send my regrets. Another said, I just bought five teams of oxen, and I really need to check them out. Send my regrets. And yet another said, I just got married and need to get home to my wife. The servant reported back, 'Master, I did what you commanded— and there's still room. The master*

> *said, Then go to the country roads. Whoever you find, drag them in. I want my house full!* (Luke 14: 16b-23)

The host is being snubbed apparently by all his acquaintances. Everyone who was invited has an excuse and refuses to come. With his honor so offended he has few choices. He orders his slave to go on the streets and bring back whomever they find to have dinner. The random gathering of persons from the roads and byways of areas produces immense diversity at the table. The Feast has a message of inclusion of those marginalized by the system.

In complexity theory the interdependent bonding of agents is defined as "*coupling*". The bonding among agents can be loose, moderate or tight. The agents that are coupled continue coupling with other clusters and form aggregates which then can connect to other aggregates. In this sense we are coupled to everything else on the planet. Most of the couplings are very loose, yet there is a sense that every part affects the whole. Tightly coupled agents display high degrees of interdependence. Moderately or loosely coupled agents have a low degree of interdependence.

In the Feast parable the host started with a tightly coupled system. Invitations had gone out to an elite circle of friends or contacts. The expectation was that they would positively respond to the invitation to dinner. When he began to lose face because his tight connections snubbed him, he changed his plan and went with a loosely coupled system. He invited whoever was available and who would appreciate his offering.

As workshop participants reflected on the concept of *Coupling* they explored the coupling pattern in their lives. Several said they were too tightly coupled to their friends in their upper class community. Others thought about their networking practices and whether they were coupled to groups and organizations with common interests.

Understanding how we are coupled to all of our stakeholders helps us move into the *Guidance* phase of the Dyke & model. Tightly coupled systems enhance close coordination and cooperation but since they are highly interdependent they are more vulnerable to disruption. Loosely coupled systems can absorb changes in the environment and assaults on its constituent parts. The most resilient systems are those that can change their level of dependence in seeking guidance, according to what is needed for survival in differing environments.

IV. Celebration—Transforming Systems And Structures: The Pearl Of Great Price

> *God's kingdom is like a trader looking for beautiful pearls. When he found one priceless pearl, he sells everything he owns and buys it.* (Matthew 13: 45-46)

To get the pearl of great price the merchant had been seeking all his life he had to sell all his possessions. To get the best he had to abandon the second best. Or perhaps his previous life was unsustainable. He may have spent years and substantial resources in his quest for the ultimate meaning of his life.

In this parable we see the possibility of moving our ethical and moral decision-making to a higher level of consciousness.

In mathematics an *attractor* is a point, a cycle, or a pattern that a system evolves to. A *"strange" attractor* is a system in continued chaotic motion that creates a subtle order. The strange attractor has been used as a metaphor to describe highly complex, but patterned, behavior in human systems. Whenever the behavior of the system is bounded, includes infinite freedom within the bounds, and generates coherent patterns over time, the human system can be metaphorically described as a strange attractor. Examples of human system aspects that fit this qualitative description include organizational culture, patterns of professional practice, or the behaviors of firms within a given industry. In each case, individual agents work within accepted boundaries in accord with patterns of behavior that are supported by the rest of the system in complex and nonlinear ways.

In the *pearl* parable the person is drawn to a strange attractor—a new pattern, a new order. His action changes the state of his life. This changed state will affect his future experiences and behavior, creating an endless series of feedback loops. Giving ourselves to the pearl of great price represents a movement to higher consciousness and wholeness. This concept has been expressed in many ways—as a search for the Holy Grail, as individuation (Jung), and as the end result of religious and philosophical paths.

In the workshop some participants reported experiences of being seized by an intuition, an inner reality, or a vision of what can be and have dedicated their lives to the pursuit of that dream—their own treasure. They found a new attractor, a new formula that makes sense of their lives. They have their own spiritual GPS[5]. Others talked about being in organizations where many people are in motion around their own new strange attractor, but each in turn has multiple dimensions and layers. As they make contact with each other, new possibilities and new attractor patterns are created

The impact a more complex system such as a pearl has on its constituent parts is referred to as *top-down causation or whole-part constraint*. In biological terms, an organism embedded in a higher-level system that includes its environment is affected by it. The universe is governed by a higher order system of natural laws and processes.

As the workshop participants considered top-down causation, they saw how their "pearl" determines how all of the parts behave. This is the opposite of reductionism in science that says that a complex system can be explained by reduction to its fundamental parts.

---

5. I am grateful to Ken Kostial for this insight.

With the holistic perspective of an attractor such as the "pearl", the *Celebration* phase of the Dyke & Wong model is realized. Shortcomings of current practices are remedied and we are able to appreciate and evaluate the next step in our evolution.

Paraplexity Method

I have named the workshop method that links parables and complexity metaphors as "paraplexity" (simply putting together parables and complexity). To apply complexity metaphors to the parables it is useful to use the *CDE model* that explains the three conditions of self-organizing (Olson & Eoyang, 2001) in facilitating change in organizations. The three conditions are:

1. **Containers** such as shared goals and objectives and constructions of reality that hold the interactions of the agents in the system to accomplish authentic and observable work.
2. **Significant Differences** such as new ideas, relevant associations, and alternatives.
3. **Transformative Exchanges** across those differences that promote learning and growth. Sharing of visions, values, and insights. Emphasis on collaboration and learning.

The steps in the method fit within the CDE model. The first two steps set out the *container* of the organization issue and the relevant parable. Steps 3 and 4 amplify the issue, the parable, and the relevant complexity concepts to identify the *significant difference*. Steps 5 and 6 encourage the participants to develop meaningful *transforming exchanges* as they apply the previous discussion to their own role and responsibility for the issue under focus.

The six steps are:

Container

1. Identify an issue, area of concern, or decision under consideration
2. Choose and present a parable to an individual or group that they will appreciate as relevant, whether because of a religious heritage, commitment to social action, or intellectual curiosity

Significant Difference

3. Encourage reflection and discussion of the meaning and ethical/moral implications of the parable in the context of the listener(s)
4. Provide a complexity explanation that enlarges the scope of the parable and its possible meaning

5. Discuss integration of the complexity perspective with the listener's previous interpretation
6. Discuss application to the issue under focus including action steps.

As people share their common concerns and their perspectives on the parables, the boundaries of separation are blurred and they move beyond comfortable conventions and seek more systemically challenging alternatives. The moral truths expressed in the parables are linked to our deepest sense of the sacred, the mysterious and unknowable, and can be the basis of a new worldview. The parables with greatest impact create a common resonance and shared experience among the participants. The experiential workshops establish ethical fractal[6] seeds that the participants carry into other aspects of their lives.

Case of the Vineyard Laborers

To illustrate how using a parable together with complexity metaphors can impact ethical decision-making I will relate a case in which I applied the paraplexity method with a group of community volunteers in southwest Florida.

1. Identify Issue

The specific issue of concern to a group of church volunteers in southwest Florida was the plight of the tomato workers in Immokalee Florida where mainly Latino, Mayan Indian and Haitian day laborers pick tomatoes bound for fast-food restaurants around the country. Like the day laborers in the parables of the Vineyard Workers, each day the farm workers gather in a local parking lot at four thirty in the morning to look for work. If they are chosen, crew leaders drive them in old school buses to tomato fields 20 to 200 miles away. There they work 10-12 hours picking at a per bucket rate that hasn't changed in more than 25 years, namely 40-45 cents per thirty-two pound bucket. At that rate, a farm worker must pick two tons of tomatoes to earn fifty dollars. They do not receive overtime or healthcare and do not have the right to organize a union.

There are many aspects of the issue, including what some would call "modern-day slavery", but we limited our focus to unjust wages.

2. Choose Relevant Parable

I chose to present the *Vineyard Laborers* parable which speaks to individual and systemic issues of fairness.

---

6. A fractal is a pattern that is repeated at different levels throughout a system. The shapes and patterns are similar at all scales.

Vineyard Laborers
(from Peterson, 2003)

God's kingdom is like an estate manager who went out early in the morning to hire workers for his vineyard. They agreed on a wage of a dollar a day, and went to work. "Later, about nine o'clock, the manager saw some other men hanging around the town square unemployed. He told them to go to work in his vineyard and he would pay them a fair wage. They went. He did the same thing at noon, and again at three o'clock. At five o'clock he went back and found still others standing around. He said, 'Why are you standing around all day doing nothing?' They said, 'Because no one hired us.' "He told them to go to work in his vineyard. When the day's work was over, the owner of the vineyard instructed his foreman, 'Call the workers in and pay them their wages. Start with the last hired and go on to the first.' Those hired at five o'clock came up and were each given a dollar. When those who were hired first saw that, they assumed they would get far more. But they got the same, each of them one dollar. Taking the dollar, they groused angrily to the manager, 'These last workers put in only one easy hour, and you just made them equal to us, who slaved all day under a scorching sun.' He replied to the one speaking for the rest, 'Friend, I haven't been unfair. We agreed on the wage of a dollar, didn't we? So take it and go. I decided to give to the one who came last the same as you. Can't I do what I want with my own money? Are you going to get stingy because I am generous?' (Matt 20: 1-15)

3. Encourage Reflection on Parable in Context of Issue

The group discussed the acts of generosity in the parable. They appreciated that the vineyard manager generously chose to pay those who worked for only a short time the same wage perhaps, they speculated, because their needs and the needs of their families would have been the same.

The group concluded that although the workers did not earn what they received, all of them received gifts from a generous benefactor. This is the usual and expected reading of the parable—that God gives grace to all, both to those who "deserve" it and those who do not.

As we reflected on the parable I pointed out that the manager of the Vineyard offered a pay system similar to the way pay is dispensed by all landowners in 1st century Palestine and in Immokalee. Landowners and managers are the arbiters of what is considered a fair wage, a system defined by the members of their ruling class. Whether in large estates or small holdings, the *fractal* pattern is repeated—those in control operate within the compensation pattern of the times—then and now.

4. Present Complexity Perspective

I explained that a *fractal* is a geometric *pattern* that can be split into parts, each of which is a reduced-size copy of the whole. The pattern is repeated at different levels—from a very small part of a system to the whole system itself. This property is called "self-similarity". Natural objects that approximate fractals include clouds, mountain ranges, coastlines, various vegetables (cauliflower and broccoli) and the coloration patterns in animals, e.g. zebras. These repeating patterns can also be seen in human cultures where subgroups or individuals mimic the overall behavior occurring in the larger community.

After acknowledging their interpretations, I asked the group, "Did any transformational change occur in the parables? If so, who was changed? Are the patterns in the parables likely to be repeated and so blocking any future change from occurring?" I suggested an alternative interpretation of the parable—viewing the manager in the Vineyard parable as disguising his control and domination of the workers by manipulating them with acts of espoused generosity.

Transformational change in a human system requires disorder and discovery, openness to uncertainty, letting go of attachments, and painful adjustments to new situations. Contacts between agents must be a transformative exchange.

As they struggled with this alternative interpretation I presented this argument by Ford (2009):

> *The parable shows how those in power disguise their domination and about how those dominated are systematically deprived of their voice. Given a market flooded with impoverished and unemployed workers, the owner knows that none of them can bargain with him. The amount he pays, the customary wage of a denarius is an amount insufficient to sustain life, given that day laborers worked so infrequently. By taking advantage of a high level of unemployment, the manager is chronically and legally stealing his workers' sustenance, the very foundation of their lives. The owner offers charity as if it were justice. By disguising his control, the owner's momentary generosity makes difficult any coherent rejoinder. The workers know something is wrong but they simply do not have the words to express what it is.*

Within the confines of the systemic injustice fractal they have always known, all the workers in the Vineyard can do is to complain that they are being treated unfairly. They are stymied because how can they mount a convincing protest if the manager remains so generously in complete control? His dominance ensures his ignorance: in this interpretation of the parable he exits convinced of his goodness.

5. Amplify Differences in Perspectives

The group struggled with seeing the possible downside of a generosity that stymied growth and development and perpetuated a fundamentally unjust pay system. I pointed out that people make unethical decisions when they are disengaged from the moral aspects. Some people use "moral justification" where individuals reconstrue harm to others in ways that make it appear morally justifiable (Bandura, 1986). Seeing themselves as generous allows them to engage in unethical behavior without apparent guilt or self-censure. The manager in the vineyard also used the mechanism of euphemistic language and reframed the situation to make it sound less egregious.

6. Apply to Issue and Action Steps

A breakthrough happened when an influential member of the group said that she could not stand to give out food and diapers to the workers any longer because "justice is not being served".

As the group held the tension between their traditional mode of weekly handouts to the workers and the new insight about the need for systemic change, the group developed a new perspective about generosity. The shift in perspective was an example of the transcendent function described by Carl Jung[7]. The parable certainly advocates generosity, love and care for others just as the group advocates for the tomato workers. But the group became aware that systemic change will not occur if everyone colludes with the current system in which the tomato growers and the end users are seen as generous by providing work to foreign immigrants, but not a suitable wage. The basic issue is that all workers have the right to a sustainable wage for them and their family.

The group saw that while their gifts and donations helped the farmworkers meet their survival needs, they were in effect subsidizing the tomato growers who were underpaying workers. Their donations helped keep the current fractal pattern in place. One told the story of how McDonald's offered money during earlier negotiations with the Coalition of Immokalee Workers (CIW), a community-based organization of immigrants working in low-wage jobs throughout the state of Florida, along with an offer to employ some people, but the Coalition said "that won't fix the problem".

We discussed how a new fractal seed could change the volunteer's role in support of the workers. They also expanded their thinking about all of the stakeholders, including the growers and the retail food corporations as to what confrontations (or support) they needed to transition to a new pattern of just wages and working conditions.

---

7. See Olson (1990) for working with the transcendent function in organization change.

The group realized that creative transformation and reconciliation is only possible if opposing parties can speak honestly to each other and share positive regard for each other. To break the human rights crisis in the tomato fields, real partnerships need to be created between the end buyer and the workers. The federal government subsidizes wheat, grain, and corn farmers, but cheap fresh fruit and vegetables are subsidized on the backs of many migrant workers.

Recently the larger sandwich restaurants chains in North America have agreed to pay an additional penny per pound for Florida tomatoes and are asking the suppliers to ensure that the additional penny be paid to the farm workers[8].

If they deliver on their promises, such partnerships can promote transformative exchanges that were missing in the Vineyard parable. Ending clever PR maneuvering to appear generous and beginning honest engagement with the workers in monitoring supplier compliance and investigating complaints and resolving disputes is a good start.

Seeing human values and economic systems as *fractal* patterns gives us an alternative way to relate to injustice. We can change patterns that are inequitable by first identifying the seed that is being iterated in the system. We can then reduce the importance of that seed or create a focus on a different seed. In the parables Jesus is unmasking the fractal of the economic system and domination attitudes that pervade human society. In the present time the fractals in the migrant farm worker systems will continue to be replicated until new fractal seeds are sown.

Further Research and Application

The polyvalency of the parables has only been partially explored in this article. The five parables I have discussed could be used in a variety of other organizational issues. Certainly organization behaviors that have led to the many global economic and environmental crises we face could be explored with any of the parables.

I have only used a handful of complexity metaphors and concepts and parables attributed to Jesus. Other studies could develop a more comprehensive list of metaphors and concepts that could be fruitfully applied.

Parables from the religious traditions of Buddhism, Islam, Judaism, Hinduism and others could also provide useful stories and narratives for application. There is also an emerging field of eco-parables and stories derived from the work of social constructionists that is very relevant.

8. See the article at http://ciw-online.org for a May 2010 agreement by Quiznos.

Conclusion

Moral consciousness emerges in response to exposure to a stimulus like a parable that challenges conventional wisdom and disorients the listener to see things differently. The chaos that is created allows a change in perspective, a shift in our worldview, and a sense that something is greater than our self-interest, even a sense of our connectedness to the sacred.

Ethical metaphors in parables can contribute to the resolution or reframing of many organization issues and practices. As a prism transforms light into a rainbow of many colors, parables transform our common understanding into something beyond ourselves when added to our evolving organization landscape.

Using the paraplexity method is useful for applying the Dyke & Wong model to ethical issues in organizations. Parables can identify shortcomings and unacceptable attributes and behaviors that need *Confession* and make organizational boundaries more permeable. Parable illustrations of the *Worship* phase help us appreciate others and tune into alternative ideas and ways of working as we let go of our needs for control and certainty. Broadening the base of participation and including others in decision-making is part of the *Guidance* phase. Parables raise questions about how tightly coupled we are to our group and organization affiliations and professional identities with their canons and strictures. Parables that stress the wholeness of life provide a holistic attractor that helps us put things in perspective in the *Celebration* phase.

Parables can be interpreted using the *CDE Model*. They enhance sharing of visions, values, and insights among stakeholders (*Exchange*). They point to important differences and bring forth new ideas and alternatives (*Difference*). They also create new constructions of reality, new directions, and motivating goals (*Container*)

Facilitative skills are required to let an ethical decision emerge from complex problem spaces (Richardson & Tait, 2010). The paraplexity method described in this article can help consultants and managers to develop skills in using both archetypal stories like parables and complexity concepts in their facilitations. Developing shared narratives and rich connections between stakeholders will develop and advance sense-making and deep learning.

References

Anderson, M. and Escher P. (2010). The MBA Oath, ISBN 9781591843351.

Banaji, M.R., Bazerman, M.H. and Chugh, D. (2003) "How (un)ethical are you?" *Harvard Business Review*, ISSN 0017-8012, 56-64.

Bandura, A. (1985). *Social Foundations of Thought and Action: A Social Cognitive Theory*, ISBN 9780138156145.

Baskin, K. (2008). "Storied spaces: The human equivalent of complex adaptive systems," *Emergence: Complexity & Organization*, ISSN 1521-3250, 10(2): 1-12.

Cilliers, P. (2004) "Knowing complex systems," in K.A. Richardson (ed.), *Managing Organizational Complexity: Philosophy, Theory and Applications*, ISBN 9781593113186, pp. 7-20.

Dodd, C.H. (1981). *The Parables of the Kingdom*, ISBN 9780060619329.

Dyck, B. and Wong K. (2010). "Corporate spiritual disciplines and the quest for organizational virtue," *Journal of Management, Spirituality & Religion*, ISSN 1942-258X, 7(1): 7-29.

Ford, R.Q. (2009). "Jesus's parable of the vineyard workers and U.S. policy on Iraqi oil," *The Fourth R*, ISSN 0893-1658, (July-August): 3 ff.

Funk, R.W., Scott, B.B. and Butts, J.R. (1988). *The Parables of Jesus: Red Letter Edition*, ISBN 9780944344071.

Galloway Seiling, J. (2010). "Knowledge generation as a complex relational process: Absorbing, combining, transfer and stickiness in the organizational context," in A. Tait & K. Richardson (eds.), *Complexity and Knowledge Management*, ISBN 9781607523550, pp. 93-108.

Gergen, K.J. (2009). *Relational Being: Beyond Self and Community*, ISBN 9780195305388.

Hill, J.A. (2010). "Fear, civic life, and the divine domain: Addressing the shadow sides of contemporary culture," *The Fourth R*, ISSN 0893-1658, (May-June): 9 ff.

Holloway, R. (2003). "The danger of sincere religion," http://homepages.which.net/~radical.faith/holloway/sermondangers.htm.

Hultgren, A.J. (2000). *The Parables of Jesus: A Commentary*, ISBN 9780802860774.

Kauffman, S.A. (2008). *Reinventing the Sacred: A New View of Science, Reason, and Religion*, ISBN 9780465003006.

Liebenberg, J. (2000). *The Language of the Kingdom and Jesus*, ISBN 9783110167337.

McCollough, C. (2008). *The Art of Parables: Reinterpreting the Teaching Stories of Jesus in Wood and Sculpture*, ISBN 9781551455631.

Olson, E.E. (1990). "The transcendent function in organizational change," *Journal of Applied Behavioral Science*, ISSN 0021-8863, 26(1): 69-81.

Olson, E.E. (2009). *Keep the Bathwater: Emergence of the Sacred in Science and Religion*, ISBN 9780615275208.

Olson, E.E. (2010). "The vital dimension in social change," *Practising Social Change*, 1(1): 20-24.

Olson, E.E. and Eoyang, G.H. (2001). *Facilitating Organization Change: Lessons from Complexity Science*, ISBN 9780787953300.

Peterson, E.H. (2002). *The Message: The Bible in Contemporary Language*, ISBN 9781600066672.

Richardson, K.A. and Tait, A. (2010). "The death of the expert?" in A. Tait & K. Richardson (eds.), *Complexity and Knowledge Management*, ISBN 9781607523550, pp. 23-40.

Weick, K.E. (1995). *Sensemaking in Organizations*, ISBN 9780803971776.

Williams, P.A. (2008). *Revealing God: A New Theology from Science and Jesus*, ISBN 9780741444868.

Edwin E. Olson, PhD has led organization change, team building, and workforce diversity initiatives for Texaco, Inc., U.S. General Accounting Office, Digital Equipment Corporation, NASA, and the FDA. He is a collegiate professor in the University of Maryland University College Graduate School and an adjunct professor in the Executive Leadership Program, the George Washington University. He is a former professor of management at Baldwin-Wallace College and professor of information service at the University of Maryland, College Park. He is the co-author of Facilitating Organization Change: Lessons from Complexity Science (Jossey-Bass, 2001). He has a B.A. in Philosophy, an M.S. in Pastoral Counseling, and a Ph. D in Government. He is a professional member of the NTL Institute and the Human Systems Dynamics Institute.

14. Institutional Fragmentation In Metropolitan Areas And Infrastructure Systems: Governance As Balancing Complexity And Linear Tasks*

Jack W. Meek
College of Business and Public Management, University of La Verne, USA

This chapter addresses governance strategies utilized in administrative planning and implementation in an institutionally fragmented metropolitan transportation environment. The strategies utilized are categorized as either "predict-and-control" and "prepare and commit" and represent two fundamentally differentiated approaches to administration in complex public environments. The sub-regional case examined demonstrates that the use of multiple—even blended—techniques (both control oriented and open systems for learning) are essential to constructively develop and manage projects in an institutionally fragmented system. These blended techniques represent a kind of complexity sensitive tools that are utilized by newly formed agencies addressing complex implementation issues in institutionally fragmented transportation systems found in metropolitan environments.

Introduction

This paper addresses the implications of institutional fragmentation within metropolitan areas for the development, construction and operation of complex infrastructure systems. Complex infrastructure systems often cut across existing government borders, thus resulting in institutional complexities which make it hard to manage these systems in an efficient way and which may even result in project failure or project delays. On the other hand, many such systems have been developed under conditions of multiple jurisdictions and

* This chapter was presented at the *1st International Workshop on Complexity and Policy Analysis* held in Southampton, England, July 21-23, 2010. The author wishes to thank Joop Koppenjan for the original orientation and framing of this research effort. A special thanks is extended to those who offered critical review and feedback and especially to Elieen Conn for her careful review and critique. An earlier version of this chapter was prepared for the panel "Complexity and Metropolitan Governance," at the Annual National Conference of the *American Society for Public Administration* held in San Jose, California, April 12, 2010.

seem to operate in an adequate way. Apparently governance methods (network governance), project management approaches and institutional arrangements can cope with these circumstances.

This condition raises the question what the implications of institutional fragmentation for the realization and operation of metropolitan infrastructure systems are, and what challenges and opportunities these implications offer for a successful development and management of these systems. These questions are investigated through a case analysis of organizational arrangements of transportation infrastructure systems in Los Angeles and an assessment of practices utilized by a sub-regional jurisdiction of the transportation system. This sub-regional jurisdiction operates across a complex array of institutional and jurisdictional boundaries in order to improve transportation infrastructure operations.

The paper begins with an overview review of the Los Angeles metropolitan transportation management system with a general description of the institutional setting and a description of the central institutional actors within the overall system. This review is framed in regard to the institutional mix associated with strategic development of transportation planning within the complex metropolitan setting. One feature of the institutional mix is the emergence of sub-regional initiatives designed to improve transportation infrastructure in one area of the overall metropolitan system. One characteristic of the Los Angeles metropolitan transportation system is a kind of 'loose-coupling' of sub-regional initiatives that benefit areas within the system where issues or problems arise that are unmet by system-wide strategic initiatives or policies.

To explore the operations of agencies within the broader institutional mix, the paper then examines current efforts at transportation and safety enhancements in one sub-region of the Los Angeles metropolitan area—the Alameda Corridor East Project (ACE)—that calls upon deeply connected stakeholder coordination and collaboration and extends governance arrangements through a transportation regime that is extending across the metropolitan region. Given the complex institutional web of transportation agencies, ACE is faced with designing, funding and implementing grade separation projects throughout a Los Angeles area sub-region (the San Gabriel Valley). The activities of ACE call upon a political and administrative will to coordinate and implement projects in numerous local jurisdictions on behalf of the region, and funded by federal, state and local resources. The activities need to be coordinated with appropriate transportation agencies, rail operators and local jurisdictions. The overarching central questions explored in this case are as follows:

1. How have complex infrastructure systems been developed under conditions of multiple jurisdictions and seem to operate in an adequate way—creating institutional arrangements that cope with these circumstances.

2. What challenges and opportunities these implications offer for a successful development and management of these systems.
3. What are the implications of institutional fragmentation for the realization and operation of metropolitan infrastructure systems?

The paper interprets this case of sub-regional efforts and decision approach by drawing upon three areas of research: (1) administrative conjunction and administrative regime development in metropolitan settings (Frederickson & Meek, 2010); (2) tools associated with the governance approaches within the context of metropolitan institutional fragmentation (Koppenjan *et al.*, 2009); and (3) characteristics of "complexity-sensitive tools" or pacticies identifiable in complex systems.

The first area of research—administrative conjunction—stresses that contemporary public administration is characterized by operations carried out by administrators working across jurisdictional boundaries in representation of a generalized form of public interest. As such, public management professionals "will engage in inter-jurisdictional administrative governance to address problems that cannot be jurisdictionally contained, and to, thereby, reduce collective uncertainty" (Frederickson & Meek, 2011).

The second area of research—governance approaches to institutional fragmentation—asserts that to be successful, administrators who engage in these cross-border activities, must balance the use of two kinds of governance approaches: administrative control (project management) and administrative flexibility (discretion). The point made here is that internal organizational relations need to be built on control to manage project costs and external relations need to be founded on discretion and openness to build support for infrastructure projects among external actors. The work of Koppenjan *et al.* (2009) provides an interpretation of administrative practices in governance settings that reflect control-orientated or open and facilitated patterns (see Table 1 below).

This framework is applied to a complex institutional setting—the Alameda Corridor East Project—where implemented projects are deeply interconnected with an array of horizontal and vertical stakeholders that include citizens. The case of ACE also represents the long-term implementation of a transportation supporting infrastructure in metropolitan Los Angeles designed for the increased transportation safety and mobility between automobiles and trains.

As a decision and implementation approach, ACE initially drew upon the original design of projects established in 1997 and developed practices that represent a blend of techniques representative of both control (predict-and-control) project management and flexible (prepare-and-commit) governance strategies. These strategies or practices can be identified as the third avenue of inquiry and practices as viewed within the framework of the "characteristics of complexity-sensitive tools" outlined for the *1st International Workshop on Complexity and*

Predict-and-control versus Prepare-and commit

	Type I Predict-and-control	Type II Prepare-and-commit
Terms of reference	Blueprint	Functional
Task definition	Narrow for best control	Broad for best cooperation
Contract	Task execution	Functional realization
Incentives	Work-task based	System-output based
Change	Limit as much as possible	Facilitate as much as needed
Steer	Hierarchical	Network
Information exchange	Limited, standardized	Open, unstructured
Interface management	Project management task	Shared task

Source Joop Koppenjan, Ernst ten Heuvelhof, Martijn Leijten, Wijnand Veeneman, Haiko van der Voort (2009)

Table 1 *Framework for Assessing Infrastructure Management*

Policy Analysis: boundary critique, pluralism, synthesis, emergence, and timeliness. In summary, an additional way to interpret these practices can be through the lens of "complexity-sensitive" tools (referred to as practices in this paper) listed and summarized in Table 2 below.

The Transportation Challenge In The Los Angeles Region

To better understand the impact of the advent of regional planning on local governmental agencies concerned with transportation, it is useful first to briefly summarize the problems that confront regional and local agencies in Los Angeles as they embark on managing transportation systems. We first look at the magnitude of the problem in Los Angeles, which is followed by an overview of the agencies involved in developing and implementing transportation solutions. We then briefly review the selected sub-regional transportation initiatives.

The Southern California Association of Governments (SCAG), the federally designated Metropolitan Planning Organization (MPO) for the Southern California metropolitan region, assists in regional planning and coordination for a vast area of over 19 million people, six counties and 189 cities. Population is expected to grow by 29% in the next 28 years creating increased demands for goods and services with consequent problems of transportation demands and congestion (see Table 3).

The SCAG metropolitan region includes numerous county-centered transportation planning and programming agencies, such as the Los Angeles County Transportation Authority (MTA); sub-regional planning agencies, such as Councils of Governments; and other governmental agencies. For a list of key stakeholders in regional transportation planning in Los Angeles, please refer to Table

Characteristics of Complexity Sensitive Tools	Character	Case Illustration
Boundary critique	All boundaries are no more than temporary patterns resulting from a filtering process (e.g., based on personal values); they are to some degree arbitrary and require ongoing review to understand how they shape our context of interest	The ACE example represents various patterns of boundary blurring in funding, operation, and implementation. The cross-boundary authority has shaped the stakeholders within limits.
Synthesis	We cannot rely completely on any one perspective	The original design of projects (Korve Engineering 1997) have been adjusted based on negotiated contexts; much of the original intent remains
Pluralism	The attempt to tailor descriptions (models) to the context of interest, rather than have the model shape the context	The original set of projects incorporated local interests within budgetary reason
Emergence	Acknowledges the emerging view of the real world, rather than the favored method / methodology; recognize that the real world will collectively conspire to respond to our design interventions in a variety of ways—some of them not considered by the "designers;	Some of the ACE projects were reconsidered in terms of dominant strategy based on negotiated practices; Agency learning is apparent
Timeliness	Control/design in complex systems is a never ending process; ongoing analysis and intervention; analyst being part of the complex system he seeks to affect	ACE projects have undergone constant review; the time-line is a function of funding (pay-as-you-go).

Table 2 *Characteristics of Complexity Sensitive Tools and Case Illustrations Source of Complexity Sensitive Tools: Richardson (2010).*

4. The list of participating agencies offers a rich representation of the kinds of stakeholders that participate in the fragmented metropolitan transportation system of metropolitan Los Angeles.

Prior to addressing the analytical assessments of cross-border governance practices, what follows is a summary description of the key institutional arrangements that operate in the metropolitan Los Angeles transportation arena: (1) the Southern California Association of Governments; (2) The Metropolitan Transportation Authority; (3) Metrolink; and (4) an example of a sub-regional agency—the Alameda Corridor East Project. The latter institutional arrangement will be the focus of the balance of governance practices under the condition of institutional fragmentation.

Key Growth Patterns 2008 Regional Transportation Plan Los Angeles Region (in millions)						
	Unit of Measure	Benchmark Year	Amount	Forecast Year	Amount	% Change
More People	Millions	2007	18.6	2035	24	29%
More Congestion	Average Speed (mph)	2003	30.5	2035	26.8	-12%
More Freight	Port Volume (20 Ft Eq Units)	2006	15.8	2030	42.5	169%
More Truck Traffic	Million Miles	2003	29	2035	50.4	74%

Table 3 *Los Angeles Region Growth Forecasts*
Source: Regional Transportation Plan, December 2000, SCAG (p. 2) http://www.scag.ca.gov/rtp2008/pdfs/outreach/2008RTPOutreach.pdf.

Southern California Association Of Governments (SCAG)[1]

The sole multifunctional agency of regional scope is SCAG, which plays a key role as the federally designated MPO charged with preparing regional growth plans. SCAG exercises influence in the area of transportation funding through the requirement under the federal transportation authorization bills, ISTEA and TEA-21[2] 9the successor to TEA-21, SAFETEA-LU), for proposed transportation projects to be included in a SCAG-prepared Regional Transportation Plan (RTP) in order to qualify for federal funds.

Although it boasts 189 member cities from the Southern California region, SCAG falls short of constituting a metropolitan government because it does not exercise authority over land use, which remains the sole domain of municipal and county government, nor does it exercise direct authority over budgeting decisions for transportation or air quality or its other issue competence areas. In addition, rather than being governed by an autonomous board, SCAG is controlled by a board composed of appointed members of the region's city councils and county boards of supervisors, accountable not to regional interests but to their local constituencies.

---

1. The material on the Southern California Association of Governments (SCAG) and the Metropolitan Transportation Authority (MTA) and the Alameda Corridor East Project (ACE) is drawn from a previously published work by the author. See Hubler & Meek (2006).
2. The Intermodal Surface Transportation Efficiency Act (ISTEA) of 1991 and Transportation Equity Act for the Twenty First Century of 1996 (TEA-21) and the The Safe, Accountable, Flexible, Efficient Transportation Equity Act: A Legacy for Users of 2005 (SAFETEA-LU).

Designated MPO	Sub-Regional Council of Governments	Local and County Governments Other Owners, Operators and Implementing Agencies
• Southern California Association of Governments (SCAG)	• Arroyo Verdugo Cities • Coachella Valley Association of Governments • Gateway Cities COG • Imperial Valley Association of Governments • Las Virgenes-Malibu-Conejo COG • Los Angeles, City of • North Los Angeles County • Orange county COG • San Bernardino Associated Governments • San Gabriel Valley COG • South Bay Cities COG • Ventura County COG • Western Riverside County COG • Westside Cities COG	• Caltrans District Offices • Airport Authorities • Port Authorities • Transit/Rail Operators
County Transportation Commissions/Transportation Sales Tax Commission		**Resource/Regulating Agencies**
• Los Angeles • Orange • San Bernardino • Riverside • Venture • Imperial		• USDOT (FHWA, FTA, FAA, FRA) • US EPA • CA DOT • CA Air Resource Board • CA EPA • Air Districts
		Other private, non-profit organizations, interest groups, and Tribal Communities

Table 4 *Stakeholders in the Development of the Regional Transportation Plan*

While SCAG is multi-functional and federally mandated to produce plans for transportation, growth management, hazardous waste management and air quality, in practice it functions primarily as an advisory planning "think tank" with strong technical expertise and as an intermediary in federal-regional transportation funding relations through ISTEA and TEA-21. SCAG does not program budgets or implement projects, leaving significant aspects of the policy-making and implementing process to the county and regional agencies charged with those functional areas.

The Metropolitan Transportation Authority (MTA)

The agency with teeth when it comes to planning and implementing transportation in Los Angeles County is the Los Angeles County Metropolitan Transportation Authority. The MTA plans, programs, builds and operates transportation systems and service in Los Angeles County, which with a population of more than 10 million constitutes more than half of the SCAG region population. As discussed in Fulton (1997), the MTA has been plagued by a deeply divided and often parochially interested governing board confronted with a struggle between bus and rail advocates, and further, between light rail and subway partisans. Indeed, the MTA is often a battleground for powerful local and regional forces brought into conflict in the struggle over spending its immense budget, $3.8 billion budget in fiscal year 2009-10 for highway, bus and rail transit. And, this conflict will likely intensify as the stakes rise with the latest "sweetening of the pot." The MTA and other local and regional authorities were granted vastly greater discretionary authority over state transportation funding for the region through a thorough restructuring of transportation funding programming in the form of California Senate Bill 45.

Metrolink (Southern California Regional Rail Authority)[3]

Metrolink is Southern California's rail system linking residential communities to employment and metropolitan areas. Formed in 1991, it is operated by a Joint Powers Authority called the Southern California Regional Rail Authority (SCRRA). This organization consists of 5 member agencies representing the county transportation authorities of Los Angeles, Orange, Riverside, San Bernardino, and Ventura. These particular counties are involved in the funding and operation of the Metrolink because this rail service runs through and connects these counties together.

The JPA was created for the purpose of developing and implementing a public transit service that would reduce highway congestion and enhance mobility throughout Southern California. On October 26, 1992, the proposed public transit service (Metrolink) began operations.

At the time of its creation, Metrolink operated on three lines of service and twelve stations, accommodating approximately 5,000 daily passengers. In the fifteen years since, the rail system has expanded to seven lines of service, fifty-five stations, and a daily ridership of over 45,000 passengers. The system's 512-mile length connects stations in Ventura County, Los Angeles County, San Bernardino County, Riverside County, Orange County, and San Diego County. Between 1990 and 1993, SCRRA member agencies had acquired track and other properties from railroad companies such as the former Southern Pacific Railroad, Santa Fe Railway, and Union Pacific Railroad. Such acquisitions were made through straight purchases or easement agreements.

The member agencies that make up SCRRA have certain roles in the operation of Metrolink. According to the JPA agreement, the extent of the member agencies' duties is to provide support in the implementation of the commuter rail system. SCRRA, as an entity separate from the member agencies, oversees the implementation of Metrolink operations. Support from the member agencies come in the form of staff support, contract and special financial support, and policy support. This means that the member agencies are bound to provide staff to assist in the design, construction, and administration of any infrastructures; to provide funding; and to either approve or respond to any recommendations relating to funding, operational decisions, fare structures, and other policy considerations.

SCRRA's governing board consists of eleven voting members selected from among the governing boards of the member agencies. Each member agency is allocated a certain number of votes. An executive staff that carries out the daily operations of the Metrolink service also supports the governing board. When it comes to the liabilities, it is clear by the JPA agreement seeks to make clear that any debts, liabilities, or obligations incurred by SCRRA have no bearing on

---

3. Thanks to Hong Kyu Lyu for developing the section of this report on the MTA and Metrolink.

the member agencies. The governing board and the employees of SCRRA have the duty to exercise ordinary care and reasonable diligence in the exercise of their power and in the performance of their duties. A summary of the three Los Angeles Regional Transportation institutions discussed here is summarized in Table 5.

Agency	Mission/Purpose	Founded	Governance	Budget/ Funding
Southern California Association of Governments (SCAG)	To facilitate a forum to develop and foster the realization of regional plans that improve the quality of life for Southern Californians.	1965	70 Member Regional Council	$38.7 Million
Los Angeles Metropolitan Authority (Metro)	Operates a regional public transportation system that consists of bus and urban rail services	1993	13 member Board of Directors	$3.8 Billion
Metrolink	Operates a commuter rail system that serves to connect sprawling regions within Southern California	1991	11 Member Governing Board	$ 168.1 Million

Table 5 *Los Angeles Region Transportation Institutions*

Sub-Regional Transportation Initiatives In The Los Angeles Region

The role of regional transportation planning played by the federally mandated MPO—the Southern California Association of Governments (SCAG)—has been interpreted in various ways. One version is that the Los Angeles area experience with regional planning is characterized as an "evolving region" or a "shadow regionalism." Under this interpretation, an emergent form of regional coordination is evolving, and avoids the direct top-down imposition of a larger governmental agenda that reduces local input and interests. In previous research, Hubler & Meek (2006) review of four sub-regional transportation initiatives—three that rely upon joint-power authority strategies and one on institutional devolution—revealed the possibilities of sub-regional cooperation not likely under a super-regional solution.

The Alameda Corridor-East Project

As to the focus of this paper, one joint-power authority, the Alameda Corridor East Project (ACE), is an example of sub-regional cooperation, where local governments joined together in order to develop a coordinated solution to ensure safety and enhance cross-traffic travel along rail lines that are increasing in use. ACE was a direct response by sub-regional leadership including the council of governments anticipating the dramatic increase in international trade coming through the ports of Los Angeles and Long Beach. The three questions outlined at the beginning of this article with regard to institutional fragmentation are addressed.

1. How have complex infrastructure systems been developed under conditions of multiple jurisdictions and seem to operate in an adequate way—creating institutional arrangements that cope with these circumstances.

The ports of Los Angeles and Long Beach are the busiest in the nation. Each day, about 35 trains laden with containers roll out of the ports along the Alameda Corridor rail expressway. With trade projected to grow from 9 million containers per year to 24 million containers a year over the next 20 years, the number of trains is expected to grow to 100 per day. Faced with those growth projections, and armed with a study projecting looming congestion and safety issues along the twin 35-mile rail lines traversing eastward through the San Gabriel Valley and connecting to the transcontinental rail network, the San Gabriel Valley Council of Governments created a joint powers authority, the ACE Construction Authority. ACE (see Table 6) was established to build a series of grade separation projects, primarily roadway underpasses, to separate the trains and vehicles at 20 key intersections, as well as to construct a series of safety improvements at other intersections.

The Alameda Corridor-East Project of the San Gabriel Valley Council of Governments was formed as a joint-powers authority in response to the projected on-

Regional Initiative	Design and Description	Objective
Alameda Corridor-East Freight Rail Congestion Relief and Safety Program	Sub-regional joint powers authority created in 1998 to enhance pedestrian and motorist safety, provide traffic congestion relief and freight rail capacity improvements.	Projects are underway, with completion extended based on continued funding

Table 6 *ACE as a Sub-Regional Transportation Initiatives*
Source: Hubler & Meek (2006)

slaught of freight trains forecast with the opening of the Alameda Corridor rail expressway from the seaports to the rail yards near downtown Los Angeles. Essentially, a new institution was established to implement the intended grade separations and safety improvements anticipated by the regional partners.

2. What challenges and opportunities under conditions of multiple jurisdictions offer for a successful development and management of these systems?

Once political will supported the establishment of the ACE joint-power authority as an administrative agency of the San Gabriel Valley Council of Government (SGVCOG), the next challenges were to design, fund and implement projects. The funding challenge was foremost as there were only limited amounts of private funding available. There was no incentive for railroad funding as infrastructure improvement anticipated by ACE was for grade separations and safety improvements and, from the perspective of rail operators, did not add value to or enhance rail operations. The Union Pacific Railroad operated trains on its two mainlines, one of which is double-tracked, largely at will across the San Gabriel Valley, with operations constrained mainly by track capacity. Its budget largely dependent on legislative earmarks of transportation funding, ACE has designed and built projects on a pay-as-you-go basis, and has been subject to the vagaries of the State of California's fiscal crisis and tightening federal budgets. Funding of the ACE projects traffic mitigation and track improvements. Funding commitments to date total $1.129 billion and incorporates are rather impressive mix of federal, state and local funding (see Table 7).

With an initial budget of $410 million for the first phase of projects, ACE has completed the safety and congestion relief projects and started construction on the first of 10 grade separation projects in Phase 1 (see Table 8).

There was no time line given to the projects, as these would be built when funding became available. In this phase, each of the projects incorporated adjustments necessary in order to achieve successful implementation. These adjustments were a function of ACE agency staff working closely with city agencies and negotiating contracts where projects were undertaken. In some cases, the original design of the project was altered in order to better meet local needs (KOA Corporation, http://www.theaceproject.org/docs). In other cases, it is com-

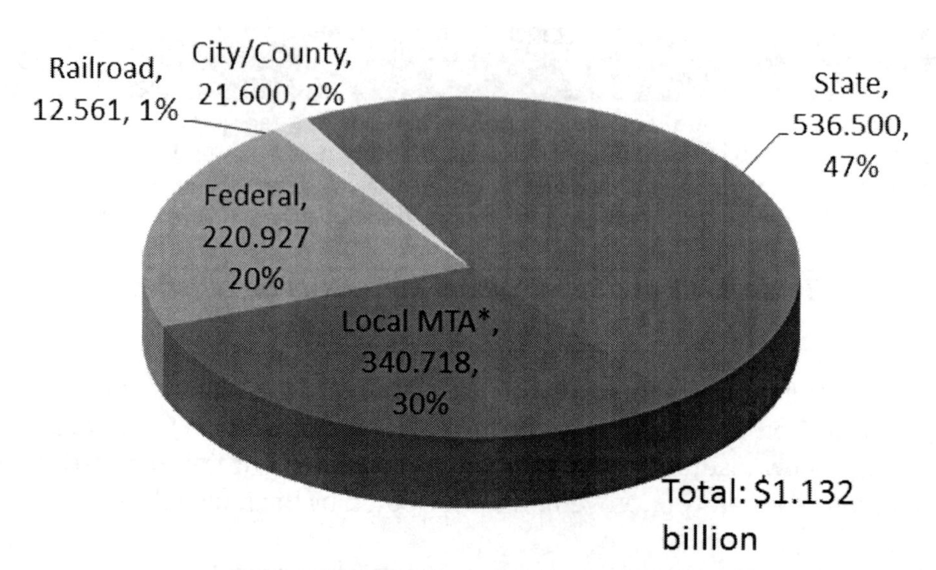

Table 7 *Public Funding Mix of ACE Projects*
Source: www.theaceproject.org/funding.htm

PROJECT	LOCATION	COST	STATUS
Corridor Safety Upgrade	39 grade crossings	$27.8 million	Completed January 2004
✓ Nogales St./Valley Blvd. Grade Separation	Industry / West Covina	$49.7 million	Completed August 2005
✓ Reservoir St. Grade Separation ✓ East End Ave. Grade Separation	Pomona	$79.1 million	Completed December 2006 Completed June 2008
✓ Ramona Blvd. Grade Separation (Cypress St. vacated)	El Monte	$51.2 million	Completed April 2008
✓ Brea Canyon Rd. Grade Separation	Diamond Bar / Industry	$67.4 million	Completed October 2008

Table 8 *ACE Completed Projects*
Table 9 Predict-and-Control Versus Prepare-and-Commit

mon that contract change orders, sometimes significant change orders that include utility redesign, are managed within the project by ACE staff.

While not all local interests in redesign are viewed as feasible because of extraordinary costs, ACE staff and leadership seek to find local solutions within budgetary constraints. These adjustments were found common in the projects and were reported to the ACE Board of Directors for review and determination. The ACE Board of Directors includes representation from cities where projects are undertaken. The ACE leadership and staff have been diligent in working with both local interests and to inform the regionally representative board.

3. What are the implications of institutional fragmentation for the realization and operation of metropolitan infrastructure systems?

After identifying need and garnering support for ACE projects, successful project implementation to date can be attributed to several practices: pay-as-you-go basis, inclusion of appropriate stakeholders, ensuring community relationships, and ensuring accountability and responsiveness through agency—board relationships. ACE leadership and staff also embraced project management practices that allowed for proper administrative control and accountability. These practices are reported to the board at regular meetings of representatives (see http://www.theaceproject.org/meetings.htm).

The implications of institutional fragmentation for the realization and operation of metropolitan infrastructure systems are that multiple administrative techniques are essential in the design and implementation of infrastructure support and improvements in complex metropolitan environments operating in fragmented institutional settings. It is clear that the use of multiple—even blended—techniques (both control oriented and open systems for learning) are essential to constructively meet the challenges offer for a successful development and management of these systems. These blended techniques address both the vertical and horizontal aspects of the complex system that characterize transportation infrastructure system. Agents in this system call upon complex sensitive approaches addressing the horizontal character of the system, such as control over their operations for accountability reasons along with practices such as the inclusion of appropriate stakeholders to ensure community relationships. At the same time agents in this system need complex sensitive approaches to demonstrate administrative control and accountability within the vertical character of the system, such as pay-as-you-go and outcome oriented results and reporting that demonstrate effectiveness. This kind of conceptualization of complex sensitive practices or approaches to both horizontal and vertical character of complex systems are revealed in practical terms mentioned earlier in this paper: to be successful agents in a complex system address both how internal organizational relations need to be built on control to manage project costs and external relations need to be founded on discretion and openness to build support for infrastructure projects among external actors. Examples of the blend of these kinds of complex sensitive practices are outlined in Table 8 below. It is apparent that this mix of strategies is part of the creative process of administrative management under conditions of institutional fragmentation and allows for the implementation of these projects as iterative and developmental while at the same time responsive and practical. The mix of strategies listed in Table 9 reveals that both kinds of management practices as outlined in the work of Koppenjan *et al.* (2009) (previously outlined in Table 1) are evident in the design and implementation of metropolitan infrastructure system illustrated in this case.

Type	Type I Predict-and-control	Type III Predict/Prepare Control/Commit	Type II Prepare-and-commit
Terms of reference	Blueprint	Blueprint & Functional	Functional
Task definition	Narrow for best control	Narrow & Broad	Broad for best cooperation
Contract	Task execution	Task execution & Functional Realization	Functional realization
Incentives	Work-task based	Work-Task and System based	System-output based
Change	Limit as much as possible	Limit & Facilitate	Facilitate as much as needed
Steer	Hierarchical	Mixed Network	Network
Information exchange	Limited, standardized	Limited and Open	Open, unstructured
Interface exchange	Project management task	Project and Shared Tasks	Shared task

Table 9 *Predict-and-Control Versus Prepare-and-Commit*

Summary

The case of ACE as a sub-regional effort represents an innovative institutional design that initiates and carries out specific tasks for a region of stakeholders. Interestingly, each of the projects coordinated by ACE, once completed, become the responsibility of the city where the project is located. Upon completion of all for the ACE projects, ACE will no longer remain in existence. ACE represents a collaborative effort that has a specific purpose. The activities of the ACE leadership and staff, working on behalf of cities within the region, represents a kind of administrative conjunction that may be characteristic of necessary operations within fragmented administrative transportation systems. As such, ACE offers important lessons in the use and application of "complexity-sensitive" tools and practices for other sub-regions in addressing unmet transportation needs within a larger metropolitan system.

References

Frederickson, G.H. and Meek, J.W. (2011). "Bureaucratie sans frontiéres: Legitimacy, authority, accountability in geo-governance systems," in E. Ongaro, A. Massey, M. Holzer and E. Wayenberg (eds.), *Policy, Performance and Management in governance and Intergovernmental Relations: Transatlantic Perspectives*, ISBN 9781848443204, pp. 62-76.

Hubler, P. and Meek, J.W. (2006). "Sub-Regional transportation initiatives: Implications for governance," *International Journal of Public Administration*, ISSN 0190-0692, 29: 31-52.

Koppenjan, J., ten Heuvelhof, E., Leijten, M., Veeneman, W. and van der Voort, H. (2009). "Balancing competing governance approaches in complex, innovative projects: The RandstadRail Project," unpublished paper.

Richardson, K.A. (2010). Conference Correspondence, *1st International Workshop on Complexity and Policy Analysis* held in Southampton, England, July 21-23, 2010.

Jack W. Meek, Ph.D. is a La Verne Academy Professor of Public Administration, MPA Chair and Coordinator of Research & Graduate Studies at the College of Business and Public Management at the University of La Verne. Professor Meek offers courses in research methods and collaborative public management. His research focuses on metropolitan governance including the emergence of administrative connections and relationships in local government, regional collaboration and partnerships, policy networks and citizen engagement. Professor Meek has published over thirty-five articles including contributions in the *International Journal of Public Administration, Public Administration Quarterly, The Journal of Public Administration Education, Administrative Theory and Practice*, and the *Public Productivity and Management Review, Public Administration Review* and *Emergence: Complexity & Organization*. Jack has coedited a book on business improvement districts and coauthored a book on governance networks. Jack serves on the editorial boards of *the International Journal of Organizational Theory and Behavior, Journal of Globalization Studies* and *Social Agenda*.

15. Community Engagement In The Social Eco-System Dance

Eileen Conn
Living Systems Research, ENG

This paper identifies two distinct types of organizational arrangements, which need to be recognized to achieve effective policies and programmes for community engagement. The differences arise from the nature of relationships in the institutional and organizational world, which are primarily *vertical hierarchical*, as distinct from the informal community world where the relationships are primarily *horizontal peer*. Lack of attention to these distinctions adversely affects the interaction of the public agencies and the community, and the community's organizational governance and working arrangements. Using a complexity perspective, the '*social eco-system dance*' model identifies some issues and new ways of thinking about them and of handling some of the practical challenges. This is leading to a set of managerial and organizational tools to develop new ways of working for policy making, managing, operating and participating in the community engagement process across all the sectors involved.

Introduction

UK Government policies over the last few decades aimed to increase civic engagement—to enhance the democratic process, and to make public services more effective and reduce their cost. On the other hand, election voting decreased, and the cost of public services has risen alongside continuing dissatisfaction. The UK Coalition Government, which took office in May 2010, has continued the focus on civic engagement by its 'Big Society' policy (Prime Minister, 2010; UK Government, 2010). Meanwhile for their own reasons, millions of people engage together in a wide variety of collective activity on a multitude of human mutual interests, from the personal to the collective and from the social to the political, called 'civic' or 'community' engagement, or 'civil society'.

Through a dual experience, as an active resident and in parallel in government (unconnected) policy making, I have developed the *social eco-system dance* model. It is rooted in practice, and informed by complexity theory which provides an additional way of thinking and of seeing the human world (Mitleton-Kelly, 2003). The model illuminates in a new way some of the nature and organization of community engagement. The conventional approach assumes there is, overall, a simple system in operation, with mechanistic attributes; but it is in reality a multiple complex dynamic system. The lack of understanding adversely affects the interaction of the public agencies and the community. Using the model is helping to develop some basic policy and organizational tools to clarify the issues underlying this, and to improve some of the arrangements. This paper outlines the model and takes a preliminary look at some of its practical applications.

The Two Systems Approach

A complex system consists of a large number of elements which interact with each other. The many interactions in a system form clusters of elements, and interact with other clusters (Cilliers, 1998). These clusters are also nested complex systems. Individuals can simultaneously be in several different systems in different roles, and the individuals can change in the systems but, as Cilliers points out, a repeated pattern of system relationships will remain. These human social activities create complex social systems. In community engagement, there are also two different dominant modes of relational behavior and dynamics. These also can be seen as clusters, as differentiated sub-systems or worlds within the overall social system, operating separately but interacting. Practitioners working at global level in international development (Dove, 2006) and in UK community development (Pitchford & Higgs, 2004) have observed these relational distinctions, and noted that they appear to be largely invisible to the academics, policy makers, decision makers and professionals in their fields. This evidence from the field is supported by social philosophy (Table 1).

Philosopher	**Ordered Authority System**	**Free Association System**
Ferdinand Tonnies 1855-1936 German sociologist	*Gesellschaft* = society groups sustained to be instrumental for members' individual aims and goals	*Gemeinschaft* = community groupings based on feelings of togetherness and on mutual bonds
Martin Buber 1878-1965 Austrian/Israeli philosopher	*The political principle* = the necessary and ordered realm, of compulsion and domination	*The social principle* = the dialogical, i.e., the realm of free fellowship and association
Jurgen Habermas b1929 German sociologist & philosopher	*The system* = institutions and governing bodies	*The lifeworld* = societal and individual-level attitudes, beliefs and values

Table 1 *Two Systems Philosophical Distinctions*

The depths of these philosophies, and the differences between them, lie beyond the scope of this paper, but their analysis provides some evidence to the existence and nature of the two relational systems. The impact of this in community engagement is also being explored (e.g., Barker, 2010b, Bakardjieva, 2009;

Jackson, 1999). The distinctions, summarized in Table 1, illuminate some of the characteristics that play out in practice. These relate to inherent differences in the nature of human relationships and behavior in the two social sub-systems:

- Where there are ordered authority hierarchies, with structured power and rule based instrumental relationships, in the world of organized activity—public, commercial and 'voluntary'[1] worlds, and;
- Where relationships are based on free and voluntary association, often referred to as the 'community'—in neighborhoods or mutual interest groups, networks and ad hoc associations.

The One System Approach

The colloquial way of referring to these two different areas of life, in the community engagement world, is to talk about 'top-down' and 'bottom-up' approaches, with the public agencies at 'the top' reaching out to engage with the community at 'the bottom' (Figure 1).

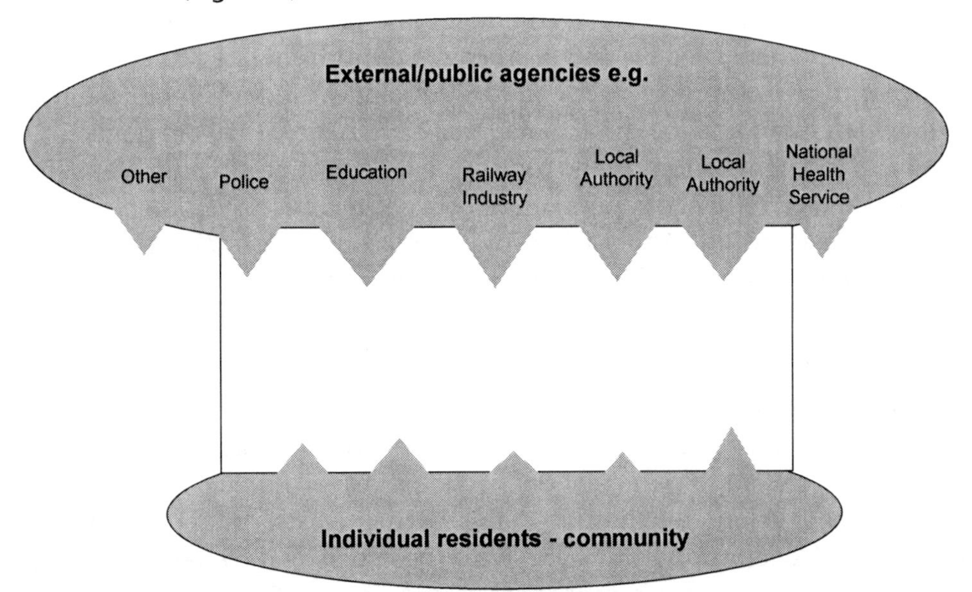

Figure 1 *The One System Approach*

This approach, used widely also in organizational management, transfers to the community the instrumental approach which is at the heart of the ordered authority system. It reflects an idea that the 'bottom-up' and the 'top-down' are like two parts of a machine to be fitted together. That often leads to an approach that the 'bottom-up' needs to behave, and have governance processes, like the 'top-down' system, to help deliver services. So the community tends to be used instrumentally by the public services, rather than treated as an independent

1. 'Voluntary' in this paper includes charities, non-profits, NGOs, social enterprises.

participant. Strenuous efforts are sometimes made to avoid this by focusing on 'bottom-up' development (e.g., Nikkhah, 2009; Larrison, 2000; citizen first, 2008), but these often fail to overcome the problems resulting from the idea that both parts are organizationally similar.

The different nature of relationships results in different organizational dynamics in the part called 'bottom-up', giving a false symmetry. The two relational systems cannot be fitted together as one machine-like system, but continue to coexist alongside each other interacting and coevolving in a shared social eco-system. There are different theories, models, views and positions, from different managerial, political and ideological perspectives, used to explain defects in the current systems and process, especially power relationships and organizational management. The two relational systems' perspective is relevant to each and all of them. Seeing it through this additional new lens can remove some of the significant blocks to effective community engagement.

Two Systems: Forms And Processes

In the authority systems of work, commerce and power, the nature of the relationships is primarily *vertical* and *hierarchical*: tightly regulated to ensure compliance with organizational policies and constraints including employment and contract laws, and financial and managerial governance. They are generally divided into segments, subjects and topics. The organization structures, and management and governance systems, have coevolved with the *vertical hierarchical* system of relationships. Other forms of more flexible, informal and boundary-spanning human relationships coexist alongside; there are strong arguments for management systems to respond to these as a more prominent form in the *vertical hierarchical* world.

However, even if there were more of these changes, there would remain the fundamental difference that civil society is not like regulated organizations, where people are recruited to particular defined jobs. Instead, individuals, when they come together voluntarily through their shared interests, connect to give each other mutual 'peer' support in some way. These personal connections are the source of nourishment for the *horizontal* relationships between *peers*. They have their roots in life and death experiences in the community, not in contractual hierarchical relationships, nor in the needs of public agencies to deliver their services. To be healthy and strong, the roots for these social relationships need to be appropriately tended. The way grass roots grow, illustrated in Figure 2, is an instructive image for this.

Grass that grows strongly and healthily, and is difficult to uproot, has a strong and intertwining mat of roots. These are like the strong interconnections in a community, all giving strength and support to the whole. If the grass is separated from its mat of roots it loses its strength and its intrinsic nature. These social

Figure 2 *Grass Roots (Miller, C. painting)*

networks, and the need to nurture them, are fundamental to resilient communities (Gilchrist, 2009; Rowson *et al.*, 2010).

The dominant *horizontal peer* nature of the community produces its own managerial and organizational challenges of a different nature from those in the *vertical hierarchical* world. My preliminary analysis, summarized in Table 2, shows some of the fundamental distinctions between the two systems.

Two Systems In Reality

Because of these differences, organizational dynamics in the *horizontal peer* world are as dissimilar from the dynamics of the *vertical hierarchical*, as different as are oil and water. This has a significant effect on how they are experienced and managed. Reflecting on our personal experiences, we can sense this. Working as an employee, in a line management chain, is not like being a resident working with neighbors, or other local residents, on a matter of mutual local interest. The way we relate to coworkers in the detailed structured regulation of the *hierarchical* world of work and business, is mainly replaced by fluid informal free association forms of *peer* relationships in the 'community' world. It is important, therefore, to name and visualize them (see Figure 3), to shine a light on the effects of the differences.

Public and Voluntary Agencies ***Vertical Hierarchical* System**	**Community Organizations** ***Horizontal Peer* System**
Organization Organizations incorporated Limited liability Command & control systems	**Organization** Organizations unincorporated Unlimited liability Free association systems
Management *Vertical hierarchical* relationships Authority/line management	**Management** *Horizontal peer* relationships Personal links
Employment Contractual Paid staff, & managed volunteers Employment law context	**Employment** Social informal Not-paid volunteers, not 'managed' General civil law
Resources Recurring annual income External sources—taxes, grants & fees Commissioned contracts Permanent physical locations	**Resources** Unpaid voluntary work Donations, ad hoc grants In-kind services Domestic + ad hoc locations

Table 2 *Two Systems: Summary of Some Different Forms and Processes*

This naming is an abstraction; the systems do not have impermeable boundaries. But to understand differences we need initially to focus on them (Cilliers,1998). The distinctive natures of the relationship patterns and associated organizational systems in *vertical hierarchical* and *horizontal peer* systems are captured as a snapshot in Table 2, but it is a snapshot that has continuing relevance to organizational, management and governance of community engagement in all forms. Differences between the nature of relationships and working arrangements in the *vertical* and *horizontal* worlds are also identified in work arenas not confined to the community engagement world—for example in communities of practice (MacGillivray, 2009, this volume), and social entrepreneurs (Goldstein *et al.*, 2008). In the *social eco-system dance* model however, the individuals in the *horizontal* world do not simultaneously occupy work roles in the directly interacting *vertical* systems, and some of the issues are therefore different.

Two Coevolving Systems

In the world of community engagement, the differences between the community and the work worlds cause familiar difficulties, as the troublesome 'community' fails to behave as the public agencies wish it would behave. The tendency is to try to cope, or to pay lip-service to community engagement, or to put it off,

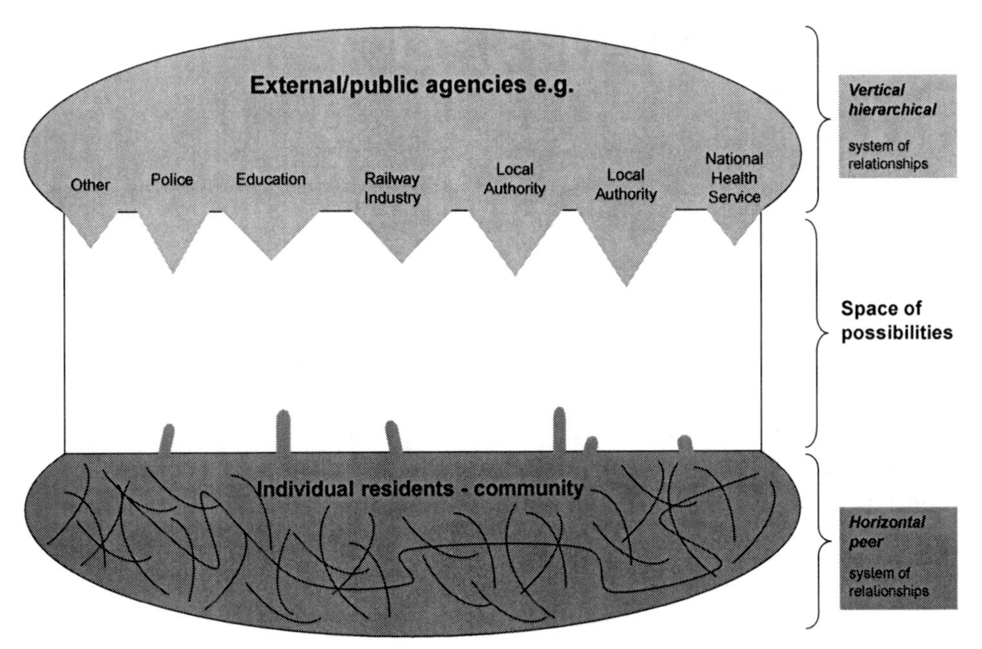

Figure 3 *Two Distinct Coevolving Systems Sharing an Eco-System*

or attempt to mould the community to be acceptable to the world of the *vertical hierarchical.* But none of these are likely to work well.

Taking a complex systems view, the two relational systems can be seen as two differentiated systems inhabiting a shared social eco-system. They are in continuous interaction, in what complexity theory calls the *space of possibilities* (Figure 3), and are coevolving. Understanding this offers more scope for appropriate governance in the *horizontal peer* and in the *space of possibilities*. Acknowledging the systemic differences is to see that the organizational forms, processes and relationships, which will work effectively cannot be just a replication and imposition of those which work in the *vertical hierarchical* world.

In this shared social eco-system the two complex systems have their very different internal dominant dynamics, for which they each need appropriate governance approaches. In addition, their continuing interaction creates further dynamics in the *space of possibilities*, which also needs its own governance. As Mitleton-Kelly points out (2003), the logic of complexity suggests that appropriate approaches for organizations need to be facilitated—nurturing *enabling environments* that facilitate learning, emergence and self organization. This needs to be applied consciously and deliberately to the other two areas of the shared social eco-system—the governance within the *horizontal peer* which has its own different needs, and also the interaction of the two systems in the *space of possibilities*.

The Social Eco-System Dance

The two systems are locked into a perpetual coupled relationship, within the shared social eco-system (Figure 3). Social entities—individuals, groups, organizations—interact in identifiable rhythms, in a continuing set of moves as in a dance, even though it is often uncomfortable. According to Kaufmann (1995), this is coevolution which "is a story of coupled dancing landscapes", "deforming landscapes where the adaptive moves of each entity alter the landscapes of their neighbors" (Kauffman in Mitleton-Kelly 2003). The idea of *fitness landscapes*, arising in the study of natural eco-systems, is described by Geyer & Rihani (2010), in applying the ideas to public policy on social issues, as "capturing well the symbiotic relationship between multiple interacting actors and units". That is, everything they do affects (deforms) the social context (environment) in some way, so the other entities adjust their actions and behavior to restore their own *fitness landscape*.

Some Dynamics

Life can be tough at the interface between these two relational systems: discordance from the different systemic dynamics; local State power meeting the citizen; and the *vertical hierarchical* system driven by a *'contagion of inwardness'* and *'organization-first'* (Crockett, 2008; Barker, 2010b) from the imperatives of organizational survival, exacerbated by short public expenditure timescales. So that system reaches out (*'outreach'*) to the 'community' to pull parts of it into its own dimension to help it achieve its own aims, objectives, programmes and service delivery. To engage, individuals in the *horizontal peer* often have to detach themselves from their natural soil and roots, defeating the purpose of community engagement. The weaker *horizontal* system can have, however, a severe negative power in hindering the successful achievement of the *vertical* system's objectives.

There are successful and positive interactions: they flow from a recognition and accommodation of the different natures. But this happens in a fragmented way, without appreciating the underlying reasons for success. So, 'successful' initiatives are 'rolled out', mechanically, often inappropriately in the *vertical hierarchical* command and control managerial style, paying little attention to the different *horizontal peer* dynamics and particular local conditions which may have been a pre-requisite for the original success.

Emergence In The *Space Of Possibilities*: A Local Case Example

In Figure 4 the *space of possibilities* is populated with emergent organizational forms. These represent some of the different forms that have emerged over a period of years in my local area of inner south east London.

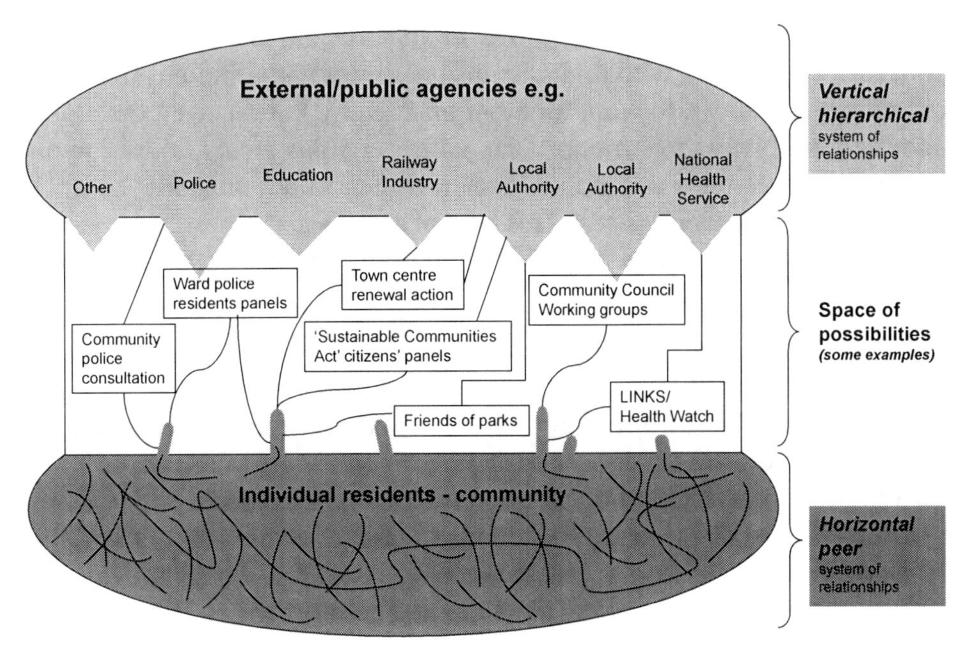

Figure 4 *Emergence in the Space of Possibilities*

They reflect a variety of degrees of openness to the *adjacent possible*, of permanence or impermanence, formality or informality, fluidity and flexibility. They provide attempts with varying degrees of success or failure for enabling citizens to be collaborative problem solvers (e.g., Olcayto, 2010). Local people become active in the *horizontal peer* in a number of ways and link by a variety of resident-led local groups and loose networks, now enhanced by web based media. The town centre renewal action for example is underpinned by a loose network of some 1500 local residents (Peckham Residents, 2011), over an urban area of about 2 to 3 miles, giving rise to an informal consortium of local interests (Peckham Vision, 2011) linked with other local groups. Figure 4 is like a map, another snapshot, of the interactions between the systems, from the *horizontal peer* perspective of one participant, of part of this particular local social eco-system dance. The experience has contributed to the development of the model and typologies outlined in this paper. Each participant in a locality can produce their own map and together they portray some of the richness and complexity of life in the civic grassroots and its interaction with the local State.

Adjacent Possible In The *Space Of Possibilities*

In *the space of possibilities* where individuals from each system come together, there are many opportunities for change by the use of the *adjacent possible* (Kauffman, 2000; Mitleton-Kelly 2003). These need small adjustments from recognized forms and processes, leading to the emergence of new ones. In the local case, for example, a local council planning officer has been collaborating with new flexible approaches in a resident-led town centre forum, to contribute to planning policy and town centre management (Southwark Council, 2011).

The small process and attitude changes, to enable this, are essential for individuals from the two systems, with their dissimilar systems with different structures, processes and dynamics, to work together effectively. In practice it becomes a collaborative problem solving approach, which enables local knowledge and expertise to complement the professional, technical and managerial contributions to the policy making process. This kind of approach is essential for solving modern complex, multi dimensional problems (Battle, 2010; Peters, 2010; Barker 2010b).

Often, the *vertical hierarchical* approach to community engagement is a tick box exercise, or dominated by their agenda and way of working that kills off real engagement, or it can be disrupted by stressful reactions from the *horizontal peer* world. Because there are significant challenges in operating in this world where the two systems interact, intentional good will from the actors in both systems, able to nurture *trust*, is essential. Trust leads to greater awareness of the 'Other' and their differences, helping to create less risky conditions for careful *adjacent possible* movement outside normal practices and comfort zones. This encourages mutual exploration of methods and approaches. But citizens' trust, and social networks and social capital that are rooted in trust, cannot be generated overnight (Savage *et al.*, 2009). Trust emerges in safer zones, which are an example of the *enabling environment* needed in managing complex systems to nurture the "coevolution that can produce new orders of coherence" (Mitleton-Kelly, 2003).

Nurturing The *Space Of Possibilities*

These safer zones need supportive culture, especially in the *vertical hierarchical* system, and they need to be sensitively nurtured. This in many instances requires a cultural change (Lurie, 2010; Morse, 2006). USA research supports the view that 'active listening by bureaucrats' and 'deliberative approaches' are needed to develop citizen trust in government which in turn can develop government trust in citizens (Cooper *et al.*, 2006). This would be an example of a positive outcome of the coevolution of systems and adaptations in their *fitness landscapes*. Action research in The Netherlands for planning water management showed that new processes to encourage *interactive governance* 'needed constant maintenance' and failed to survive, because of the resistance of some elements of the *vertical hierarchical* system (Edelenbos *et al.*, 2009).

Keys to the success of relationships, and organizational processes, in the *space of possibilities*, and the nurturing of trust, can be found in values and practices that are intrinsically in tune with how people can live and work together humanely and constructively. These include encouraging and nurturing cooperative human relationships where there are different perspectives or different modes of operation or there is potential for conflict. The knowledge and skills for this can be found in abundance in the community development approach (Gilchrist, 2009), deliberative dialogue (Carcasson, 2010; Battle, 2010; London, 2010) and also in others such as community organizing, mediation, group relations, peer mentoring and so on.

All of these methods and skills are relevant for those who navigate the *space of possibilities*, in the engagement process between the *vertical* and *horizontal* systems. Their training should reflect the distinct natures of the two systems, and their interactions. Table 3 indicates some of the topics to include in developing policy for creating and nurturing an enabling environment, and for training participants from both systems.

Encouraging *adjacent possible*	**Nurturing *Horizontal Peer system***
• Moving beyond habitual routines • Resisting fixed design in advance • Enabling new structures to emerge	• Social gardening • Facilitating networking • Nurturing relationships
Collective efficacy	**Facilitating the process**
• Creating potential connections • Enabling loose connections • Sustaining connections	• Nurturing informal fluid processes • Growing from the roots • Creating support systems

Table 3 *Facilitating an* Enabling Environment *to Nurture and Navigate the* Space of Possibilities. *Starter Checklist for Policy and Training Development*

Hybrid Sub-Systems

The political sector and faith/religious sector each have large numbers of organizations intertwined alongside very large numbers of volunteers. They are therefore hybrids of the two sub-systems, *vertical hierarchical* and *horizontal peer*: see Figure 5. Neither of the two relational systems is dominant.

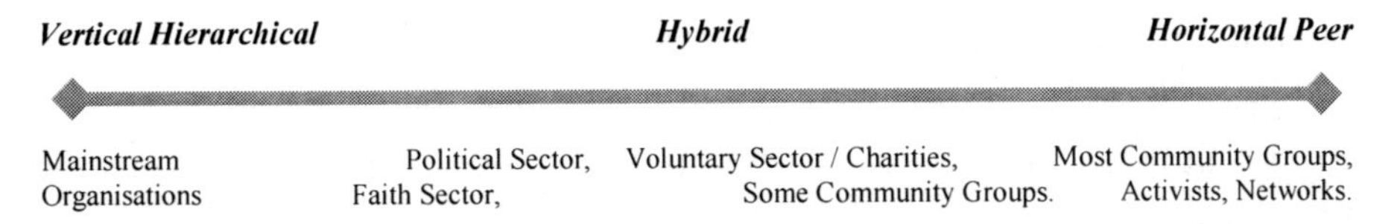

Figure 5 *Continuum of Organizations in the Vertical and Horizontal Systems*

A combination of numerous political organizations and thousands of volunteer activists constitutes the political sector. It has major intentional and unintentional effects on the macro and micro details affecting the *fitness landscapes* within the overall social eco-system for community engagement. The faith sector, containing all religious faiths, is increasingly engaged in civic activities, and it too is a hybrid through the combination of its religious structures, and the very large numbers of volunteers who freely associate with each other in their chosen religious or faith group. The sectors interact in significant ways in the *space*

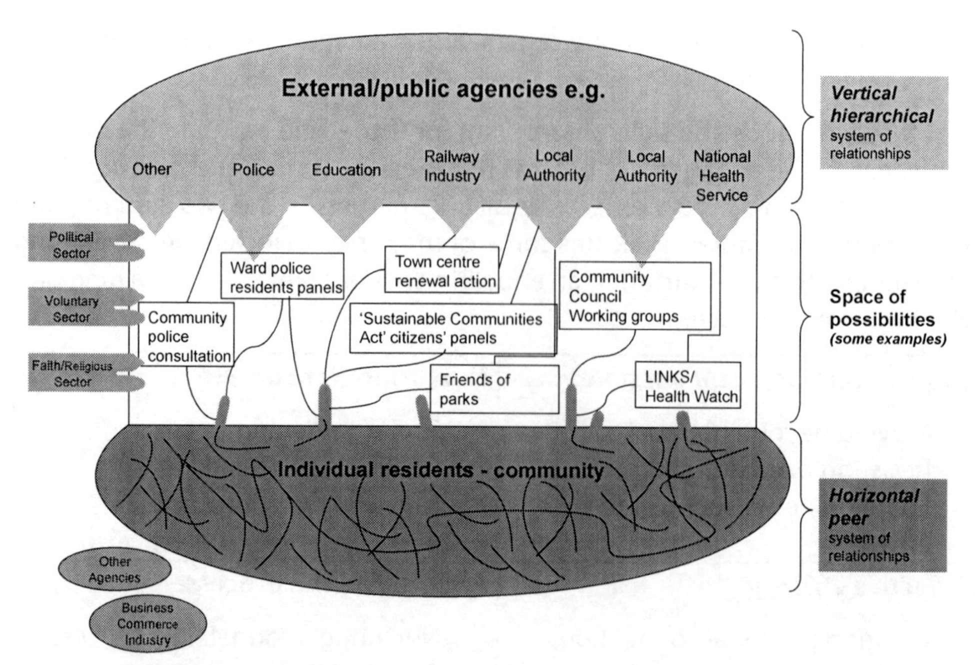

Figure 6 *Hybrid Systems in the Shared Social Eco-system*

of possibilities alongside and between the public agencies and the community, as indicated in Figure 6.

There is not the space here to detail this further, but because of their hybrid nature the two systems lens can illuminate many aspects of the roles of these sectors, in the working of community engagement.

It is worth noting that aspects of the *vertical hierarchical* and *horizontal peer* distinctions may find some expression also in the world beyond community engagement, with some similar and some dissimilar effects:

- Interactions between commercial organizations and their customers.
- The fluid free association networks at *horizontal peer* levels within corporate organizations in both public and commercial sectors.
- Communities of practice, transcending organizational boundaries, as noted earlier.

Vertical And *Horizontal* Systems In The Voluntary And Community Sectors

The professional 'voluntary' sector is also a hybrid of the relational *vertical* and *horizontal* systems because of its historical roots. But the *vertical hierarchical* system is now dominant in the organized professional 'voluntary' world with vertical and rule-based relationships, where the thousands of volunteers contribute their time and skills to structured roles in those organizations. However, the informal community sector, comprising a sub-system of small and informal groups of local activists and active citizens, is primarily a *horizontal peer* social system based on free association: "Voluntary sector players have different interests from them [community sector], usually function in different ways, and often fail to understand these differences..." (Pitchford & Higgs, 2004).

The nature of the *vertical hierarchical* system in the voluntary sector results in a mismatch in policies and programmes for the community sector (Barker, 2010b), and "though many civic organizations… use the language of civic engagement, their routines appear to be misaligned with citizens who seek a sense of agency" (Barker, 2010a). The misalignment reinforces the failure to perceive the difference between *vertical* processes for service delivery and those for strengthening *horizontal* citizen capacity. The summary in Table 2 shows some of the underlying reasons for these differences in the two systems which lead to different managerial and organizational challenges.

There is increasing awareness that the formal voluntary sector is just 'the tip of a large iceberg' with most of the community sector 'under the water', 'below the radar (BTR)', or to use a living image—the teeming micro-life revealed under a stone in a garden. This out of sight and poorly understood activity is the missing element named in this paper as the *horizontal peer* system. The scale of the issue is shown by the estimate that there are 600,000 to 900,000 'below the radar' groups in the UK, three quarters of the total in the 'voluntary & community' sector (Phillimore & McCabe, 2010). Yet they are generally grouped together as if they were largely the same, reinforcing the inability to see the differences.

Community Organizations' Spectrum

The distinction between the relational sub-systems within the voluntary & community sectors can be seen in my preliminary analysis of the variety of organizational forms (Figure 7). The diagram focuses on the small end of the spectrum. It magnifies and shines a light on some of the organizational variety in what is out of sight. Some small community groups that employ no staff and are very informal are aspiring to become formal staff employing groups, and need traditional 'capacity building' support for this. They are however the minority of groups in the informal community sector, yet there is a tendency in the *vertical hierarchical* world to assume incorrectly that all groups should aspire to such growth and focus their support for that (Foster, 2010): another misalignment.

But small community groups whether wanting to grow or not, employing staff or not, whether formal or informal, all need governance support tailored specifically for their *horizontal peer* needs. Some of the characteristics that make up their various structures are indicated in the Notes in Figure 7 which is the beginning of a set of diagnostic tools to distinguish between different forms, to match the different forms of appropriate support. Recognizing the distinctions between the *vertical* and the *horizontal* is essential to see clearly enough the informal community sector—the *horizontal peer* world—to achieve a better

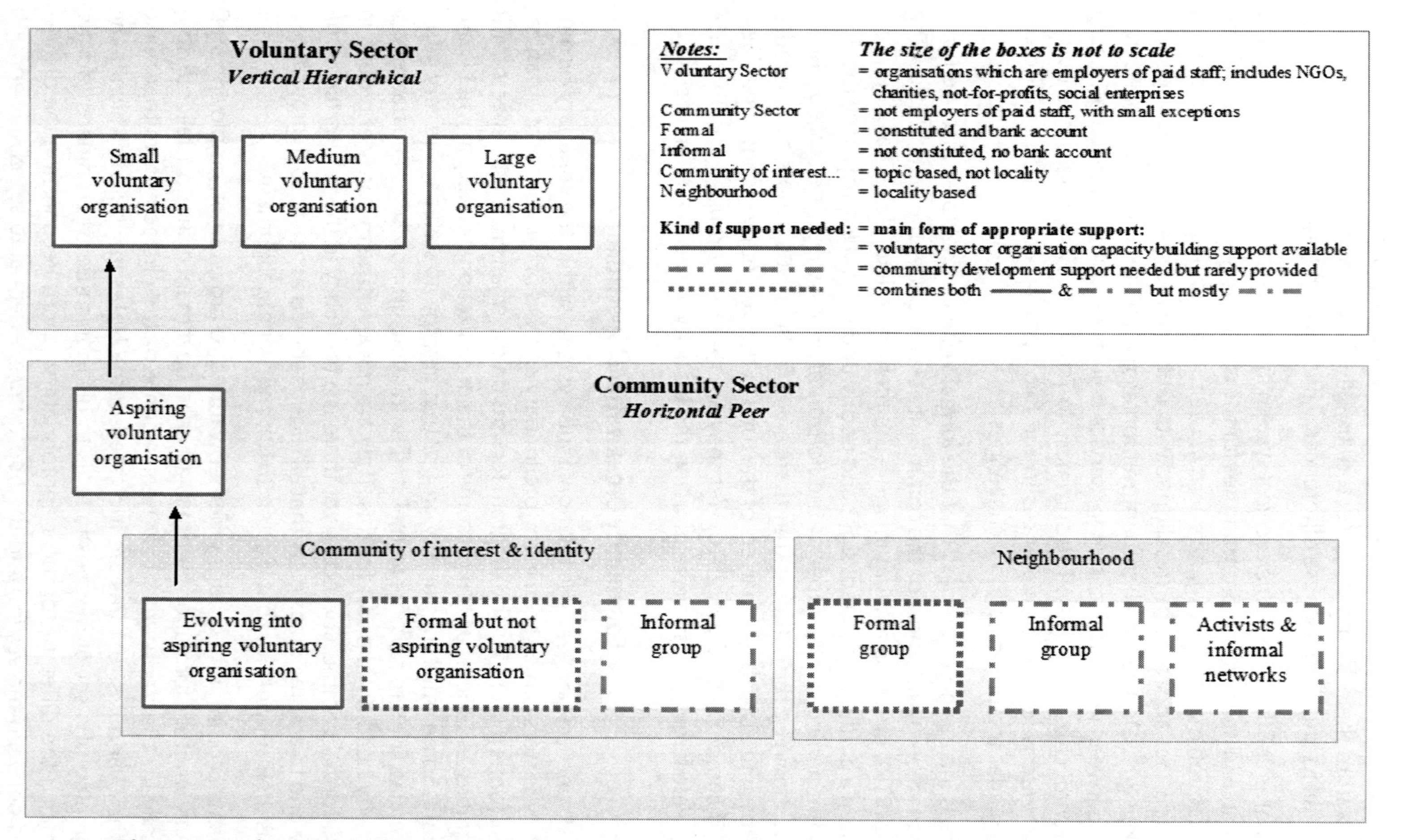

Figure 7 *Voluntary and Community Sectors—Organization Types: Size And Formality Spectrum Chart*

balance between support systems for the organized voluntary sector and those to meet the needs of the much larger but out-of-sight community sector, as recommended recently (NSG, 2010).

Typology Of Active Citizen Roles

Thousands of volunteers work in the voluntary sector alongside employed professional staff. These volunteers occupy well defined roles within the structured voluntary organizations (Figure 8, type 2). However there are several other different kinds of volunteer shown in the typology in Figure 8.

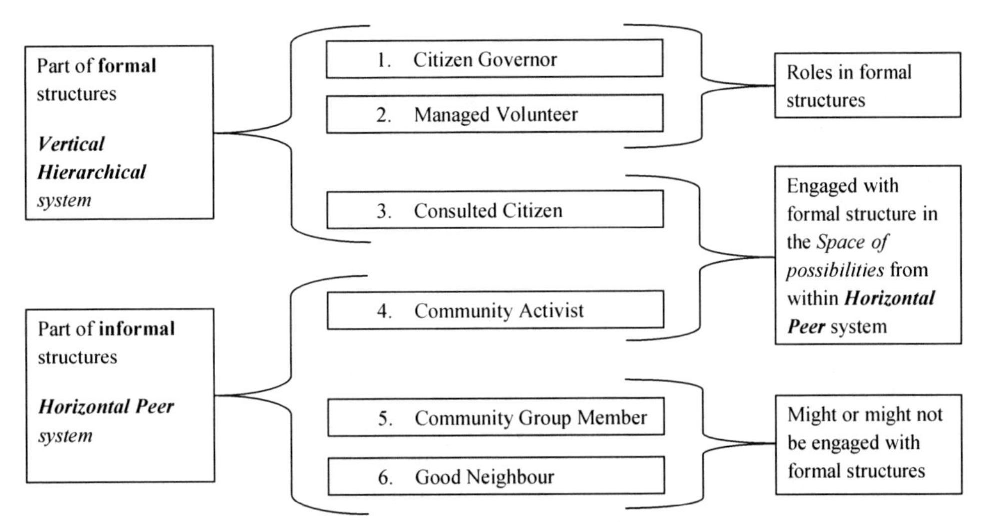

Figure 8 *Typology of Different Active Citizens Roles. These roles can be fluid and without fixed boundaries, and the same individual may switch between roles and fulfill more than one role at once, while some individuals may carry out only one role at a time. The types do not indicate progressive pathways in activism. The original typology was devised within the Southwark Active Citizens Hub (SACH) Steering Group. It was further developed to show relationships with the vertical and horizontal systems. For type descriptions see leaflet (SACH, 2006).*

Some are part of the *vertical hierarchical* system, and others interact with it from the *horizontal peer* system. Each role has different kinds of support needs. Using the typology, an analysis of the needs of the different types, and the existing support, begins to indicate where the gaps are, and how they might be met. In my experience, gaps relate particularly to the roles identified in Figure 8 as activists, community group members and 'consulted citizens' and their related small groups and networks (see Figure 7). This London experience is supported by reports from elsewhere, for example in NW England (NWCAN, 2010).

Strengthening The *Horizontal Peer* System

The voluntary sector already provides a broad range of support in local areas for voluntary activity. This is an essential underpinning for informal community action, but traditional ways of providing it need adapting to give a more appropriate response to the different needs of the *horizontal peer* world. The NSG (2010) study recommends some ways of achieving this. Distinctive characteristics include peer-to-peer relationships creating their own support, and strengthening citizen capacity as distinct from developing alternative forms of 'service delivery' (Barker, 2010a, 2010b). Gaps need to be identified with local activists. If something is already provided for the voluntary sector, an adjusted form may need to be tailored to meet the *horizontal peer* needs. Starting with existing forms is an example of collaborative *adjacent possible* work between the voluntary sector and local activists. Preliminary indications of some adapted and new forms of local systems and structures, reflecting the nature of *horizontal peer* relationships which need to be explored, are summarized in Table 4.

Management & Organization	**Peer support systems**
• New advice, support systems and training on management and organization reflecting *horizontal peer* dynamics. • Reflecting needs of informal groups, residents' networks, and community activists.	• Nurturing of local residents' networks. • Action learning sets for activists. • Peer and mentoring support systems for activists. • Using digital social media.
Local neighborhood based systems	**Citizen collaborative problem solving**
• New organizational systems for access to local physical practical resources. • Advice & information systems for active residents, incorporating local peer advice with local voluntary sector systems.	• Creation and support of activists' reference panels for engagement with public agencies. • Enhancing community engagement processes through linking informal residents' networks in the *horizontal peer* system.

Table 4 *Some Preliminary Indications of Horizontal Peer Forms of Local Systems and Structures*

There is more to be uncovered to provide appropriate support systems for the invisible community life. Greater understanding is needed of the links between personal life and experiences and peer-to-peer networks (Battle, 2010) that are part of what Mary Parker Follett called 'neighborhood consciousness' (Nielsen, 2010; Morse, 2006). Recent UK research (Ockenden, 2008; Phillimore, 2010; Hutchison, 2010) is beginning to shine light on this micro life in the informal civic sector, confirming its differences from the organized voluntary sector and the need for more detailed research. A current research project (Pathways, 2011) is investigating why people are involved in different forms of social and civic participation. Another is exploring the 'Civic Commons' as a model for supported citizen-led contributions to problem solving (Norris & McLean, 2011). The practical tools in the *social eco-system dance* model, and typologies outlined in this paper, can help in such research to understand the informal world whose processes have been in the shadow.

The UK Coalition Government's Big Society policy is unfolding (Prime Minister, 2011), and will have significant effects on the *horizontal peer* world (Chanan, 2011; McCabe, 2011), affecting its *fitness landscape*. There will be a loss, from public expenditure cuts, of some local support systems, and the policy to train 5000 new 'community organizers' (Urban Forum, 2010; UK Government, 2011) will have a direct impact. It will be essential that the policy and training illuminate the distinction between the *vertical* and *horizontal* systems. Otherwise, the new programmes risk repeating the mistakes of relating not to the real needs from within the informal community sector, but to the needs as perceived by those outside it to deliver public services.

In the US, attention is being drawn to the increasing technical and professional natures of public and civic organizations *(vertical)* which have created a growing gap between them and citizens *(horizontal)* and the loss of citizen contributions to shaping their collective future (Barker, 2010a). The consequence of this gap at its extreme was seen in Cairo in January 2011 where *horizontal peer* informal civil networks, developing slowly out of sight, strengthened by digital social media, found their voice (Alexander, 2011). A similar phenomenon is reported in Russia where civil networks, supported by digital social media, emerged from the free association of individuals with shared personal experiences (Loshak, 2010). These examples display organizational dynamics and relationship patterns familiar to those with inside experience of the *horizontal peer* world in liberal democracies. This indicates that the two systems' relational experience and impact may be a social phenomenon in modern technocratic societies across the global complex social system.

Conclusion

Human social affairs create a complex evolving system. Within that, for community engagement, there are two differentiated sub-systems, each embodying distinct relational experiences. Naming these sub-systems *vertical hierarchical* and *horizontal peer* helps to identify them as different interacting systems. Perceiving this reality can enable more effective policies and programmes for encouraging community engagement, and breathing new life into democracy in the global technocratic age. Attention must be paid to the ways the two systems interact and engage with each other, and how the community system can be strengthened and supported for that engagement.

Complexity theory's understanding of the nature of complex systems has a liberating effect in thinking about some of the repetitive problems in community engagement. The ideas of *eco-systems, coevolution, fitness landscapes*, the *space of possibilities* and the *adjacent possible* have a particular resonance with the reality of the experiences in this world of community engagement. The *social eco-system dance* analogy captures the close coupled repetition of systemic behavior patterns that need to be recognized and accommodated.

These ideas provide the basis for further research to apply complexity theory to community engagement, as well as developing management and organization practical tools:

- To help strengthen resilience in the *horizontal peer* world, and;
- To nurture the *space of possibilities* for effective interactions and community engagement between *horizontal peer* civic society and the *vertical hierarchical* institutional world.

The issues to be tackled span a variety of worlds—research, policy, management, and front-line workers in public sector, voluntary sector and community development, and activists and other active citizens. The model has been used in many discussions with practitioners in these various worlds, and the model and tools evolve and develop through this continuing interaction. This paper, using the model, indicates some of the areas that need to be addressed to develop new ways of thinking about the *vertical—horizontal* interactions, and new ways of managing and nurturing them, so they can work together more constructively for the greater whole. There is rich material here for complexity analysis, and still much work to do. I invite those interested in that, or community engagement from these perspectives, to contact me: complex.community@gmail.com

References

Alexander, A. (2011). "Internet role in Egypt's protests," BBC News Middle East, February 9th, http://www.bbc.co.uk/news/world-middle-east-12400319.

Bakardjieva, M. (2009) "Subactivism: lifeworld and politics in the age of the internet," *The Information Society*, ISSN 0197-2243, 25(March): 91-104.

Barker, D. (2010a). "Citizens, organizations, and the gap in civil society," *Kettering Connections 2010*, http://www.kettering.org/media_room/periodicals/connections.

Barker, D.W.M. (2010b). "The colonization of civil society," *Kettering Review/Fall 2010*, ISSN 0748-8815, http://www.kettering.org/media_room/periodicals/kettering_review.

Battle, D. (2010) "Uncovering organic community politics: A View from the Inside," *Kettering Connections 2010*, http://www.kettering.org/media_room/periodicals/connections.

Carcasson, M. (2010). "Developing democracy's hubs: building local capacity for deliberative practice through passionate impartiality," *Kettering Connections 2010*, http://www.kettering.org/media_room/periodicals/connections.

Chanan, G. and Miller. C. (2011). "The big society and public services: complementary or erosion?" PACES Publications (Public Agency and Community Empowerment Strategies), http://www.pacesempowerment.co.uk.

Cilliers, P. (1998). *Complexity & Postmodernism: Understanding Complex Systems*, ISBN 9780415152877.

Citizen First (2008). "Top-down or bottom-up services: How best to put the citizen first," EU North West Europe Programme, Guidelines, http://www.citizen-first.net.

Cooper, T.L., Bryer, T.A. and Meek, J.A. (2006). "Citizen-centered collaborative public management," *Public Administration Review*, ISSN 1540-6210, 6(Supp 1).

Crockett, C. (2008). "Public work vs. organizational mission," *Kettering Connections 2008*, http://www.kettering.org/media_room/periodicals/connections.

Dove, M. (2006). "Little people and big resources," *Carbon Trading: A Critical Conversation on Climate Change, Privatization and Power*, ISSN 0345-2328, p. 133, http://www.greenchoices.cornell.edu/downloads/development/Carbon_Trading.pdf.

Edelenbos, J., van Schie, N. and Gerrits, L. (2009). "Organizing interfaces between government institutions and interactive governance," *Policy Sciences*, ISSN 0032-2687, 43(1): 73-94.

Foster, J. (2010). "Finding a different path," *Kettering Connections 2010*, http://www.kettering.org/media_room/periodicals/connections.

Geyer, R. and Rihani, S. (2010). *Complexity and Public Policy: A New Approach to Twenty-First Century Politics, Policy and Society*, ISBN 9780415556637.

Gilchrist, A. (2009). *The Well-Connected Community: A Networking Approach to Community Development*, ISBN 9781847420565.

Goldstein, J.A., Hazy, J.K. and Siberstang, J. (2008). "Complexity and social entrepreneurship: a fortuitous meeting," *Emergence: Complexity & Organization*, ISSN 1532-7000, 10(3): 9-24.

Hutchison, R. (with Eliza Buckley, Becky Moran and Nick Ockenden) (2010.) "The governance of small voluntary organizations," Institute for Voluntary Action Research (IVAR) with Institute for Volunteering Research (IVR), http://www.ivr.org.uk/evidence-bank.

Jackson, N. (1999). "The Council Tenants' Forum: A liminal public space between life-world and system?" *Urban Studies*, ISSN 0042-0980, 36(1): 43-58.

Kaufmann, S.A. (1995). *At Home in the Universe: The Search for Laws of Self-Organization and Complexity*, ISBN 9780140174144.

Kauffman, S.A. (2000). *Investigations*, ISBN 9780195121049.

Larrison, C.R. (2000). "A comparison of top-down and bottom-up community development interventions in rural Mexico," 12th National Symposium on Doctoral Research in Social Work, https://kb.osu.edu/dspace/bitstream/1811/36912/1/12_Larrison_paper.pdf.

London, S. (2010). *Doing Democracy: How a Network of Grassroots Organizations is Strengthening Community, Building Capacity, and Shaping a New Kind of Civic Education*, ISBN 9780923993320, http://www.kettering.org/media_room/publications/doing-democracy.

Loshak, A. (2010). "Parallel worlds: How connected Russians now live without state," Open Democracy 22 November, http://www.opendemocracy.net/author/andrei-loshak.

Lurie, P. (2010). "Governmental agencies as civic actors," *Kettering Connections 2010*, pp. 28-29, http://www.kettering.org/media_room/periodicals/connections.

MacGillivray, A. (2009). *Perceptions and Uses of Boundaries by Respected Leaders: A Transdisciplinary Inquiry*, PhD Dissertation: Fielding Graduate University, AAT 3399314.

MacGillivray, A. (2011). "The application of complexity thinking to leaders' boundary work," this volume.

McCabe, A. (2011). "Below the radar in a Big Society? Reflections on community engagement, empowerment and social action in a changing policy context," Third Sector Research Centre Briefing Paper 51, and Working Paper 51, http://www.tsrc.ac.uk/LinkClick.aspx?fileticket=OMbpEZaAMKI%3d&tabid=500.

Miller, Curt. "I am grass roots", http://www.clothespinreferee.com/p22.htm.

Mitleton-Kelly, E. (2003). "Ten principles of complexity and enabling infrastructures," in E. Mitleton-Kelly (ed), *Complex Systems and Evolutionary Perspectives of Organizations: The Application of Complexity Theory to Organizations*, ISBN 9780080439570.

Morse, R.S. (2006). "Mary Follett, prophet of participation," http://www.iap2.org.au/sitebuilder/resources/knowledge/files/452/journal_issue1_folletbymorse.pdf, adapted from "Prophet of participation: Mary Parker Follett and public participation in public administration," *Administrative Theory and Praxis*, ISSN 1084-1806, 28:1-32.

Nikkhah, H.A. and Redzuan, M. (2009). "Participation as a medium of empowerment in community development," *European Journal of Social Sciences*, ISSN 1450-2267, 11(1): 170-176.

Nielsen, R. (2010). "How the 'body politic' thinks and learns: the role of civic organizations," *Kettering Connections*, http://www.kettering.org/media_room/periodicals/connections.

Norris, E. and McLean. S. (2011). "The Civic Commons: A model for social action," *RSA Projects*, http://www.thersa.org/projects/citizen-power/civic-commons.

NSG (2010). "NSG infrastructure report," Neighborhoods Sub Group of the London Regional ChangeUp Consortium, http://www.londoncivicforum.org.uk/wordpress/wp-content/uploads/downloads/2011/07/NSG-report-April-2010.pdf.

NWCAN, (2010). "More about us: report: NWCAN/NWTRA conference 23rd March 2010," North West Community Activists Network and North West Tenants & Residents Assembly, http://nwcan.org/index.php?page=report-23-march-2101.

Ockenden, N. and Hutin, M. (2008). "Volunteering to lead: a study of leadership in small, volunteer-led groups," *Institute for Volunteering Research*, http://www.ivr.org.uk/evidence-bank.

Olcayto, R. (2010) "Localism can, and does, improve the quality of the built environment," *Architects' Journal*, ISSN 0003-8466, 29 July, http://www.peckhamvision.org/blog/?p=463.

Pathways through Participation (2011). "What creates and sustains active citizenship?" NCVO (National Council for Voluntary Organizations), Involve, IVR (Institute for Volunteering Research) research project, http://pathwaysthroughparticipation.org.uk.

Peckham Residents Network, (2011). http://peckhamresidents.wordpress.com.

Peckham Vision, (2011). http://www.peckhamvision.org.

Peters, S.J. (2010). "Relating reason and emotion in democratic politics," *Kettering Connections 2010*, http://www.kettering.org/media_room/periodicals/connections.

Pitchford. M. and Higgs. B. (2004). "Community catalyst report: Closing of operations, lessons from capacity building work 1995-2004," internal report, unpublished, pp. 40-43 in Appendices, NSG infrastructure report (2010), http://www.londoncivicforum.org.uk/wordpress/wp-content/uploads/downloads/2011/07/Appendixes-NSG-report-April-2010.pdf.

Phillimore, J. and McCabe, A. (2010). "Understanding the distinctiveness of small scale, third sector activity: the role of local knowledge and networks in shaping below the radar actions," Third Sector Research Centre Briefing Paper 33, and Working Paper 33, http://www.tsrc.ac.uk/LinkClick.aspx?fileticket=iBB6cFBtNYU%3d&tabid=500.

Prime Minister. (2010). "Big Society speech 19 July 2010," http://www.number10.gov.uk/news/big-society-speech/.

Prime Minister. (2011). "Big Society speech 15 February 2011," http://www.number10.gov.uk/news/pms-speech-on-big-society/.

Rowson, J., Broome, S. and Jones, A. (2010). "Connected communities: How social networks power and sustain the Big Society," RSA Projects. http://www.theRSA.org/projects/connected-communities.

SACH (2006). "How to be an active citizen," Southwark Active Citizens Hub, http://www.volunteercentres.org.uk/how_to_guides.htm.

Savage, V. with O'Sullivan, C., Mulgan, G. and Ali, R. (2009). "Public services and civil society working together: An initial think piece," *The Young Foundation*, http://www.youngfoundation.org/files/images/yf_publicservices_nov09_v2.pdf.

Southwark Council (2011). "Town center strategies update, item 7," Regeneration and Leisure Scrutiny Sub-Committee 2 February 2011, http://moderngov.southwarksites.com/mgConvert2PDF.aspx?ID=16020.

UK Government. (2010). "Building the Big Society," *Cabinet Office*, http://www.cabinet-office.gov.uk/news/building-big-society.

UK Government. (2011). "Government names new partner to deliver community organizers," *Cabinet Office*, http://www.cabinetoffice.gov.uk/news/government-names-new-partner-deliver-community-organisers.

Urban Forum, (2010). "Community organizers briefing," http://www.urbanforum.org.uk/briefings/community-organisers-briefing.

Eileen Conn, MA (Oxon) FRSA MBE, worked for many years in central UK Government Whitehall policy-making on the organization, management and development of government systems and subsequently in developing systems of business corporate social responsibility. In the 1990s she established 'Living Systems Research' as an umbrella for her study of '*Social Dynamics & Complex Living Systems*', and her work in the field for *'Sustainable & Cohesive Communities'*. As an RSA Fellow she founded the RSA Living Systems Group in 1994 looking at companies and other human social systems as complex living systems. In parallel, she has been an active citizen in London community organizations. She was Southwark Citizen of the year in 1998, and in 2008 was Community Activist of the year, Active Citizen of the year, and Southwark Woman of the year. In the UK's New Year's Honours of 2009 she was awarded the MBE for services to the community. She has had a long-term interest in the dynamics of communities and the emergence of community organizations which interact with the structures of public agencies and commercial companies. She has found that complexity theory provides a rewarding approach to understanding these complex social systems. Eileen is an informal associate in the LSE Complexity Research Programme, and an Associate Fellow of the TSRC (Third Sector Research Centre). She is facilitating and studying the emergence of community networks and other community engagement processes in London. She was coeditor and a coauthor of *Visions of Creation* (1995), and is currently working on a new book with the Living Systems Group. She received an MA in philosophy, politics and economics from Oxford.

16. Language, Complexity And Narrative Emergence: Lessons From Solution Focused Practice

Mark McKergow
The Centre for Solutions Focus at Work, ENG

This chapter examines the case for viewing conversations as emergent phenomena, and the practical consequences for complexity practitioners and others engaged in 'talking cures'. Post-structural thinking from Wittgenstein onwards is connected to the school of Solution-Focused practice, which has made explicit use of these ideas in a practical, pragmatic and effective form of psychotherapy and coaching. These fields can be connected by the idea of 'narrative emergence', which casts light on the ways in which new narratives are formed within apparently everyday conversations.

Introduction

> *A living language is in a state far from equilibrium. It changes, it is in contact with other languages, it is abused and transformed. This does not mean that meaning is a random or arbitrary process. It means that* meaning *is a local phenomenon, valid in a certain frame of time and space.... Above all, language is a system in which individual words do not have significance on their own. Meaning is only generated when individual words are caught in the play of the system.* (Cilliers 1998: 124)

This first international workshop on complexity and real world applications has brought together participants from a wide variety of areas and disciplines, with a wide range of views on the salient aspects of complexity. It seems to me that we have seen broadly three ways of connecting complexity with real world applications:

1. Modelling complex real world issues with agent-based models, to help decision making in unfolding real-time situations (for example routing of oil tankers or taxis—as seen in the work of George Rzevski (see for example Glaschenko *et al.*, 2009).
2. Using complexity ideas as a new way of engaging with an ever-changing world, particularly in business and organizational work (see for example the notable work of Eve Mitleton-Kelly and her group at the London School of Economics). This involves making people aware of the key terms and distinctive features of complexity. Some have referred to this as 'metaphorical complexity'.

There is, however, a *third* strand. This was not highlighted as such during the workshop, but has become much clearer to me in the intervening period.

3. Taking seriously the idea that the world IS a complex place (in the strict meaning of the term). How are we to act when the world is always changing and unfolding in ways which are unknowable even within our understanding of the laws of physics? The contribution of Joseph Pelrine on 'On understanding software agility—from a social complexity point of view' (Pelrine, 2011) falls into this category, as does the present chapter. This is *not* metaphorical, in my view.

Acting In A Complex World—Molecules And Meanings

In seeking to grapple with a complex world, we might start by looking at two fundamentally different types of interaction—the physical and the semantic. This distinction has been memorably made by social psychologist Rom Harré as 'molecules and meanings'.

As far as anyone knows, we are all subject to the laws of physics. People can interact physically, as in one boxer striking another with gory consequences. Likewise, an oak tree grows from an acorn using comprehensible, if amazing, mechanisms. These are both molecular interactions.

Humans and other creatures also have various levels of semantic interaction—including our language. This has a physical/molecular component—vibrations in the air or marks on a page. However, the meaning of such an interaction cannot be found by merely examining the physical components—there have to be two parties who have, by some presumably social process, learned to (mis) understand each other.

It is important not to confuse these two levels. To attempt to understand a social phenomenon at a physical level would be a muddle—like trying to find what makes traffic stop at a red light by taking the light off to the lab and examining it.

This chapter is an attempt to grapple with language and conversation as an emergent phenomenon, and to see what practical conclusions might be drawn. It seeks to connect these somewhat philosophical issues with a well-established yet not widely known field of conversational practice. 'Solution-focused' work in the form of therapy, coaching, OD and many other fields is practised around the world. Its development, stemming from the Brief Family Therapy Center in Milwaukee in the mid-1980s, has been very practitioner-led. One aspect of this school is to avoid theorizing about generalities—and therefore most practitioners are not explicitly aware of the connections with complexity and emergence.

This chapter will set out:

- Challenges to conventional thinking from complexity—'lessons from Life', connecting Conway's famous Life cellular automaton format with everyday thinking and practice;

- How conventional psychological thinking fails these challenges;
- A brief description of Solution-focused practice, as applied in the field of organizational coaching and change;
- How conversations can be treated as emergent, and what this looks like in practice (with connections to Wittgenstein, Stacey and Cilliers), and;
- How the 'interactional view' of SF practice intersects with constructionist thinking and complexity in the form of a paradigm of 'narrative emergence'

The Challenge Of Complexity—Lessons From Life

Introducing ideas about complexity is always a challenge, and I have found great utility in illustrating these with Conway's Life. This might be thought of as 'a very simple complex system'—the rules of the cellular automaton can be understood in seconds, and yet the ramifications of those rules are counter-intuitive and curious. For the purposes of this chapter, I will not give a detailed description of either the general characteristics of a complex system (see Cilliers 1998 for at least one good version of this), or the way in which Conway's Life works (see Poundstone, 1987, for one illuminating account).

Richardson *et al.* (2001) give four key properties of complex systems, which give a handy way to start thinking about our lessons from Life.

Diversity of behaviors: Complex systems display a rich diversity of qualitatively different operating regimes. This is the result of the non-linear nature of the interactions. These behaviors cannot be predicted by examination of the interactions themselves—they are said to emerge. In Life terms, this means that simply knowing the interactions between the cells—even completely—does not allow us to extrapolate upwards to the behavior of groups of cells other than in the most trivial cases. However great the understanding of ONE cell, the future behavior of the whole system is elusive—even if all the cells are the same!

Lesson from Life: Even total understanding of one agent/unit/person is not at all the same as knowing where the interactions can take the whole system.

Large sensitivity to initial conditions and tiny changes: Since the 1960s, mathematicians have become aware that the emerging behavior of complex systems depends on arbitrarily small differences in the initial conditions, or known starting points, of the system. This means that tiny differences introduced, either at the start or as perturbations, will introduce unknowable effects over time. These effects may be large, small or zero. And since the original conditions are not knowable with total precision, the 'status quo' future without perturbations cannot be known accurately either. The idea of designing interventions to achieve a known result—for example the propagation of best practice across an organization—is therefore a misleading distraction in a complex system.

Lesson from Life: Tiny changes and differences can matter a great deal. Applying the 'same' intervention cannot be relied upon to produce the same result—as indeed it cannot be precisely the 'same' intervention.

Robust self-organization: Auyang (1999) points out that alongside this extraordinary sensitivity to tiny perturbations comes a broad robustness to large disturbances. While the future of a complex system cannot be known with certainty, it is usually recognizable as being the same 'sort' of system as before. This is down to self-organization—the agents interact to adapt to the disturbance and find some way of continuing their interaction. As large numbers of agents are involved, such systems are capable of exploring very large 'possibility spaces' and allowing the viable futures to emerge.

Lesson from Life: The future is unknowable, but recognizable. There is no chance of the dots in Life all turning into pieces of cheese.

Incompressibility: Complex systems are incompressible (see for example Cilliers, 1998), so that it is impossible to have a complete account of the system which is less complex than the system itself without losing some of its aspects. And, as tiny changes can perturb the systems in unknowably large ways, no description (either mathematical or linguistic) can offer a complete look at the system. This is 'probably the single most important aspect of complex systems when considering the development of any analytical methodology, or epistemology, for coping with such systems' (Richardson *et al.*, 2007: 27).

Lesson from Life: Any model or description of a complex system must, at some level, be inadequate, and the inadequacy itself is not easily knowable. Alternative ways to work with complex systems should be sought.

Taking these four properties together, it can be said that complex systems display emergence; they develop over time in unknowable yet robust ways, and show properties which cannot be predicted, or even modelled, from knowledge of the individual agents and their interactions. Indeed, Conway's 'Game of Life' shows remarkably that even complete knowledge of the position of each agent, and complete knowledge of all the interactional rules, fails to lead to the ability to predict the system behavior in advance. The system is mathematically 'NP hard' (non-deterministic polynomial-time hard) (see for example Barrow 1998 for a non-technical account).

Broadly speaking, such systems have more possible futures than it is possible to assess mathematically, and so the actual future state of the system cannot be determined in advance. This is very distinct from quantum uncertainty as defined by Heisenberg in his famous Uncertainty Principle. Complex uncertainty is not a result of an inability to measure the state of the system accurately (the position and momentum of an electron, in Heisenberg's case), but of the lack of a computational approach to sort through the vast number of possible futures of the system to decide which will pertain.

The concept of emergence is not new, of course, and has roots and threads from the ancient philosophy of Aristotle and John Stuart Mill (1843) amongst many others. However, the broad conclusions about ways to deal with complex, incompressible and emergent systems seem to have gone unnoticed by both the general academic community and even by generations of 'complexity researchers' (mainly mathematicians) investigating the 'properties' of complex systems within the safe boundaries of their computers. There are of course exceptions to this, see for example Clayton and Davies (2006). I hope to lay out an overall view of the consequences of incompressibility and emergence for the investigation of social systems here.

Are Conversations Emergent? Are All Conversations Emergent?

We can examine conversation through George Rzevski's criteria for complexity (as given in his chapter herein).

- Interaction of agents—clearly present in a conversation, by definition (taking a person as an 'agent');
- Autonomy of the agents (but not too much)—each party has some freedom of choice in the next conversational turn, within the structure of grammar in use;
- Far from equilibrium—this is an interesting point—what would an equilibrium conversation look like? Perhaps stuck in a rut?;
- Non-linearity—conversations have a non-linear aspect to them, they do not always evolve steadily;
- Self-organization—a conversation self-organizes, within a grammatical structure (which might be seen as providing the container for emergence), and;
- Coevolution—the conversation (and the participants) coevolve as the conversation progresses—each turn depending on the turns preceding it and therefore the context at the time.

It may even be that emergence is the most 'honest' description of the development and evolution of families, organizations and so on. Conversation and discourse, apart from a few exceptional cases such as ritual, are surely emergent—in terms of one utterance being a response to the totality of previous utterances rather than being governed by some overarching structure.

Let us examine a simple conversation and see how it relates to emergence. This is a very simple conversation between two people.

Turn 1 "Hello Eileen!" (has lots of possible grammatical responses, from which the other participant picks one).

Turn 2 "Good afternoon Mark, how are you?" (which has lots of possible responses from which the first participant picks one).

Turn 3 "Fine thanks. Have you seen Kurt lately?" (which has lots of possible responses, etc.).

Notice that each turn is taken in the context of the previous turns, so a whole set of possible futures is present at each step, some of which are cut off by the choice of the next turn. However, new possibilities also enter at each turn.

'Conversation' here includes all the non-verbal gestures, nodding, etc. Conversations which generate new meaning are certainly emergent. These can be very everyday conversations—indeed, the potential for emergence is always present, even if not realized. To quote from Ralph Stacey (2007):

> *the thematic patterning of conversation is iterated over time as both repetition and potential transformation at the same time. However, this potential need not always be realized... Change can only emerge in fluid forms of conversation. However, it is important to understand that fluid conversation is not some pure form of polar opposition to repetition.* (Stacey, 2007: 283-284)

This being the case, it is not necessary to use such special methods as Open Space, 'deep' discussion, etc to work with complexity (though such methods may be useful—and having many conversations going at once in a group may provide useful synergy). Conversation is already emergent.

The next step is to take this idea seriously and see what it means. Let us first examine how some common conversational therapeutic and coaching practices fail to deal with the challenges of complexity.

Psychology, Therapy And Coaching Conversations Through The Lens Of 'Complicated'

There is a great deal of writing and experience about people and change in the fields of psychology and psychotherapy. The latter dates from the time of Sigmund Freud in the early 20th century, and it would be surprising if this featured a complexity-friendly framework. It doesn't. (We will see later, however, how Freud's contemporary Wittgenstein comes much closer to the mark.)

Conventional psychotherapeutic work has a focus on:

- Looking beneath the symptoms and the presenting problem to the 'real underlying problem;
- Seeking the cause of this underlying problem—often in childhood relationships and trauma;
- Talking through the emotional impact of this cause—and thereby releasing the pressure which, if ignored, may 'burst out' elsewhere;
- Diagnosing the particular condition and carrying out appropriate treatment, and;

- Coming to terms with these factors over an extended period and gaining insight into them, allowing life to carry on in a better way.

This seems to me to be an example of applying thinking about 'complicated' systems—the cause of the problem must be found, requiring a highly skilled (and expensive) expert and a great deal of time. Only if the cause is dealt with can something better happen. The way of talking about emotions, as hydraulic factors which press here and burst there, is instantly understandable in the mechanics of Freud's day. Equally, the idea of diagnosis, categorizing the condition as necessary information for knowing what to do about it, results from a mechanical paradigm. This is a huge simplification of a complex field into a couple of paragraphs and there are exceptions (such as Jung's work on active imagination). However, the mechanical paradigm lives on strongly today in the latest developments to the The Diagnostic and Statistical Manual of Mental Disorders (relating symptoms, disorders and cures) and positive psychology.

A conclusion to be drawn here is to be very aware of the difference between stability-focused language—the pretence that things are mechanical, causal, inevitable—with progress-focused language—where smalls signs of useful change (possible and actual) are the focus.

An Alternative: The Interactional View

Since the 1950s an alternative view has been put forward by therapists and systems thinkers. Starting in 1952, Gregory Bateson's research project sought to look at mental illness not as something which the patient 'had' (like the measles), but rather the results of interactions with those around them—in chief their families. Initial work in the 1960s connected with the then-current system dynamics traditions. The Mental Research Institute, Palo Alto, developed these approaches into a brief therapy method (see for example Watzlawick, Weakland & Fisch, 1974). This approach was further developed by Steve de Shazer and Insoo Kim Berg of the Brief Family Therapy Center, Milwaukee during the 1980s and 1990s, moving towards an even more fluid and emergent form of practice, 'solution-focused' therapy, which I wish to highlight here. This tradition has elements in common with social construction theory (see for example Weick, 1995 or Gergen, 1999)—however, it predates or is at least concurrent with that development, and is for some reason rarely cited by social construction theorists.

Solution-focused brief therapy was initially developed by a group of therapists and academics associated with the Brief Family Therapy Center (Milwaukee, WI, USA) in the 1980s and 1990s. It emerged within a practical clinical context concerned with fostering effective and efficient change in the lives of a socially diverse population of clients (Miller, 1997). The precursor to the approach was a variant of strategic therapy (de Shazer, 1982) and, later, it evolved into contemporary solution-focused brief therapy (de Shazer, 1988, 1991). The history of solution-focused brief therapy is a process of moving from a systemic perspective to a discursive orientation emphasizing how problems and solutions are organized within clients' use of language. These changes were accompanied by evolving descriptions of observable interactional processes in therapy (Miller, 1997, 2001).

A central principle of solution-focused practice is:

Change is happening all the time—the simple way to change is to find useful change and amplify it.

The discursive emphases of solution-focused brief therapy became clear with the ascension of a Wittgensteinian (1958) perspective as the dominant descriptive language of the early solution-focused brief therapists (de Shazer, 1988, 1991). Wittgenstein's view of the significance of language-in-use, with meaning being continually defined and redefined, draws attention away from fixed concepts of meaning. Wittgenstein is, in this author's reading, the originator of meaning as an endlessly-emerging local property—well ahead of his time, and of the later social constructionists and post-modernists.

De Shazer draws upon Wittgenstein in developing an interpretive framework for giving language to practice (Mattingly & Flemming, 1994), that is, for seeing and talking about the otherwise unnoticed aspects of their interactions with clients. The therapists' interest in discourse went beyond the questions and other 'techniques' that are often associated with solution-focused brief therapy. They also included the 'philosophy' of solution-focused brief therapy, which consists of the assumptions and concerns that organize solution-focused brief therapy interactions. These are now in use in organizational change settings around the world, with a thriving international community of practitioners who have gathered at the SolWorld conferences and summer university events since 2002, and more recently a professional body, SFCT, with its own journal, InterAction (ISSN 1868-8063).

The SF approach has been shown to be effective in therapy in several randomized controlled trials and there are over eighty peer-reviewed studies in various contexts (see Macdonald, 2007). Interestingly, the success of the approach does not seem to depend on the diagnosis of the client, nor on their class or culture. This is practical evidence that the diagnose-and-categorise notion of therapy is flawed or at least unnecessary.

Since SF practice was developed in the 1990s, further connections have appeared to other fields. It is becoming a popular approach in the world of coaching and facilitation (see for example Berg & Szabo, 2005). In addition there is continuing success in practical organizational settings (see for example McKergow & Clarke, 2007).

Six Simple Principles

The art of being wise is the art of knowing what to overlook.

—William James

I present the Solution Focused model here for several reasons. Firstly, it is almost unknown to complexity practitioners. Secondly, it is very well researched across many fields. Thirdly and most importantly, it offers a rare direct example of complexity in action. By taking the world and its conversations as active and emergent, huge and lasting change can be produced with very modest effort. This is in stark contrast to conventional Freudian psychotherapy, which takes many years to learn and even more (we are told) to be effective.

In their book *The Solutions Focus*, the present author and Paul Jackson defined SF practice in terms not only of simplicity but also in six SIMPLE principles. These principles are a guide to how to stay simple, by knowing what to focus on and what, in the words of William James, to overlook. Bearing in mind the difficulties of trying to clearly define a field where definitions are treated sceptically, we present them here. These principles will prepare you for the nature of SF practice in the cases that follow. The principles are:

- Solutions—not problems;
- Inbetween—not individual;
- Make use of what's there—not what isn't;
- Possibilities from past, present and future;
- Language—clear, not complicated, and;
- Every case is different—beware ill-fitting theory

Solutions—Not Problems

In a field calling itself Solutions Focus this may seem obvious. However, this principle is often misunderstood. The 'solution' here is not something to do next (though that will emerge) but what is wanted. This is made clear in our Albert Model (named after Albert Einstein and his famous quotation 'Things should be made as simple as possible, but no simpler').

At the heart of the SF approach is this distinction between *narratives* relating to the problem—i.e., what's wrong—and the solution—i.e., what's wanted. Most approaches to change seek to discover what to do next by examining the problem and seeking to address is. This works well for broken motor cars and washing machines, but less well for people and organizations. The insight of SF is that these two narratives are not related. Take five people with the same 'problem', ask them about signs that things are improving and you get five different answers. There is much more relevant information in the details of what's wanted than in the story of what's wrong.

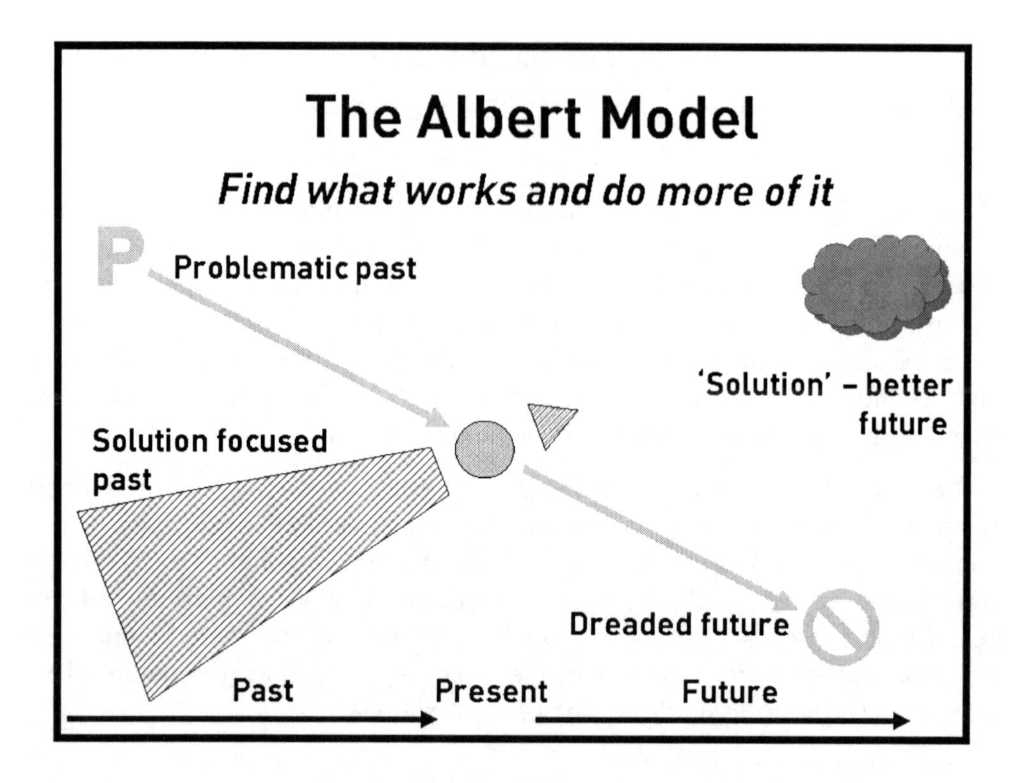

Figure 1 *The Albert Model*

At the same time, there is a (sometimes thin and ignored) narrative of elements or enablers of the 'solution' which have happened in the past, and may well be continuing to happen in the present. SF practitioners start to connect up these elements and look for small steps to build on what is working. The small steps, often taken as some kind of experiment, start to move things forwards and give even more information about what helps in this case. Then what works can be used and built on with even more confidence and so on.

Bear in mind that statements of 'what's wanted' and so on are formed as part of emergent conversations, are subject to change over time—they are emergent themselves. This is therefore not a 'goal' in the usual sense. A goal is something we set today to be achieved at some date in the future—and we are then judged by whether we meet it. Taking the more fluid complex lens, a goal may well make sense today and then make different sense tomorrow. This is not normal business thinking—goals must be achieved! It is the formation of these narratives through complex self-referential processes that makes this such a relevant field for the present context. The field of Appreciative Inquiry (see for example Cooperrider & Whitney, 2005) has arrived at somewhat similar conclusions, though via a different route.

Looking through this lens of emergent conversations, the idea of goals and targets seems to become problematic

In the cases that follow you will find many examples and methods of this shift from looking at the problem axis to the solution axis. It can take skill and patience. However, the advantage of leaving the problem and its diagnosis behind are considerable, both in terms of motivation and time.

Inbetween—Not Individual

This is the most conceptual of the six SIMPLE principles. A key element of SF practice is the Interactional View, originally deriving from the work of Gregory Bateson's research project into communications which developed into family therapy and then brief therapy in the 1960s. Following the development of the simpler SFBT, Steve de Shazer drew connections with the philosophy of Ludwig Wittgenstein, whose later work forms a basis for post-structural linguistics. We are continuing to develop these connections today, including connections with discursive psychology and the science of emergence.

Conventional wisdom has it that individual behavior is driven from within—by values, beliefs, motivations and many more. The 'inbetween' view is that trying to explain behavior by reference to internal states is misleading. In practice people develop their faculties interacting with others, and so learn to speak and act during conversations. Even when they are thinking 'alone', they are doing so in the context of their prior interactional experience.

The practical upshot of this is to greatly enhance simplicity. Just don't use 'mentalistic' language about beliefs, values and so on. Instead, keep it simple by using interactional language relating to observable signs and activity. During the cases you will find many references to 'signs that things are better' or 'what would be the first sign your colleagues noticed'. These are examples of questions to prompt interactional language.

Why is this a good idea? Because the findings from research into the effectiveness of SF practice shows results that are as good or better than other approaches, in less time. It therefore seems the case that such mentalistic language is not necessary to produce change. Rather, it seems to be connected with mechanical/causal thinking. Indeed, it may be serving to muddle the issue and delay change. This is another example of the benefit of taking simplicity seriously.

Make Use Of What's There—Not What Isn't

SF is about finding what works in a particular given context. Therefore, anything that seems to be connected with things working, going better or even going less badly that normal may be worth utilizing. This includes personal strengths, positive qualities, skills and cooperation as well as examples of the 'solution' happening already.

This may seem obvious. However, it's quite different to the commonly used idea of 'gap analysis', where the *differences* between the present and the desired future are the most important aspects. It's not unusual to find managers who quickly nod towards what's already going well before getting onto the allegedly more relevant topic of what is not. This is not to say that organizations should not be aware of these differences. It's simply that the building blocks for change are much more likely to come from what's already going well. You will find the practical application of this principle in every case in the book.

Possibilities From Past, Present And Future

A sense of future possibility, of hope and optimism and positive expectations, is of course vital in any change effort. Where there is no hope, things are very difficult. The idea of possibilities in the past, however, is perhaps a more novel idea. Some think that the past has happened, is gone, is immutable. We take a different view. The past is a useful source of stories of success, examples of resources in action, even tales of adversity overcome.

The key to bringing these stories into the present, where they can influence the future, lies in identifying and drawing out their connections with the 'solution'—what's wanted. It is much easier in practice to identify relevant examples if you have some idea what you are looking for.

The idea of possibilities is also different from certainties. As soon as you become certain of what's going to happen, or indeed what happened in the past, it's much harder to start to identify the alternative possibilities. In some cases people use explanations of the past to create a firm (and unhelpful) view of the future. In SF practice we tend to steer away from explanations towards experiments and a sense of 'seeking to find what works'—particularly if the explanation is about why the problem happens and is difficult to change.

Language—Clear, Not Complicated

One of the working principles of SF is the idea that "$5 words may be worth more than $5000 words". $5000 words are usually long, complicated and abstract, and sound very impressive. They are often used in explanation, theorizing and academia. $5 words, on the other hand, are short, concrete and detailed. They may not sound very impressive. They are, however, usually more connected to action, movement and everyday life.

In the cases described here, you will find many examples of people gently seeking to simplify the language being used. The focus will be on the language used by the people involved in going about their everyday business, rather than on abstract and generalizing language imposed by observers. This turns out to a surprisingly practical idea—it not only shows respect to those involved by taking their words seriously, it is also a good way to avoid getting into dubious explanations and generalizations.

Every Case Is Different—Avoid Ill-Fitting Theory

Most approaches to organizational change seek to simplify the messy everyday world by establishing some classification scheme. If the issue is of type X, then action A is advised. This is also true of science, medicine and many other fields, and has served humanity very well.

SF works in a different way. Simply approach each case with a completely clean and open mind, seek to identify what the people want and what works. If you 'know' what should work, then you will undoubtedly try to find it and will therefore miss important clues. Not knowing what should be done is the easiest way to see clearly. In the words of Zen writer Shunryu Suzuki, 'In the mind of the expert, there are few possibilities. In the mind of the beginner there are many.' So it may be more helpful to have the mind of the beginner than the mind of the expert.

Of course there are many occasions when theories may be helpful. In these cases, we urge people to go on using them. When pre-conceived ideas are failing to fit your experience, however, you may find it helpful to let go of the idea and examine your experience again with fresh eyes.

Conversations And Organizations As Emergent

Richardson *et al.* (2001) examine the challenges posed by the 'inbetween' nature of complexity. Stacey (2007) discusses the difference between conversations which remain trapped in reproducing the same patterns of talk, and those which may foster the emergence of new knowledge.

We saw earlier that no special kind of language is necessary for the emergence of new knowledge. Rather, the fluidity, spontaneity, and 'good enough holding of anxiety' of the conversational interaction itself (Stacey, 2007: 285) seem important in encouraging potentially transforming themes. The deliberate utilization of the clients' everyday language in solution-focused brief therapy and the conversational focus on small details show how this emergence of knowledge is encouraged without every referring to it in such terms.

Solution-focused brief therapists' interest in initiating change through small and often provisional steps is also consistent with complexity theorists' appreciation of modest depictions of social realities (Cilliers, 2005). Modest proposals (made in the knowledge of an incomplete and uncertain context) acknowledge and accept the limitations of our understandings of and control over the complex processes of life—in stark contrast to the booming certainty of many organizational change frameworks. They also orient to the unpredictability and transformative potential that are built into complex interactional processes. McKergow and Korman (2009) explain that the interactional emergence of modest proposals in solution-focused brief therapy occurs 'inbetween'—neither inside (stemming from inner drivers, urges, motivations or other mentalistic explanatory mechanisms) nor outside (determined by external systems, narratives or other mechanisms). There are also links to the 'second cognitive revolution' of discursive psychology (see for example Harré, 1995; Potter & Wetherall, 1987).

Narrative Emergence

The phrase 'narrative emergence' has been coined by Miller and McKergow (2010) as a way to summarize what we have been discussing here. We use this term to call attention to several interrelated aspects of solution-focused brief therapy as a distinctive form of discursive therapy and complex systems. The concerns include recognition that while the future is unknowable, it is an ever present possibility in the present. We continuously create and discover the future by engaging in self-organizing activities (particularly social interactions) that are, at least partly, improvised, and potentially transformative. Thus, the narratives emergent in our everyday lives are always under construction. They exist in our ongoing 'work' to make sense of and manage the exigencies of life.

What helps these narratives to emerge is somewhat beyond the scope of this chapter. Some initial guidelines might include:

- Taking whatever the client says very seriously—assume it all makes good sense, in some version of reality at least;
- 'Stay at the surface'—don't listen for causes, motivations and inner drivers, but focus on responding to whatever is said to amplify connections with what's wanted and what's working;
- Get the detail—once a useful strand is established, asking 'what else?' again and again can help add much further substance, and;
- Give space—gently encourage thoughtful reflection

When narratives emerge which connect possible better futures or better lives with 'evidence' from the past (stories, experiences, observations from others, etc), lasting and meaningful change results with astounding frequency. Research from SF therapy shows a success rate of 60%-80% within a few sessions, across a wide range of diagnoses, age ranges and social class.

Conclusions: Lessons For Complexity Practitioners

I propose that SF practice has important lessons for complexity practitioners, This form of practice has important parallels with the work of Wittgenstein (1958), the first post-structural thinker and linguist, and shows one way of thinking about, and participating in, conversations leading to useful and sustainable change. In so doing it challenges many tacit assumptions about therapy (McKergow & Korman, 2009).

I would like to point to three aspects of language and complex systems here, which may be relevant in applications of complexity in the real world:

1. The difference between *stability-focused language* (explanations, reasons, causes) and *progress focused language* (signs of change in both future and past). Experience in SF practice shows that focus on the latter seems to be a good way to get away from causal thinking and into something more useful;
2. The difference between *internal mentalistic language* (beliefs, thoughts, feelings) and *interactional language* (descriptions of ongoing everyday life from a variety of perspectives). The former echo Freud in terms of hydraulic causes of behavior, the latter are about rooting change in our everyday lived experience;
3. The difference between *goals* (targets set in the future) and *small steps* (things to do quickly to move in a desired and defined direction. This may be the single most important conclusion for progress in a complex world. If things are ever-shifting, then what's clear today may not be so clear tomorrow. Focus on small steps, to be done quickly in order to find out more about what works (rather than in the expectation that everything will be instantly resolved) seems to give people a sense of control and purpose in a confusing and muddled world. There is a clear parallel here with the 'agile' school discussed by Pelrine (2011).

References

Auyang, S.Y. (1999). *Foundations of Complex-System Theories in Economics, Evolutionary Biology, and Statistical Physics*, ISBN 9780521778268.

Barrow, J.D. (1998). *Impossibility: The Limits of Science and the Science of Limits*, ISBN 9780195130829.

Berg, I.K. and Szabo, P. (2005). *Brief Coaching for Lasting Solutions*, ISBN 9780393704723.

Cilliers, P. (1998). *Complexity and Postmodernism: Understanding Complex Systems*, ISBN 9780415152877.

Cilliers, P. (2005). "Complexity, deconstruction and relativism," *Theory, Culture & Society*, ISSN 0263-2764, 22: 255-267.

Clayton, P. and Davies, P. (2006). *The Re-Emergence of Emergence*, ISBN 9780199287147.

Cooperrider, D. and Whitney, D. (2005). *Appreciative Inquiry: A Positive Revolution in Change*, ISBN 9781576753569.

De Shazer, S., Dolan, Y., Korman, H., Trepper, T., McCollum, E. and Berg, I.K. (2007). *More Than Miracles: The State of the Art of Solution Focused Brief Therapy*, ISBN 9780789033987.

Glaschenko, A., Ivaschenko, A., Rzevski, G., Skobelev, P. (2009). "Multi-agent real time scheduling system for taxi companies," Proc. of 8th Int. Conf. on Autonomous Agents and Multiagent Systems (AAMAS 2009), Decker, Sichman, Sierra, and Castelfranchi (eds.), May, 10-15, Budapest, Hungary, http://www.rzevski.net/09%20Scheduler%20for%20Taxis.pdf.

Gergen, K. (1999). *An Invitation to Social Construction*, ISBN 9780803983779.

Harré, R. (1995). "Discursve psychology," in J.A. Smith, R. Harré and L. van Langenhove (eds.), *Rethinking Psychology*, ISBN 9780803977358.

Jackson, P.Z. and McKergow, M. (2007). *The Solutions Focus: Making Coaching & Change S.I.M.P.L.E.*, ISBN 9781904838067.

Macdonald, A. (2007). *Solution-Focused Therapy: Theory, Research and Practice*, ISBN 9781412931175.

Mattingly, C. and Flemming, M.H. (1994). *Clinical Reasoning: Forms of Inquiry in a Therapeutic Practice*, ISBN 9780803659377.

McKergow. M. and Clarke, J. (2007). *Solutions Focus Working: 80 Real-Life Lessons for Successful Organizational Change*, ISBN 9780954974947.

McKergow, M. and Korman, H. (2009). "Inbetween-neither inside nor outside: The radical simplicity of solution-focused brief therapy," *Journal of Systemic Therapies*, ISSN 1195-4396, 28(2): 39-49.

Miller, G. (1997). *Becoming Miracle Workers: Language and Meaning in Brief Therapy*, ISBN 9780202305714.

Miller, G. (2001). "Changing the subject: Self-construction in brief therapy," in J.F. Gubrium and J.A. Holstein (eds.), *Institutional Selves: Troubled Identities in a Postmodern World*, ISBN 9780195129274, pp. 64-83.

Miller, G. and McKergow, M. (2010). "From Wittgenstein, complexity, and narrative emergence: Discourse and solution-focused brief therapy," in A. Lock and T. Strong (eds.), *Discursive Perspectives in Therapeutic Practice*, to be published.

Pelrine, J. (2011). "On understanding software agility: A social complexity point of view," in A. Tait and K. Richardson (eds.), *Moving Forward with Complexity: Proceedings of the International Workshop on Complex Systems Thinking and Real World Applications*, pp. 63-74.

Potter, J. and Wetherall, M. (1987). *Discourse and Social Psychology: Beyond Attitudes and Behavior*, ISBN 9780803980563.

Poundstone, W. (1984). *The Recursive Universe: Cosmic Complexity and the Limits of Scientific Knowledge*, ISBN 9780688039752.

Richardson, K. A., Cilliers, P. and Lissack, M. (2001). "Complexity science: A 'gray' science for the 'stuff in between,'" *Emergence*, ISSN 1521-3250, 3: 6-18.

Stacey, R. (2007). *Strategic Management and Organizational Dynamics: The Challenge of Complexity*, ISBN 9780273708117.

de Shazer, S. (1982). *Patterns of Brief Family Therapy: An Ecosystemic Approach*, ISBN 9780898620382.

de Shazer, S. (1988). *Clues: Investigating Solutions in Brief Therapy*, ISBN 9780393700541.

de Shazer, S. (1991). *Putting Difference to Work*, ISBN 9780393334708.

Watzlawick, P., Weakland, J. and Fisch, R. (1974). *Change: Problem Formation and Problem Resolution*, ISBN 9780393011043.

Weick, K. (1995). *Sensemaking in Organizations*, ISBN 9780803971776.

Wittgenstein, L. (1958). *Philosophical Investigations*, ISBN 9780631146704.

Mark McKergow is codirector of **sf**work—The Centre for Solutions Focus at Work. He is an international consultant, speaker and author. He is known for his energetic and accessible style, cutting edge ideas presented with his inimitable blend of scientific rigour and performance pizzazz. Mark's PhD in the physics of self-organization and ordering of metal hydrides has subsequently transferred to a career in management consulting and development.

Mark is a global pioneer applying Solutions Focus ideas to organizational and personal change. He has written and edited three books and dozens of articles; his book 'The Solutions Focus: Making Coaching and Change SIMPLE' (coauthored with Paul Z Jackson) was declared one of the year's top 30 business books in the USA, and is now in nine languages. He is now developing new metaphors for leadership, notably around leader as host.

Mark is a Board Member of SFCT, a new worldwide association for SF consultants, coaches and leaders (www.asfct.org), and edits the academic journal InterAction. He is also a member of the Association for Management Education and Development (AMED) and the Association for MBAs (AMBA). Find out more about Mark at www.sfwork.com.

17. Bells That Still Can Ring: Systems Thinking In Practice

Martin Reynolds
Communication and Systems Department, The Open University, ENG

Complexity science has generated significant insight regarding the interrelatedness of factors and actors constituting our real world and emergent effects from such interrelationships. But the translation of such rich insight towards developing appropriate tools for improving real world situations of change and uncertainty provides a further significant challenge. Systems thinking in practice is a heuristic framework based upon ideas of boundary critique for guiding the use and development of tools from different traditions in managing complex realities. By reference to five systems approaches, each embodying more than 30 years of experiential use, three interrelated features of the framework are drawn out—contexts of systemic change, practitioners as change agents, and tools as systems constructs that can themselves change through adaptation. The 'bells that still can ring' refer to tools associated with the Systems tradition which have demonstrable capacity to change and adapt by continual iteration with changing context of use and different practitioners using them. It is in the practice of using such tools whilst being aware of significant 'cracks' associated with traps in managing complex realities that enables systems thinking in practice to evolve. Complexity tools as examples of systems thinking can inadvertently invite traps of reductionism within *contexts*, dogmatism amongst *practitioners*, and fetishism of our *tools* as conceptual constructs associated with ultimately undeliverable promises towards achieving holism and pluralism. The heuristic provides a guiding framework on monitoring the development of tools from different traditions for improving complex realities and avoiding such traps.

Introduction

Ring the bells that still can ring
Forget your perfect offering
There is a crack in everything
That's how the light gets in

(Cohen, 1993)

I first read this verse written by Canadian songwriter and poet Leonard Cohen in a small book called *Inside Out* (Huston, 2007: 8) written by an experienced systems practitioner, Tracy Huston. The book has the sub-title *Stories and Methods for Generating Collective Will to Create the Future We Want*. It is about planning for the future and in particular generating meaningful organizational change drawing upon our existing untapped *internal* human resources rather than continually seeking *external* answers. With the insightful revelations of complexity science on the nature of reality, the book prompts thinking about how such insight may interface with practitioners wanting to effect meaningful change. What is it about complex situations of systemic change that may inform our use and development of tools—as internal systems constructs—for improving such situations? The question signals a point of departure between two closely affiliated traditions—Complexity thinking and Systems thinking. How might the more explicitly purposeful orientation of systems thinking provide offerings to the development of complexity tools?

The words in Cohen's verse capture for me something of the importance behind five systems approaches chosen by myself with a team of academics for a publication *Systems Approaches to Managing Change: A Practical Guide* (Reynolds & Holwell, 2010). The approaches chosen are *System Dynamics* (SD), the *Viable Systems Model* (VSM), *Strategic Options Development Analysis with Cognitive Mapping* (SODA), *Soft Systems Methodology* (SSM), and *Critical Systems Heuristics* (CSH). They were chosen from a vast array of systems approaches because they shared three qualities: (i) adaptability to variable complex situations, (ii) an appeal to different practitioner communities, and (iii) an underpinning constructivist mindset enabling different conceptual use of the tools dependent on different complex situations of use and different practitioner communities using them.

The five approaches are drawn from three philosophical traditions underpinning systems thinking: SD and VSM from the cybernetics tradition (primarily dealing with feedback interrelationships and interdependencies) which also informs much of complexity thinking; SSM and SODA from the interpretivist tradition (primarily focusing on multiple perspectives); and CSH from the tradition of American philosophical pragmatism and European critical social theory (primarily addressing issues of ethics and politics). The five approaches cover the fundamental concepts of systems thinking and the essential elements of the different perspectives across the main theoretical strands of systems thinking in practice.

Box 1 *Five Systems Approaches Described*
(adapted from Reynolds and Howell, 2010: 18-21)

1. System dynamics was founded in the late 1950s by Jay W. Forrester of the MIT Sloan School of Management with the establishment of the MIT System Dynamics Group (Forrester 1961). It is an approach to understanding the behavior of complex systems over time. It deals with internal feedback loops and time delays that affect the behavior of the entire system. What makes using system dynamics different from other approaches to studying complex systems is the use of feedback loops and stocks and flows in displaying nonlinearity.

2. Viable systems model was developed by the cybernetician Stafford Beer (1974a, 1972). It describes the necessary and sufficient conditions for the viability of systems in order to keep an independent existence. To do so it needs to be organized in such a way as to meet the demands of surviving in a changing environment. The principles of recursion (whereby a viable system itself can be seen as either part of a wider system or constitutive of many viable systems), and Ashby's law of requisite variety (capacity to exhibit diversity) are central to VSM.

3. Strategic options development and analysis (with cognitive mapping) is an approach developed in the 1970s by Colin Eden—an Operational Researcher—for revealing and actively shaping the mental models, or belief systems (mind maps, cognitive models) that people use to perceive, contextualize, simplify, and make sense of otherwise complex problems. Whilst being appropriate at the individual level in clarifying thoughts around a particular issue, work on SODA encompasses much wider contexts of strategic thinking; neatly encapsulated through the software acronym JOURNEY making (JOintly Understanding Reflecting and NEgotiating strategY). SODA is the methodology used for cultivating organizational change through attention to and valuing of individual perspectives in a concerted manner (Ackermann *et al.*, 2005)

4. Soft Systems Methodology is an approach to process modelling developed in England by academics lead by Peter Checkland at the University of Lancaster Systems Department through a program of action research (Checkland ,1981; Checkland & Scholes, 1990). The primary use of SSM is in the analysis of complex situations where there are divergent views about the definition of the problem—'soft problems' (e.g., how to improve health services delivery, or what to do about homelessness amongst young people?). In such situations even the actual problem to be addressed may not be easy to agree. The soft systems approach uses the notion of a 'system' as an interrogative learning device that will enable debate amongst concerned parties

5. Critical systems heuristics represents the first systematic attempt at providing both a philosophical foundation and a practical framework for critical systems thinking (Ulrich 1983). CSH is a framework for reflective practice based on practical philosophy and systems thinking, developed originally by Werner Ulrich. The basic idea of CSH is to support boundary critique—a systematic effort of handling boundary judgments critically. Boundary judgments determine which empirical observations and value considerations count as relevant and which others are left out or are considered less important. Because they condition both 'facts' and 'values', boundary judgments play an essential role when it comes to assessing the meaning and merits of a claim.

The five approaches collectively provide significant tools in systems thinking. Each approach embodies at least 30 years of experiential use—30 years of road-testing. They are the 'bells that still can ring', with a pedigree of time and experience. Over that period, challenges have exposed new offerings, insightful ways on how better to use these approaches in the light of invaluable experience. They are presented not as new tools to replace old tools, but as composite tool sets that have been adapted to deal with different contexts and changing circumstances. Box 1 provides a brief description of each approach.

The purpose here is not to present these five approaches as tools for supporting complex decision-making activities. That task is fulfilled through the Reynolds and Howell publication in which each approach is updated by originators and/or experienced practitioners using a common simple template (what it is, how it's done, and why it is important).

The 'tool' being offered in this paper is a framework for guiding the use of tools more generally in supporting decision making for improving complex realities. I call it a framework for systems thinking in practice—the namesake of the UK-based Open University (OU) postgraduate programme to which the book contributes as a Reader on the core module *Thinking Strategically: systems tools for managing change* (Open University, 2010).

In what follows, I'll briefly explain the heuristic framework for systems thinking in practice relating to three constituent activities and associated entities. Features of each entity and the framework as a whole are then examined.

What Is Systems Thinking In Practice?

Systems thinking in terms of promoting a more holistic perspective is not new. In emphasizing the integral relationship between human and non-human nature, systems thinking can be traced back to spiritual traditions of Hinduism (e.g., through ancient texts like the Upanishads and Bhagavad Gita), Buddhism (oral traditions of the Dhama), Taoism (basis of acupuncture and holistic medicine), sufi-Islam (in translations of the *Kashf al-Mahjûb* of Hujwiri, and the *Risâla* of Qushayri), and ancient Greek philosophy (particularly Hericles and Aristotle). It is also prevalent through the oral traditions of many indigenous tribal spiritual traditions which have existed for tens of thousands of years.

Since the early 20th century when Bertalanffy published his first papers on systems theory, there has grown a multitude of systems approaches, many of which, like the traditions in complexity sciences, deal with the essential ontological challenge in developing more holistic understandings of reality. Systems thinking in the later part of the 20th century took on more the epistemological challenge of dealing with multiple perspectives on reality, and the ethical and political challenge of confronting power relations associated with different realities. These constitute what have been called the 'soft' and 'critical' systems traditions respectively (Jackson, 2000).

Bringing these different traditions together and appreciating systems as conceptual constructs, systems thinking *in practice* involves three interrelated activities: (i) stepping back from messy situations of complexity, change, and uncertainty, and *understanding* key interrelationships and perspectives on the situation; (ii) *practically* engaging with multiple often contrasting perspectives amongst stakeholders involved with and affected by the situation, and (iii) *responsibly* directing joined-up thinking with action to bring about morally justifiable improvements. Elsewhere I have described these activities as being supported by three (sub)frameworks respectively—framework for understanding (fwU), framework for practice (fwP), and a framework for responsibility (fwR)—constituting an overall critical systems framework (Reynolds, 2008a). The activities can be represented as a triadic interplay of making judgements associated with boundary critique (Ulrich, 2000). This involves continual revising of boundary judgements (systems thinking) with judgements of 'fact' (observing) and value judgements (evaluating) (see Figure 1).

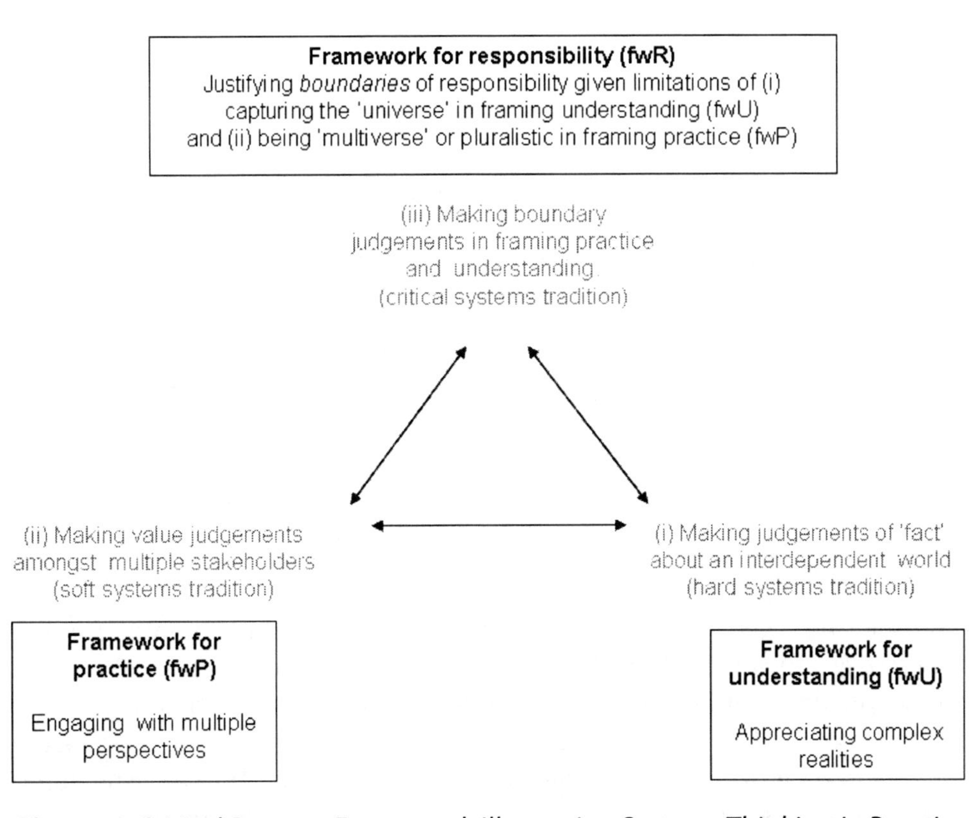

Figure 1 *Critical Systems Framework Illustrating Systems Thinking In Practice Activities (adapted from Reynolds, 2008a: 386)*

In developing this into a broader heuristic for systems thinking in practice, three complementary entities can be added: (1) real-world *contexts* of change and uncertainty, (2) people or *practitioners* involved with making change, and (3) the ideas and concepts—including systems—as *tools* for effecting change. Figure

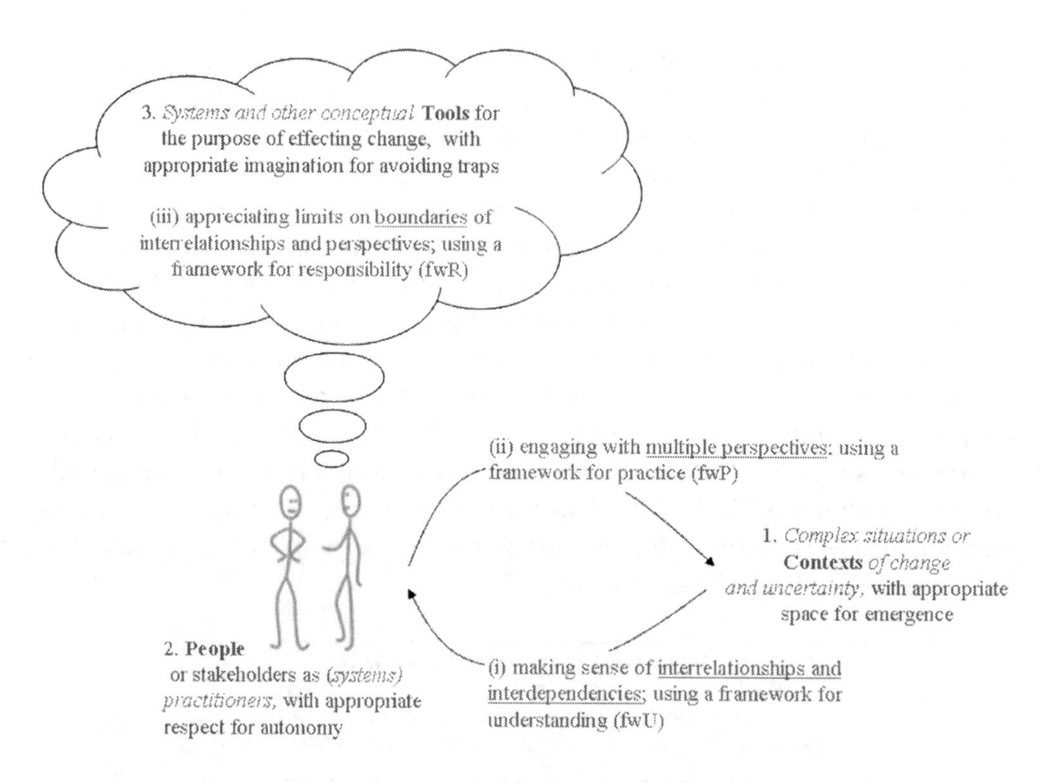

Figure 2 *Heuristic Framework Of Systems Thinking In Practice*

2 illustrates the constituent activities and entities of the heuristic framework for systems thinking in practice.

The heuristic provides a benchmark for gauging effective action in managing change. Whilst some tools may have a particular focus on one of the three activities and associated entities, the effectiveness of use in supporting decision making can be gauged according to how well all three entities are dealt with. The five systems approaches described in Box 1 each qualify with a particular strength in one of the three activities, but they each also have a track record of adaptation towards addressing all three domains.

At present the heuristic tool resides in an OU distance learning module *Thinking Strategically: Systems Tools for Managing Change* (Open University, 2010). The module provides a framework for students to engage with each of the five systems approaches but using their own chosen area of professional practice and developing their own particular life experiences and skills. The heuristic framework is used in the module to gauge the competence of a practitioner in systems thinking in practice (see Appendix). Here though I want to illustrate how the tool might be used to appreciate the value of complexity thinking and complexity sciences in general, the potential in conversation between complexity and systems thinking in practice, and the limitations of claims made by complexity and systems traditions. The following section examines each entity and associated activity of the heuristic tool in turn.

What Matters In Systems Thinking In Practice

The rich history and current variety of systems tools prompt questions as to how they may relate to each other and what emphasis is given to the context of use, the users or practitioners, or the actual tools being used. The tools used in systems thinking in practice need not be exclusively recognized as being derived from what some recognise as the Systems tradition. They may derive from traditions ranging from Complexity science to Performance arts such as puppetry. Any tools that attempt to (i) make sense of a context of complex realities whilst (ii) enabling amongst practitioners different perspectives on such realities to flourish in order to (iii) enable systemic improvement in the real world, qualify to be exemplars of systems thinking in practice. What matters in systems thinking in practice are the expression of these three entities, but also the interplay amongst all three entities and associated activities, and the resultant dynamics of change that emerge.

Context Matters

It's confusing, but we have a right to be confused. Perhaps even a need. The trick is to enjoy it: to savor complexity and resist the easy answers; to let diversity flower into creativity. (Mary Catherine Bateson, 2004, "Afterword: To Wander and Wonder", p. 410)

You cannot step twice into the same river. (Heraclitus of Ephesus, c.6th Century BC)

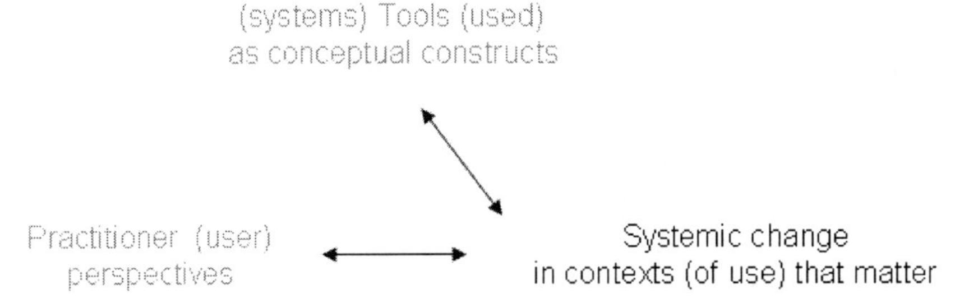

Figure 3 *Contexts In Relation To Practitioners And Tools*

These depictions of context capture important notions of systemic change implicit in complexity thinking. As an ontological point of departure from Complexity science, in Systems thinking complexity resides not in systems but the situations to which systems speak. To use a well-worn though significant adage amongst systems practitioners, a system is merely a *map* of a situation or territory, not to be confused with the actual *territory*. Arguably the prime purpose of systems thinking is to make simple the complex—that is, to bound the unbounded ontological complex realities variously referred to by systems thinkers as messes (Russell Ackoff), the swamp (Donald Schön), or wicked problems (Horst Rittel). Drawing on the signal-to-noise ratio used in the language of communications engineering (cf. Richardson, 2010: 2), systems as conceptual constructs provide purposeful ways for generating meaningful 'signals' or patterns of abstracted data sets from the cacophonous 'noise' of reality.

Real world complexities represent something that exists outside of any one conceptualization of context. Whereas complexity science has made valuable and intriguing strides in capturing real world complexity, particularly through computational modelling (see both Richardson and Rzevski contributions in this volume), systems thinking prompts a more cautionary note against achieving some ultimate understanding of reality.

One significant reference system for depicting contexts generated in complexity sciences is offered by the Cynefin framework (Kurtz & Snowden, 2003). The framework demarcates between simple, complicated, complex and chaotic contexts. A situation is regarded as complex when there is no evident central controlling element but there are strong connections between elements. A complicated situation also has strong connections between elements but is regarded as more knowable and predictable then complex situations because of there being a central controlling element. Simple situations have a very strong controlling element with little interconnections, and chaotic situations have no controlling element and little interconnections between elements.

A similar reference system used by systems practitioners for appreciating the importance of context is total systems intervention (TSI) (Flood & Jackson, 1991a). TSI draws upon a system of system methodologies (SOSM) typology to classify situations into six different types. SOSM maps 'appropriate' systems approaches that might be suitable for implementing change in different situations (Jackson, 1990). Table 1 illustrates the SOSM classification along two dimensions—level of complexity (simple or complex), and the degree of shared purpose amongst stakeholders (unitary, pluralist, coercive) along with some typical alignment of systems approaches (including my own guess of where users might likely align VSM and SODA according to the traditions from which they have arisen) relating to perceived realities.

		Stakeholder perspectives		
		Unitary 'hard' systems based on mechanistic metaphor	*Pluralist* 'soft' systems based on organismic metaphor	*Coercive* 'critical' systems based on prison metaphor
Systems view of problem situations	*Simple*	Simple unitary: e.g. systems engineering	Simple pluralist: e.g. Strategic assumption surfacing and testing	Simple coercive: e.g., critical systems heuristics
	Complex	Complex unitary: e.g., system dynamics, viable systems model	Complex pluralist: e.g. soft systems methodology, strategic options development and analysis	Complex coercive: (non available!)

Table 1 *System Of Systems Methodologies (adapted from Jackson, 2000: 359)*

A significant difficulty with TSI as with Cynefin is in assuming from the outset that a problem situation can somehow be easily identified as constituting one of the 'problem situation' or 'context' types. Both Cynefin and TSI make assumptions about knowing whether a situation can be type-cast from the outset. The Cynefin framework does acknowledge possibilities of differing perspectives on the situation amongst stakeholders involved in the situation but there appears little acknowledgement that the expert practitioner doing the typecasting may also have a skewed perspective. Contexts that are initially regarded through expert intervention as unitary or simple may often turn out to be very complex. A further difficulty with TSI is in the 'fixing' or pigeon-holing of particular systems approaches as being only suitable for specific types of situation. Such pigeon-holing, dependent on the root paradigms of intellectual tradition to which they are perceived to belong, denies the potential for systems approaches to themselves adapt and develop through different contexts of use. It also detracts from

Box 2 *Systems Approaches In Different Contexts*

SD started with work on servo-mechanism devices to control radar in the late 1950s, before moving into the field of industrial relations, and later modelling global resource depletion (Meadows *et al.* 1972, 1992). System dynamics provided the crux of the systems approach to organizational development advocated as the Fifth Discipline in the celebrated book of the same title (Senge, 1990).

VSM has been used in contexts ranging from promoting efficiency in small organizations and communities to large corporate bodies (Hoverstadt, 2008). It has been deployed for organizing national economies (Beer, 1974b) and guiding major environmental policy at national and regional levels (cf. Espejo, 1990; Espinosa & Harden, 2008)

SODA has been used in various contexts ranging from dealing with individual decision making to small and large enterprises (Eden & Ackermann, 1998; Ackermann *et al.*, 2005). It has also been recommended for dealing with wider international interorganizational relationships (Robinson *et al.*, 2000) and environmental planning (Open University, 2006)

SSM has been used to examine organizational change in large multinational corporations, with several hundred participants in the study; it can be used by an individual to manage, for example, personal recovery from substance abuse; it has been used to research Inuit fishing in Labrador; by an NGO volunteer to engage local people in mine clearance after war in the Middle East; by members of a women's forum in Japan to make sense of the impacts of societal changes on their lives; by consultants working on information systems planning in the NHS. (Checkland & Poulter, 2006)

CSH has been deployed in environmental management (Reynolds, 1998; Midgley & Reynolds, 2004; Ulrich & Reynolds, 2010) health care planning, city and regional planning, and energy and transportation planning (Ulrich, 1987: 276), enhancing prison service support (Flood & Jackson, 1991b) promoting an alternative lens for corporate responsibility (Reynolds, 2008a) and informing international development initiatives (McIntyre-Mills, 2004; Reynolds, 2008b).

opinions on where different systems approaches 'fit' based upon actual experiences of using the approach.

Box 2 gives a few examples of some different contexts in which our five systems approaches have been used through the passage of time.

As a general rule, any context of use is best regarded as being complex from the outset. From a Systems perspective (described below) this means a context with variable perspectives on what needs to be done. Systemic failure in intervention can often be attributed to the sidelining of such perspectives. Another rule is that tools—whether derived from Systems or other traditions such as Complexity sciences—are adaptable to different contexts of use depending on different users' experiences.

Practitioner Matters

A systems approach begins when first you see the world through the eyes of another. (Churchman, 1968: 231)

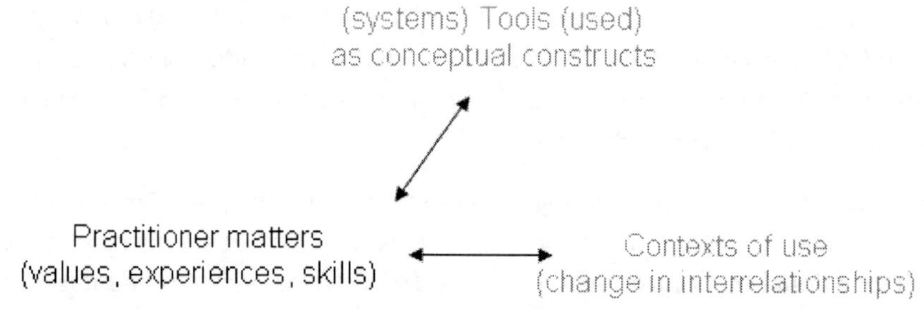

Figure 4 *Practitioner Matters In Relation To Contexts And Tools*

For West Churchman systems thinking not only requires 'building a bigger picture' of the situation—for which he described a process of unfolding increasingly more variables from the context of use—but also appreciating other conceptual constructs or perspectives on the situation. The transition speaks of two worlds; one, the holistic ontological real-world 'universe' of interdependent elements, encapsulating complex interrelationships; another, an epistemological socially constructed world of 'multi-verse' (cf. Maturana & Poerksen, 2004: 38), encapsulating differing constructs on reality.

Whereas Complexity science regards complexity as residing in the ontological features of dynamic *interrelationships* in the situation (e.g., see Rzevski's criteria of complexity in this compilation—interactions, non-linearity, emergence, disequilibria etc.), complexity as understood in the Systems thinking in practice tradition presented here resides on the layering of differing *perspectives* on the dynamic interrelationships in the situation.

People are pivotal to the systems thinking in practice heuristic framework. As described in the anthology, *Systems Thinkers* (Ramage & Shipp, 2009), our own individual experiences, competencies, skills, as well as weaknesses, shape how we engage with any particular context of change. Part of my own academic and practical experience for example is situated in a context of life-science education and international development. The conceptual tools derived from these disciplines, along with my experiences in using them, have helped me value different tools differently, and to reshape and mould them accordingly in different contexts of use.

In shifting emphases from explicating tools according to contexts of use, towards practitioner experiences and influences as users of tools, Ison and Mait-

Box 3 *Systems Approaches Derived From Different Experiences*

Jay Forrester was influenced by his practical problem-solving upbringing in a rural agriculture and cattle ranching context before starting work on servo-mechanism devices to control radar in the late 1950s. He then significantly moved into the field of, first, industrial relations, and later modelling global resource depletion through invitations to construct 'world systems models' on sustainability from the influential Club of Rome.

Stafford Beer's ideas arose out of a synthesis of Eastern and Western thought. His time in India as a very young man and subsequently his interest in Eastern thought, particularly Indian cultural traditions, was a very important factor in the emergence of the VSM. Beer's own engagement with practicing VSM was most notably carried out under invitation to Allende's Chile in the early 1970s before the military coup. Beer effectively founded management cybernetics - now known as Organizational Cybernetics.

Colin Eden worked as an Operational Researcher in the engineering industry followed by a period as a management consultant specializing in small business problems before focussing interest on University teaching and research. Eden's ideas developed originally from an interest in Kelly's psychological work on 'personal construct theory' (Kelly, 1955). He received support from institutions ranging from British Telecom to the Northern Ireland Office to help processes of making strategy with cognitive mapping and the practice of 'action research'.

Peter Checkland was interested in applying Systems Engineering (SE) to management issues. After 15 years as a manager in the synthetic fibre industry Checkland joined Lancaster University in what became a thirty-year programme of action research in organizations. The 'failure' of the early work on SE highlighted a different direction that ultimately yielded SSM as an approach to tackling the multi-faceted problems which managers face.

Werner Ulrich, like Checkland, was influenced by the ethical systems tradition promoted through the works of the American systems philosopher C. West Churchman. Ulrich's own work in developing CSH as a means of supporting social planning was also influenced by traditions of American philosophical pragmatism and European critical social theory.

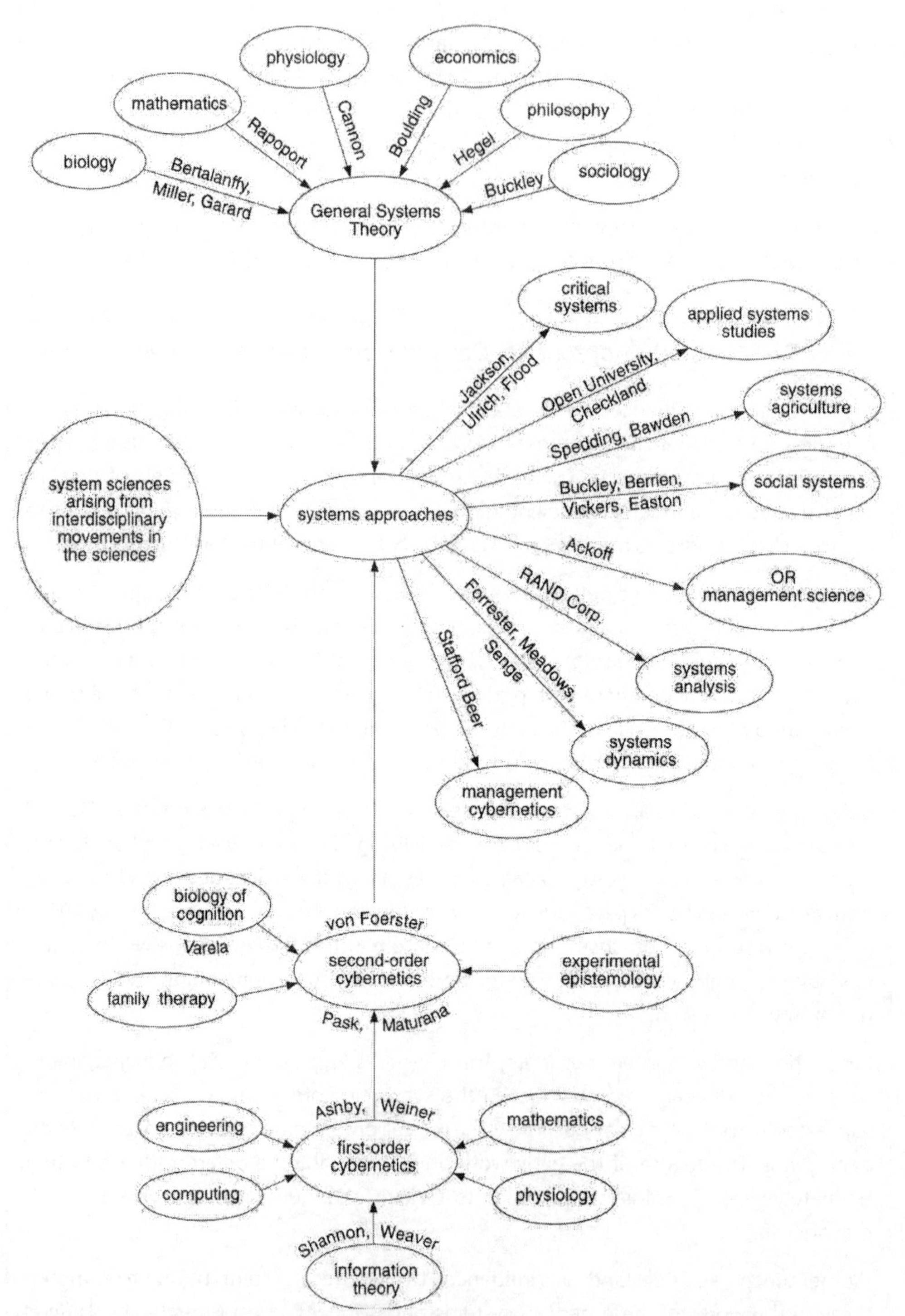

Figure 5 *An Influence Diagram Of Different Systems Traditions And Some Key Practitioners Which Have Shaped Contemporary Systems Practice (Maiteny & Ison, 2000)*

eny captured some of the wider influences and cross-fertilization that generates innovative development of systems approaches. The aim was to broaden the understanding and practice of spheres of influence both with respect to other tools and approaches outside the traditional systems toolbox, and to other contexts in which such approaches were evident (Figure 5).

Box 3 provides brief biographical sketches of original authors to the five systems approaches.

The importance of simple 'conversation' and language is key to improving situations of change (see McKergow, Dalmau & Tideman, and Michiotis in this volume). The tendency for practitioners belonging to a community of practice to become self-referential and insular applies as much to some systems practitioners and complexity thinkers as other communities. The message here is to avoid seeking some methodological purism in testing out any one approach, but rather to explore its validity and adaptation in conjunction with other approaches familiar to the user. A particular feature of the five systems approaches referred to in this paper is the sought-after working relationships and dialogues with communities of practice outside of the practitioner community associated with any one approach. Such interactions enhance not only the practice but also serve to strengthen the theoretical underpinning associated with each approach. They also serve to protect against the risk of becoming trapped in 'group-think' that can sometimes be a feature of long-standing communities. The increasing dialogue between complexity theorists and policy makers provides a healthy check against such insularity (cf. Boulton, 2010)

Systems matter

To a man with a hammer, everything looks like a nail. (Mark Twain)

True scientific simplicity is never reductive; it is always a relevant simplicity that is a creative achievement... The true grandeur of science is not power but the demanding quest for relevance... How to learn? How to pay attention? How to acquire new habits of thinking? How to concentrate or explore other kinds of experiences? Those are questions that matter. (Michael Lissack interpreting Stengers, 2004: 92)

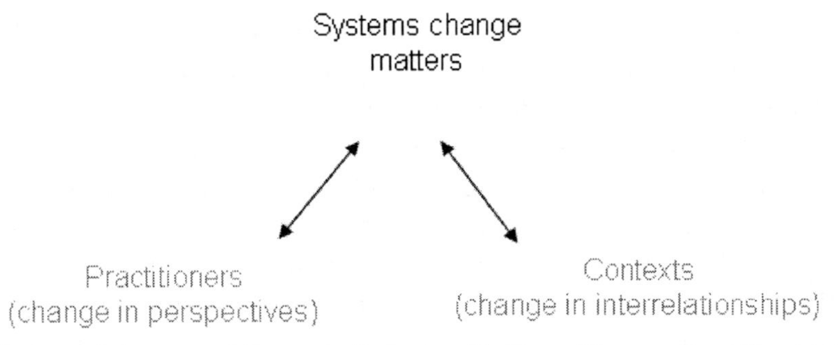

Figure 6 *Systems Matter In Relation To Practitioners And Contexts*

Our tools and models, including cognitive frameworks as systems tools, can often be sub-consciously overpowering in determining how we approach issues. Similarly, adopting 'new' systems runs the risk of elevating the notion of 'a system' to a fetish status; the panacea for resolving a crisis. Here I use the term 'system' generically, referring both to an ontological construct representing a real world situation (i.e., a 'complex system'), and as an epistemological tool for inquiry into reality.

The trap of systems maintenance, or being obsessive with the tools we construct, lies in reifying and privileging the 'system'—whether it's old or new—as though it has some usefulness, existence and worth outside of the user and some status beyond its value in a context of use. Perhaps the most pervasive example of an implicit system resilient to change is a conventional model of management hierarchically imposed and indiscriminately applied across all parts of an organization, regarding stakeholders as objects rather than subjects. It is a pervasive way of thinking that continues to hold a widespread grip on management practice. There are many other 'systems' that similarly entrap our understanding and practice. A generic descriptor for these is 'business as usual' (BAU)—frameworks for understanding and practice that stifle innovation. For example, think of the annual cycles of organizational planning, target setting, budgeting, the development of performance indicators and performance related pay incentives etc. BAU models maintain existing 'systems' principally because of a fear for change. But the fear is not evenly distributed amongst all stakeholders. Some fear change more than others simply because the system works in a partial manner. The system works for some and not for others.

All systems are partial. They are necessarily partial—or selective—in the dual sense of (i) representing only a section rather than the whole of the total universe of interrelationships in any context that matters, and (ii) serving some stakeholder parties including practitioners—or interests—better than others (cf. Ulrich, 2002: 41). As described elsewhere, no proposal, no decision, no action, no methodology, no approach, no tool, no system can get a total grip on the situation nor get it right for everyone (Reynolds, 2008a). In using and designing systems we need to keep an eye on changing contexts and practitioner matters.

With an eye on appreciating matters of context and changing complex realities, there is an imperative to continually ask questions of 'systems'; to appreciate them as *judgements* of fact rather than *matters* of fact. For example, when confronted with situations that appear simple or even complicated, we should be wary of disregarding unvoiced perspectives that may reveal complexity or even chaos. Or when confronted with arguments of an iniquitous 'economic system' generating continual social and ecological impoverishment, or an 'education system' that systematically continues to marginalize particular sectors of our community, as systems practitioners we have a responsibility to create space for, and help support the framing of, better systems, rather than perpetuating the myth that these are some God-given realities that we need to simply live with.

With an eye on appreciating practitioner matters, the risk of systems obsession is akin to moralism. Humberto Maturana makes a relevant point distinguishing between being moralistic and ethical. Moralists, he suggests, "lack awareness of their own responsibility. People acting as moralists do not see their fellow human beings because they are completely occupied by the upholding of rules and imperatives; that is a particular systems design. They know with certainty what to be done and how everybody else has to behave" (Maturana & Poerksen, 2004: 207). Being ethical, in contrast requires giving legitimacy to people, and particularly those who may disagree with the rules (see also MacGillivray in this volume).

Systems matter not because they provide some ultimate reification of complex realities, but rather because they provide a cross-disciplinary and transdisciplinary literacy for identifying traps in conventional thinking. SD and VSM arising from a holistic cybernetics tradition are particularly good for countering traps of *reductionism* (focusing on parts rather than the whole). SODA and SSM coming from a pluralist interpretivist tradition counter tendencies towards *dogmatism* (privileging one particular perspective). CSH addresses similar aspirations but also takes a step back in reminding practitioners of the need to be both modest in making holistic claims—seeing the *whole* big picture (trap of *holism*)—and cautious about claims of being multi-verse—taking in *all* perspectives equitably (trap of *pluralism*). Figure 7 illustrates these traps through a causal loop diagram.

Whilst the five systems approaches have traditional strengths in springing particular traps, all five have evolved with a capacity for dealing with each trap. This evolution and ongoing development of each approach has been a function of the variety of contexts of use and the different users through processes of iteration.

Iteration Matters

> *Thinking through the triangle means to consider each of its corners in the light of the other two. For example, what new facts become relevant if we expand the boundaries of the reference system or modify our value judgments? How do our valuations look if we consider new facts that refer to a modified reference system? In what way may our reference system fail to do justice to the perspective of different stakeholder groups? Any claim that does not reflect on the underpinning 'triangle' of boundary judgments, judgments of facts, and value judgments, risks claiming too much, by not disclosing its built-in selectivity.* (Ulrich, 2002: 42)

Systems thinking in practice might be seen as an expression of Ulrich's eternal triangle of boundary critique described above. All five approaches assume that

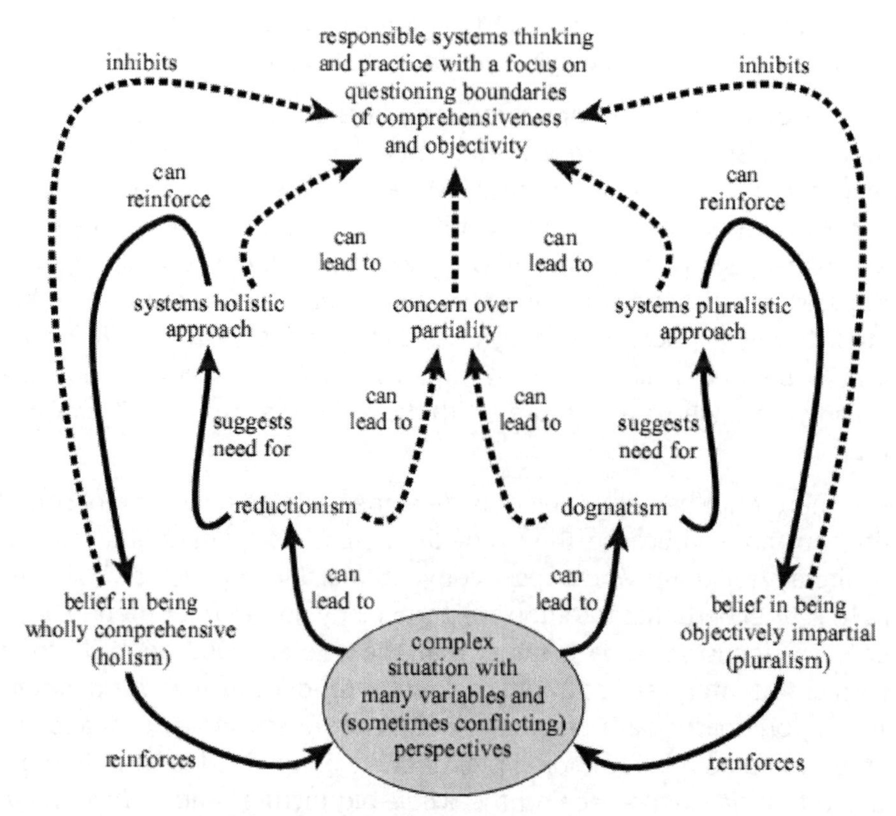

Figure 7 *Reflective Systems Thinking In Practice (From original by Martin Reynolds in Open University, 2010: 133)*

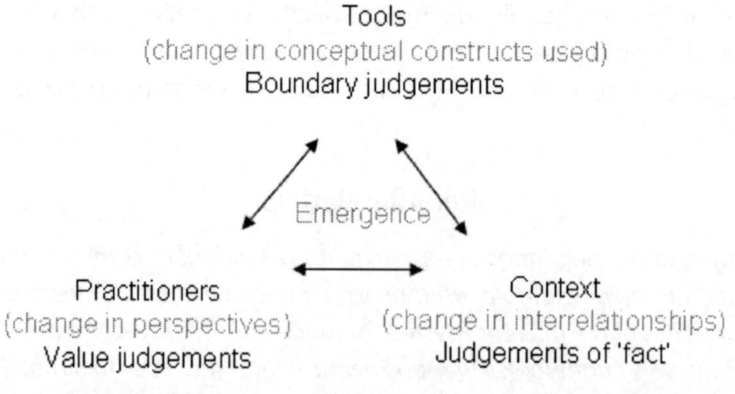

Figure 8 *Iteration In Relation To Change In Context, Practitioners And Tools*

complex realities in the form of messes cannot be resolved or improved upon without engaging in a process that is cyclic and iterative; recognizing for example that changes in perspective reveal new insights that require continual revisiting of earlier judgements of the context, and refinement of the conceptual tools with which we use to frame our understanding of, and practice in, contexts of change and uncertainty. There is an ongoing dynamic between ideas (tools), the situation (context), and the practitioner for any given approach.

This iterative quality is akin to the artistic practice of improvisation; a quality associated with the works of Donald Schön:

> *...Schön, who stresses reflection in the midst of action ... frequently used jazz as an image of reflection-in-action: the process of improvisation in the moment based on a response to the situation (what other musicians are playing, the audience's response etc), to the established rhythm and melody of the piece, and also on one's own abilities and enthusiasms.* (Ramage & Shipp, 2009: 292)

The notion of improvisation is helpful in grasping some of the nuances of systems thinking in practice as a literacy—a form of communication amongst scientists, systems practitioners and others, in dealing with complex realities.

Summary

> *Systems literacy is not just about measurement. The learning journey up the ladder of complexity—from quarks, to atoms, to molecules, to organisms, to ecosystems—will be made using judgment as much as instruments. Simulations about key scientific ideas and visualizations of complex knowledge can attract attention—but the best learning takes place when groups of people interact physically and perceptually with scientific knowledge, and with each other, in a critical spirit. The point of systems literacy is to enable collaborative action, to develop a shared vision of where we want to be.* (Thackara, 2005)

The name—systems thinking in practice—suggests an important interplay between understanding and practice; systems thinking continually being informed, moulded and (re)shaped by ongoing practice. It provides a tool for nurturing the type of systems literacy alluded to by John Thackara. It is this interplay between conceptual tools and practice that resonates for me the idea of 'bells that still can ring'.

An approach or tool of any kind of itself cannot guarantee success in managing and improving complex realities. Whilst we may discuss different tools in their abstract sense, any claims towards their value in improving situations are dependent on the context of use and the practitioner's purpose, skill and insights. The systems thinking in practice heuristic presented here supports three intentions behind complexity thinking:

1. Making sense of, or simplifying (in *understanding*), relationships between different entities associated with a complex situation. The prime intention is not to get some thorough comprehensive knowledge of situations, but rather to acquire a better appreciation of wider dynamics—to counter *reductionism*—in order to improve the situation.

2. Surfacing and engaging (through *practice*) contrasting perspectives associated with complex situations. The prime intention here is not to embrace all perspectives on a predetermined problem so as to solve the problem, but rather to allow for possibilities in reshaping a problem-situation—to counter *dogmatism*—for improved possibilities of resolution.
3. Exploring and reconciling (with *responsibility*) ethical issues, power relations, and boundary issues associated with inevitable partial understandings of a situation and partiality amongst different stakeholders. The aim here is not to provide yet another ready-to-hand matrix to offer clients through a consultancy, but rather to gently disrupt and unsettle patterns of thinking—including claims of *holism* and *pluralism*—thereby prompting innovative critical thinking in practice.

Figures 1 and 2 provide graphic illustrations of the heuristic. The Appendix provides a tangible expression of an assessment device guiding the development of skills in systems thinking in practice. This paper argues that an effective systems approach to managing real world complex situations embodies all three aspects of systems thinking in practice—the entities and associated processes. It presents a departure from Total Systems Intervention (TSI) where systems approaches tend rather to be regarded as fixed externalized artefacts suitable for different well-defined contexts.

None of the five approaches has developed out of use in restricted and controlled contexts of *either* low *or* high levels of complicatedness. Neither has any one of them evolved as a consequence of being applied only to situations with *either* presumed stakeholder agreement on purpose, *or* courteous disagreement amongst stakeholders, *or* stakeholder coercion. The paper is not a celebration of abstract 'methodologies', but of theoretically robust approaches that have a genuine pedigree for supporting real world decision-making activities. Taxonomic devices like TSI and particularly Cynefin can provide important spaces for exploring the nuanced dynamics of complex realities, but they are maps of the territory and should not be confused with the actual territory. As with any systems construct the value lies in their respective adaptability towards changing contexts of use and changing users.

The 'bells that still can ring' refer to all tools, whether traditional systems based or belonging to other traditions of professional practices. Behind the concepts and techniques constituting the tools, there are the bell ringers. Not only do they have the experiences that they bring to bear on the skill of bell-ringing but also the uniquely human qualities that determine how and why they do it as they do, and that allow them to enjoy and appreciate it. But you and I are also bell ringers, perhaps as novices wanting to cross between professional traditions and academic boundaries. We should never expect 'perfect offerings' in systems thinking in practice. But the offerings should allow for the joy in further cultivating our approaches—critically appreciating the cracks—in order to meet the challenge of improving complex situations of change and uncertainty.

Acknowledgements

The author wishes to thank participants from the *1st International Workshop on Complexity and Real World Applications Using the Tools and Concepts from the Complexity Sciences to Support Real World Decision-making Activities* (Southampton, England, July 21-23, 2010) for their feedback on an earlier draft. Appreciation is also given to colleagues working on the postgraduate Systems Thinking in Practice programme at The Open University for nurturing ideas formulated in this paper.

References

Ackermann, F., Eden, C. and Brown, I. (2005). *The Practice of Making Strategy*, ISBN 9780761952251.

Beer, S. (1972). *Brain of the Firm*, ISBN 9780471948391.

Beer, S. (1974a). *The Heart of the Enterprise*, ISBN 9780471948377.

Beer, S. (1974b). "The real threat to 'all we hold most dear'," in S. Beer (ed.), *Designing Freedom*, ISBN 9780471951650, pp. 1-34.

Boulton, J. (2010). "Complexity theory and implications for policy development," *Emergence: Complexity & Organization*, ISSN 1532-7000, 12(2): 31-40.

Checkland, P.B. (1978). "The origins and nature of 'hard' systems thinking," *Journal of Applied Systems Analysis*, ISSN 0308-9541, 5: 99-110.

Checkland, P.B. (1981). *Systems Thinking Systems Practice*, ISBN 9780471279112.

Checkland, P.B. and Scholes, J. (1990). *Soft Systems Methodology in Action*, ISBN 9780471927686.

Checkland, P.B. and J. Poulter. (2006). *Learning for Action: A Short Definitive Account of Soft Systems Methodology, and its use Practitioners, Teachers and Students*, ISBN 9780470025543.

Churchman, C.W. (1968). *The Systems Approach*, ISBN 9780440384076.

Espejo, R. (1990). "The Viable Systems Model," *Systems Practice*, ISSN 0894-9859, (special edition on VSM) 3.

Espinosa, A. and Harden, R. (2008). "A complexity approach to sustainability: Stafford Beer revisited," *European Journal of Operational Research*, ISSN 0377-2217,

Flood, R.L. and Jackson, M.C. (1991a). "Total Systems Intervention: A practical face to Critical Systems Thinking," in R.L. Flood and M.C. Jackson (eds.), *Critical Systems Thinking*, ISBN 9780306454516.

Flood, R.L. and Jackson, M.C. (1991b). "Critical Systems Heuristics: Application of an emancipatory approach for police strategy towards carrying offensive weapons," *Systems Practice*, ISSN 0894-9859, 4: 283-302.

Forrester, J.W. (1961). *Industrial Dynamics*, ISBN 9780915299881.

François, C. (ed.) (1997). *International Encyclopaedia of Systems and Cybernetics*, ISBN 9783598116308.

Hoverstadt, P. (2008). *Fractal Organization: Creating Sustainable Organizations with the Viable System Model*, ISBN 9780470060568.

Jackson, M.C. (1982). "The nature of soft systems thinking: The work of Churchman, Ackoff and Checkland," *Journal of Applied Systems Analysis*, ISSN 0308-9541, 9: 17-28.

Jackson, M.C. (1990). "Beyond a System of System Methodologies," *Journal of the Operational Research Society*, ISSN 0160-5682, 41: 657-668.

Jackson, M.C. (2000). *Systems Approaches to Management*, ISBN 9780306465062.

Kurtz, C.F. and Snowden, D. (2003). "New dynamics of strategy: Sense making in a complex complicated world," *IBM Systems Journal*, ISSN 0018-8670, 42: 462-483.

Maiteny, P. and Ison, R. (2000). "Appreciating systems: Critical reflections on the changing nature of systems as a discipline in a systems-learning society," *Systemic Practice and Action Research*, ISSN 1573-9295, 13: 559-586.

Maturana, H. and Poerksen, B. (2004). *From Being to Doing: The Origins of the Biology of Cognition*, ISBN 9781932462159.

McIntyre-Mills, J. (2004). *Critical Systemic Praxis for Social and Environmental Justice: Participatory Policy Design and Governance for a Global Age*, ISBN 9780306480744.

Meadows, D.H., Meadows, D.L., Randers, J. and Behrens, W.H. (1972). *The Limits to Growth: A Report for the Club of Rome's Project on the Predicament of Mankind*, ISBN 9780876639016.

Meadows, D.H., Meadows, D.L. and Randers, J. (1992). *Beyond the Limits of Growth*, ISBN 9780930031626.

Midgely, G. and Reynolds, M. (2004). "Systems/operational research and sustainable development: Towards a new agenda," *Sustainable Development*, ISSN 1571-4780, 12: 56-64.

Open University (2006). Making Environmental Decisions and Learning from Them, ISBN 9780749202668.

Open University. (2010). Thinking Strategically: Systems Tools for Managing Change (Study Guide), ISBN 9781848731325.

Ramage, M. and Shipp, K. (2009). *Systems Thinkers*, ISBN 9781848825246.

Reynolds, M. (1998). "'Unfolding' natural resource information systems: Fieldwork in Botswana," *Systemic Practice and Action Research*, ISSN 1573-9295, 11: 127-152.

Reynolds, M. (2008a). "Getting a grip: A critical systems framework for corporate responsibility," *Systems Research and Behavioral Science*, ISSN 1092-7026, 25:383-395.

Reynolds, M. (2008b). "Reframing expert support for development management," *Journal of International Development*, ISSN 0954-1748, 20: 768-782.

Reynolds, M. and Holwell, S. (eds.) (2010). *Systems Approaches to Managing Change: A Practical Guide*, ISBN 9781848828087.

Richardson, K. (2010). *Thinking about Complexity: Grasping the Continuum through Criticism and Pluralism*, ISBN 9780984216451.

Robinson, D., Hewitt, T. and Harriss, J. (eds.) (2000). *Managing Development: Understanding Inter-Organizational Relationships*, ISBN 9780761964797.

Senge, P. (1990). *The Fifth Discipline*, ISBN 9780385260954.

Stengers, I. (2004) "The challenge of complexity: Unfolding the ethics of science—In memorian Ilya Prigogine," *Emergence: Complexity & Organization*, ISSN 1532-7000, 6(1-2): 92-99.

Thackara, J. (2005). *In the Bubble: Designing for a Complex World*, ISBN 9780262201575.

Ulrich, W. (1983). *Critical Heuristics of Social Planning: A New Approach to Practical Philosophy*, ISBN 9780471953456.

Ulrich, W. (2000). "Reflective practice in the civil society: The contribution of critically systemic thinking," *Reflective Practice*, ISSN 1462-3943, 1: 247-268.

Ulrich, W. (2002). "Boundary critique" in H.G. Daellenbach and R.L. Flood (eds.), *The Informed Student Guide to Management Science*, ISBN 9781861525420, pp. 41f.

Ulrich, W. and Reynolds, M. (2010). "Critical systems heuristics," in M. Reynolds and S. Holwell (eds.), Systems Approaches to Managing Change, ISBN 9781848828087, pp. 243-292.

Martin Reynolds, Ph.D is a Lecturer in Systems Thinking at The Open University, UK. His previous work at the Institute for Development Policy and Management at Manchester University involved developing a critical systems approach to evaluating participatory approaches in natural resource management. He has since been involved with numerous collaborations including work with the Centre for Systems Studies at the University of Hull, the University of Guyana, the University of KwaZulu-Natal, and with internationally based colleagues associated with the American Evaluation Association and the European Evaluation Society. He has produced Open University distance learning material for postgraduate courses on Global Development Management, Environmental Decision Making, and Systems Thinking in Practice. He is one of the founding members of the Open Systems Research Group based at the OU, and provides workshop support for professional development in systems practice and critical systems thinking. Martin has published widely in the fields of systems studies, professional evaluation, international development, and environmental decision making (many publications of which are freely downloadable on Open Research Online http://oro.open.ac.uk/view/person/mdr66.html). In 2009 he was lead editor of, and contributing author to, *The Environmental Responsibility Reader* published by Zed Books. In 2010 he was lead editor of, and contributing author to, *Systems Approaches to Managing Change: A Practical Guide* published by Springer Books.

Appendix: Assessing a Systems Thinking in Practice Practitioner

Adapted from assessment overview for OU students undertaking TU811 module (Open University, 2010)

	Broad systemic characteristics of *being* a systems thinking in practice practitioner	**1. Understanding situations**: *contextualising* interrelationships and interdependencies	**2. Practicing systems design**: *engaging* with multiple , often contrasting perspectives	**3. Reflective practice 1**: *managing* ethical and political boundaries	**Reflective practice 2**: *complementing* with other (non-Systems) traditions
Weighting:	(100%)… *but* whole is more than sum of parts!	c. 28%	c. 28%	c.28%	c.16%
'Distinctive' systems practitioner *c.80% and above*	Can apply a range of key systems tools with ideas from other traditions in a coherent way for supporting strategic thinking.in situations of change and uncertainty. Can vary approach in line with change in context.	Systems used for imaginatively and creatively relating to realities. Identifies reductionism	Appreciates importance of developing perspectives in shaping strategy. Identifies dogmatism.	Reflects on design of methodology - what's at stake and whose stakeholdings are affected. Identifies holism and pluralism.	Open to relevance of other ideas and traditions outside of systems approaches. Can fine-tune strategies based upon such ideas.
Systems practitioner with 'merit' *c. 60-80%*	Has a solid grasp of systems tools covering all three traditions, which can be applied over a wide range of contexts - but without the innovative ability to reflect imaginatively as with a 'distinctive' practitioner.	Systems used for imaginatively grasping dynamics and feedback, but less confidence with implications on variety and autonomy	Appreciates need for facilitating perspectives in making strategy. Has a good grasp of systems tools in mapping perspectives	Reflects need for developing ethical and political awareness in strategic thinking amongst stakeholders but with less confidence in practice	Aware of relationship between Systems and other disciplines and traditions.
'Fairly good' systems practitioner *c.50-60%*	Has gist of some systems tools and has some comptetence in a few approaches, but not wholly confident about being a systems practitioner.	Systems used more as direct (residual) representations of the world – rather than as (relational) learning devices	Appreciates multiple perspectives but mostly on a tokenistic level	Ethics and politics seen largely as an interference with systems design.	Limited appreciation of other traditions relevant to making strategy.
'Barely adequate' systems practitioner *c.35-50%*	Has gist of a few systems tools and demonstrates limited comptetence in a few approaches, but not confident about being a systems practitioner.	Situations regarded only at levels of complicatedness in a world reduced to component parts.	Limited if any appreciation in importance of different perspectives.	Ethics and politics of situtaions and systems ignored	Has little time or inclination to value other ideas.
'Clearly not' a systems practitioner *c. 35% and below*	Has little or no appreciation of the range of tools in a systems approach. Either not understanding or flouts the principles of systems thinking in practice through being egoistical instead of humble, dogmatic rather than open, and cynical rather than being creatively critical .	Complexity is someone else's fault.. Either wilfully ignores the bigger picture or is not able to grasp simple ideas of dynamics and feedback and emergence.	Not responsive to values, beliefs and circumstances outside practitioners own sphere. Either abuses others' perspectives or ignores them.	Unable to reflect on context and practice. Either ignores and/or does not recognise political and ethical issues amongst stakeholders.	Unable to make cross disciplinary connections. Either a tyrannical belief in only one approach and/or closed to being influenced

18. A Consilient Approach: Supporting Leaders To Manage And Sustain Successful Change In Complex, Emergent And Contingent Environments

Brian Lawson
Consilient Change, ENG

This paper proposes an approach to change with leadership teams in the context of an understanding of the dramatic failure rate of most change programmes. It explores how this failure rate can be understood in relation to complexity approaches to management and leadership and their practical application. A Consilient approach to change is then outlined, exploring how this approach can address some of the failures of change programmes as well as supporting a complexity thinking approach to organizations. The application of this approach is then explored with a Local Authority leadership team. In a year they faced: political uncertainty; a 30% reduction in their budget and the transformation of the policy context for service delivery. Evidence is provided of the impact during the first year. The paper concludes by identifying next steps in both the development of the work and the approach. This work was developed and facilitated with Steve Marshall a Director of Meus.

Setting The Context: Managing And Leading Change

In 1995, John Kotter published one of the key works in change management suggesting that only 30% of change programmes are successful and identifying eight key errors made in change programmes of all sizes. Over ten years later Isern and Pung (2006) reporting results from a McKinsey survey of 1,500 business executives reported that still only 30% agreed that they considered their change programmes successful or partially successful. The IBM study of Global Chief Executives in 2008 identified a growing need to manage change along with a growing gap (22%—a gap which had tripled in three years) in their perceived ability to manage change successfully in relation to their expected need for it.

Keller and Aiken (2008) identified that change programmes fail because they fail to change the key things they are trying to transform: employee attitudes and management behavior. They cite Price and Lawson's (2003) holistic approach to influencing employee attitudes and management behavior as providing a rational, yet flawed, basis for a psychology of change management. Keller and Aiken argue that human behavior is not rational:

> *In the same way that the field of economics has been transformed by an improved understanding of how uniquely human, social, cognitive and emotional biases lead to seemingly irrational decisions, so too the practice of change management is in need of a transformation through an improved understanding of the irrational (often unconscious) way in which humans interpret their environment and choose to act* (p. 3).

They then proceed to identify ten inconvenient truths about the irrationality of human behavior which should be taken into account in order to improve the odds of leading successful change.

Higgs and Rowland (2005) conducted a number of research studies into three core questions in relation to leadership and change, relating to concerns that the root cause of many change problems is leadership behavior and an inability to learn from previous experiences. They asked three key questions:

- What approach to change management is likely to be most effective in today's business environment?
- What leadership behaviors tend to be associated with effective change management?
- Are leadership behaviors related to the underlying assumptions within different approaches to change?

Their key finding was that change approaches built on the assumptions of complexity were most successful and those classified as emergent change were found to be the most successful across most contexts. If change is perceived as complex and emergent then Wheatley (2000) argues leaders must be brought to a transformational edge so that they can work differently.

In assessing the leadership factors associated with each change approach Higgs and Rowland found that emergent change emphasizes creating capacity followed by framing change. Creating capacity focuses on creating individual and organizational capabilities, understanding the power and significance of informal networks which promote understanding and communicating and making connections. Framing change creates an overall container for the whole of the change process which includes: establishing starting points for the journey; designing and managing the journey and communicating guiding principles in the organization. Two of their key findings were:

> *In high magnitude change, that which impacts on a large number of people and entails changes to multiple parts of the system, an emergent approach is the most effective. The leadership factor accounting for the highest variant in success in this context was framing change...*

In short term change which needs to be implemented in under 12 months and will impact on a large number of people in the organization, leadership behaviors are critical to success. The set of behaviors encompassed within the factor framing change appear to be those most likely to lead to success (p. 8).

Finally they draw on the work of Meyer & Allen (1990) who identified three types of commitment: affective; continuance and normative. They demonstrated that levels of affective commitment in an organization are positively related to organizational performance. Using this work Higgs and Rowland established a significant relationship between emergent change and affective commitment, leading them to conclude that: "this approach may not only lead to change success but also to individual performance within the change."

The Application Of Complexity Theory To The Management And Leadership Of Change In Organizations

Overview

The most recent IBM survey of Chief Executive Officers (2010) noted the following development:

In our past three global Chief Executive Officer (CEO) studies, CEOs consistently said that coping with change was the most pressing challenge. In 2010, our conversations identified a new primary challenge: complexity. CEOs told us they operate in a world which is substantially more volatile, uncertain and complex (p. 8).

One of their key findings was that: 'Today's complexity is only expected to rise and more than half of CEOs doubt their capacity to manage it—a 30% gap.' The report concludes that:

The effects of rising complexity calls for CEOs and their teams to lead with bold creativity, connect with customers in imaginative ways and design their operations for speed and flexibility to position their organizations for 21st Century success (p. 9).

So what does complexity theory have to offer managers and leaders in dealing with this situation? Smith and Humphries (2004) undertook a critical evaluation of complexity theory as a management tool. They cite Tetenbaum's (1998) assertion that seven trends help explain why complexity theory helps to understand the dynamic context of organizations: technology, globalization, competition, change, speed, complexity and paradox. She concludes:

The new world is full of unintended consequences and counterintuitive outcomes. In such a world, the map to the future cannot be drawn in advance. We cannot know enough to set forth a meaningful vision or to plan productively (p. 24).

Marion and Barnes (2000) contrast complexity theory with the classical scientific management view which assumes linear causality and encourages reductionist approaches to management. They specify three characteristics of complex systems:

- The whole is more than the sum of individual behaviors;
- Complex organizations stimulate outputs that cannot be predicted simply by understanding all of the inputs, and;
- Organizations can create behavior that is neither definitively predictable nor unpredictable-they can exist on the 'edge of chaos' where there is enough order to ensure functionality and also enough chaos to preclude all prediction.

The Edge Of Chaos, Emergence And Change

The 'edge of chaos' can be seen, in organizational terms, as a balance between structure and chaos; (Brown & Eisenhardt, 1997) as a chaordic state which contains elements of order and chaos at the same time (Fitzgerald & Van Eijnatten, 2002). As Hock (2005) describes in his account of the creation of VISA such chaordic states can lead to the transformation of organizations and the creation of new forms. This obviously has implications for the management of change. Lissack (1999) notes that emergent behaviors are typically unanalyzed and may contain at least as much risk as opportunity—"In the study of emergence, complexity science and organization converge".

Smith and Humphries conclude their review of complexity theory as a practical management tool as follows:

> *Complexity theory is therefore best seen as a device for thinking about and for encouraging managers to cultivate and foster the environment that facilitates emergence...The danger facing managers is that applications of complexity thinking become reduced to another simplistic recipe for success* (pp. 103-4).

Leadership And Complexity Thinking

For Dooley (2008) complexity leadership focuses on the dynamics of leadership as it emerges over time in all arenas of an organizational system. Each interchange and every connection provide opportunities for leading, as peers individually and collectively learn and grow and engage in the continual process of organizing. This perspective on leadership may offer new insights into the emergence of innovation, the creation of order and the dynamics of performance in 21st Century networks and organizations.

Marion and Uhl-Bien (2001) argue that complex leaders enable interactions but do not control them, recognizing the importance of interactions, correlation and unpredictability among ensembles or aggregates of individuals. Instead they allow them to emerge through engaging in nonlinear processes.

Lichtenstein *et al.* (2006) situates complexity leadership within the framework of the idea of a complex adaptive system where relationships are primarily understood as interactions between agents rather than being defined hierarchically. Leadership is understood as an emergent event rather than a person-a complexity view suggests a form of distributed leadership and the creation of a collective identity:

> *According to the adaptive leadership perspective, this identity formation occurs over time, as participants together define 'who we are' and what we are doing with our interactions. In this way, the emergence of a social object occurs through the" in-forming" of a joint social identity* (p. 5).

Complexity leadership theory suggests that participants need to be made aware of this dual process of identity creation and projection, in order to take back ownership of their role in the identity formation process:

> *Complexity leadership theory provides a clear and unambiguous pathway for driving responsibility downward, sparking self organization and innovation, and making the organization much more responsive and adaptive at the boundaries. In turn, significant pressure is taken off formal leaders, allowing them to attend more directly to identifying strategic opportunities, developing unique alliances and bridging gaps across the organizational hierarchy* (p. 8).

Complexity Thinking And Local Government

Battram (1996) produced a 'Learning From Complexity Pack' for use by Local Government in the United Kingdom arguing that:

> *Local Government needs new approaches to learning and change which recognizes the complex characteristics of local government; approaches that offer a new language to facilitate dialogue and flexibility* (p. 4).

David Henshaw of the Society of Local Authority Chief Executives (SOLACE) noted in the forward: "Attempting to deal with this ever increasing complexity requires new approaches... the simple issue is that Local Authorities need to transform themselves in their thinking."

A Consilient Approach To Change

Introduction

A complexity science based approach encourages, supports and legitimizes the abandonment of rational, linear reductionist approaches to planning. It also supports managers and leaders to engage with and understand their particular, unique, contingent history. This constitutes the sensitive dependence to initial conditions on which to focus their perceptions as the external environments and drivers shape and reshape this in a continually unfolding process.

Complexity thinking alerts facilitators to attend to the networking and relationships, both formal and informal, of the group they are working with, as well as supporting and trusting that things will self organise at a higher level of complexity if we attend to them well enough and for long enough.

The tools that we require are those which allow us to take a group and hold them for long enough, well enough, in an uncertain, anxiety provoking, risky space to enable them to reframe their identity and their organizational boundaries. But tools, as Seddon (2007) reminds us, can be dangerous things in the wrong hands without the understanding or the knowledge of how they should be used. Seddon is particularly critical of tools which are a codification of method:

> *From codifying methods it is a short step to choosing those 'tools' that appear to be making the big difference and describing them as a series of tasks or steps to be undertaken. Codification itself suits the command and control culture. Tools can be taught, directed at problems and reporting on progress can be institutionalized through the hierarchy* (p. 1).
>
> *The danger with codifying method as tools is that by ignoring the all important context it obviates the first requirement to understand the problem, and, more importantly, to understand and articulate the problem from a systems perspective* (p. 7).

Resistance To Change And Disrupting Patterns Of Organization And Communication

What is it that makes organizations so resistant to change? Maturana & Varela (1987) developed the notion of autopoiesis to account for what is distinctive about living systems. For Stacey (2005) autopoietic systems have three characteristics: identifiable components, a boundary and internal mechanisms and communications. Boundaries are not imposed from outside but determined by internal relationships, and in determining its own boundaries an autopoietic system establishes its own autonomy and therefore its own identity. An autopoietic system interacts with its environment in order to preserve its homeostasis and its identity, and therefore to resist change. For Stacey:

> *Organizations can be seen as a self contained entity functioning according to the principles of its own identity, as a living autopoietic system. It is an organizationally closed system but it is perturbed by changes in other organizations. These perturbations trigger change, but the change itself proceeds according to its own internal dynamics, its identity. Organizations coevolve reflecting the history of their structural coupling* (p. 146).

For Battram (1996) this has profound implications for communication and change in organizations:

- We don't experience the world directly by receiving incoming data;
- We preserve ourselves: we are conservative, we resist change, and when forced to change, we respond in such a way as to maintain our unbroken sense of self, and;
- We are only interested in what we are interested in, and not much else. So we will only learn what we want to learn, and we will fit it into our existing view of the world.

He concludes: "These points lead us to a healthily pessimistic view of human communication as innately difficult and human behavior as both self determined and resistant to external changes" (Complexicon: 233).

For any change process to be successful it must acknowledge: the way in which systems seek to preserve their identity; understand how this process operates and can be interrupted and then support an organization to emerge into a different identity.

For Luhmann (1986) social systems maintain their autopoiesis by focusing on continual communications and the ongoing flux of events rather than coordinated action and reflective processes. Price & Shaw (1998) suggest that by slowing down and becoming aware of the patterns we are caught up in we can 'shift the patterns':

> *Evolution addresses itself to change, to the change in such patterns, which, over time, has led to such richness of life and living forms on the earth...We and not the pattern and its replication, nor the vagaries of blind evolutionary processes, have the capacity to take the lead. That capacity has been granted, interestingly, by virtue of cultural evolution. As a species, we, alone have grown beyond our biological inheritance. We can also grow beyond our cultural inheritance* (p. 313).

For Distin (2010):

> *Human culture is built by human agents on the basis of cultural information which they are able to create and acquire by virtue of cognitive mechanisms that discretize cultural information in ways that match the discretizing methods of the cultural language within which this information is shared* (p. 232).

Discrete units of cultural information are linked to the notion of the meme, a term first coined by Dawkins(1976) and taken on into the field of organizations by Price & Shaw (1998):

> *An organization is coded via 'ideas and images of the mind', abstract strands of thinking, perception and language, the smallest units of which may be thought of as memes which may be interpreted as: the smallest element capable of being exchanged, with an associated sense of meaning and interpretation, to another brain* (p. 160).

It is also important to understand the impact of contingency on our ability to consistently apply tools and to explain change. For Gould (2004):

> *The central importance of contingency as a denial of reductionism in the sciences devoted to understanding human evolution, mentality, and social or cultural organization strikes me as one of the most important, yet least understood, principles of our intellectual strivings* (p. 225).

Consilience

Consilient Change is a transition focused consultancy based in Sheffield in the United Kingdom. We provide personalized and bespoke support and solutions for individuals, teams, companies, organizations and partnerships facing crisis and transition to create successful and sustainable futures. Consilience, with reference to William Whewell (1840) and Steven Jay Gould (2004), describes:

> *The act of bringing together separate experiences, areas of knowledge, skills and expertise to create a new whole which is more than the sum of its parts.*

The best example of this 'consilience of inductions' in action is the creation of the theory of evolution. Whewell was a mentor of Darwin.

The Process And Key Elements Of A Consilient Approach To Supporting And Sustaining Change Based On Complexity Thinking

It is important to understand why things stay the same and with this understanding to work with people to help them make the changes that are important for them. It is important to assist people to get into a defended space where they will not be subject to the distractions noted by Luhmann (1986) earlier. It is also important to be clear with people where they feel they are stuck and to try and be explicit with them about which patterns they would like to change and gain a shared commitment to that purpose.

The next stage of the work is to build a community of practice and through social learning to develop the social and intellectual capital of the group. Within this it is important to retain a focus on the unique contingency of their circumstances, context and configuration and to pay attention to what emerges from the process of trying to shift the patterns within this environment. A range of techniques are described which support this emergent process.

It is important to pay attention to the emotional states and responses within the group, particularly in relation to the management of anxiety and uncertainty and to try and stay internally grounded and anchored so that their reactions don't impact too much. A mindful awareness is important.

As the group becomes more reflective they should become more system aware and start to see the patterns underneath their reactions to the patterns. As identity starts to shift this may provoke a crisis in the group which needs to be supported, contextualized and framed.

The group then needs to be assisted to move from knowing to doing, to avoid being paralysed from acting. The group can then reflect on the journey travelled and whether or not the patterns have been interrupted. Cycles of action learning and action research are important in helping people make the transition from one identity to another and this is taken forward by developing the dialogue and assisting the group to move from breakthrough to transformation (Scharmer, 2000).

This process of engagement, informed by the complexity thinking about change and leadership described above, is linked to three key elements described below to form a Consilient approach: framing, supporting and shaping the flow of emergence; integral hosting and mindful awareness.

Framing And Designing Emergence

In our approach to assisting groups in framing and designing emergence we use the following conceptual frameworks to support them in thinking about framing supporting and shaping emergence as it arises from their unique contingent circumstances (I will discuss how these conceptual frameworks are deployed as part of the discussion of the case study in the next section):

- Stacey's work (2007) on how we understand the development of strategy and leadership in situations which are uncertain, incoherent and emergent, juxtaposing this with the expectation that organizations are stable, rational, linear bureaucracies and how we move management teams from one construct of experience to the other. This is used to introduce groups to a different nonlinear non-reductionist way of thinking about the circumstances in which they find themselves.
- Mintzberg's (1987) work on emergent strategy. This supports groups to focus on thinking about strategy as an essentially emergent experience. It acts as a frame for thinking about how they create strategy from the current emergent and contingent circumstances in which they find themselves.
- The work on situational awareness (Endsley, 2000) defined as: 'The perception of the elements in the environment within a volume of time and space, the comprehension of their meaning and the projection of their status in the near future' is used to support the focus on understanding the current environment in detail. This is linked to the work on error reduction and decision making under stress (Flin *et al.*, 1997) to assist them to slow down in the process and to pay attention to seeing the systems and circumstances they are caught up in as they are unfolding.
- Wenger's work on developing communities of practice (2002) is used to assist groups to understand the importance of forming communities of prac-

tice in relation to social learning, the difficulty of changing identity in groups and as a method of managing knowledge collectively based on shared experience.

- Scharmer's (2009) work on the 'U' journey through change and the processes of 'presencing'—letting go, letting come, crystallizing and prototyping which support this theory of change is used to help situate groups in a process of change which is focused on leading from the future as it emerges, helping to contextualize and frame change in the present with a focus on the future, rather than seeking to repeat the old patterns of the past.
- Whole system approaches (Mumford, 2003) assist in supporting bringing together a whole leadership team and help to frame the conversation dialogue and graphic facilitation.
- Dialogue approaches based on Bohm (2004), Scharmer (2000) and Gunnlaugson (2007) help to deepen the conversation once a community of practice has been established.

Integral Hosting

Integral hosting is used as part of the Consilient approach to groups to work with them in a variety of ways and to use space in different ways to support a journey of emergence, based on the exploration of the conceptual frameworks identified above. This process is linked to the art of hosting which is described by Corrigan as:

> *An emerging set of practices for facilitating group conversations of all sizes, supported by principles that: maximise collective intelligence; welcome and listen to diverse viewpoints; maximise participation and civility and transform conflict into creative cooperation.*

Hosting is also linked into an integral philosophy and to an explicit intention to support the development of integral thinking as identified by Wilber(2001) who argues for the dawning of 'an integral age at the leading edge' based on a full spectrum of knowledge in four quadrants: the cultural; the conscious; the behavioral and the systemic.

We use the following techniques to facilitate an integral approach:

- Open space methodology (Harrison, 2008) is used to help the important questions for individuals emerge into the group and to help to shape both shared significance and the community of practice.
- Scenario planning (De Geus, 1997) is used to help groups explore a range of different potential futures as they might emerge from the situational awareness of the present.

- World Café approaches (Brown & Isaacs, 2005) help to share significance in the group; to build communities of practice and to create and develop the social and intellectual capital of the group.
- Graphic facilitation (Sibbet, 2006) helps to capture material as it emerges in a dynamic, memorable and integrated way reflecting material generated as part of a community of practice.
- Kinaesthetic, auditory and visual approaches to engagement and learning help us to engage all learning styles and a full range of intelligences.
- Done on the day digital capture and harvesting means we get the material generated from the day to the participants in electronic format the following working day.

Mindful Awareness

Jon Kabat-Zinn (1990) defines mindful awareness as: 'paying attention in a particular way: on purpose; in the present moment and non judgementally'. For Corrigan being present means 'showing up, undistracted, prepared, clear about the need and what your personal contribution can be.' This adapted quote summarizes for me the state I aspire to in facilitating events:

> *Can you love and respect the people and respect their inquiry without imposition of your will?*
>
> *Can you intervene in the most vital matters and yield to events taking their course?*
>
> *Can you attain deep knowing and know you do not understand?*
>
> *Conceive and give birth and nourish without retaining ownership?*
>
> *Trust action without being guided by outcome?*
>
> (Interpretation of words attributed to Lao Tzu, c. 550 BC)

Interventions are designed based on Kolb's (1984) theory of learning and Prochaska & Di Clementi's (1982) *Cycle of Change* to assist groups with blocks to the process and flow of learning and change.

Isabel Menzies-Lythes's (1960) work on managing anxiety is used to help people understand and work with the relationship between structure and relationship as part of the management of anxiety and uncertainty in risky situations.

Fineman's (2003) work underpins our approach to managing and containing the emotion and anxiety raised by the contemplation of the personal implications and impacts of addressing major budget reductions, loss of services and redundancies of staff.

Applying The Consilient Approach To A Real World Situation: The Case Study

Background And Overview: A Year Of Change, Emergence And Uncertainty

We began work with a Local Authority in the UK in December 2009 as it was beginning to face up to the likely implications of significant budget reductions for the Public Sector following the General Election in May 2010, whatever the outcome. There was awareness in the organization that previous attempts to manage change had not been successful and that the current programme of Leadership Development was not delivering the desired results.

The development programme initially began as an inquiry with the Chief Executive and his leadership team of around forty Senior Managers focused on managing a budget reduction of between 20 and 30% over the next three years. Our aim was to help them to create a narrative and options for the budget reduction which could be presented to local politicians following the Local Elections in May 2010. The initial commission for the work ran from February to June 2010.

This work was extended through to March 2011 in two further phases. From July to November we worked on moving the group from knowing to doing, helping them cope with the emergent and uncertain nature of the budget settlement as it progressed from the emergency budget in July to the proposed settlement in November. From December 2010 to March 2011 we helped them with the implementation of the proposals as well as coping with a final settlement which was much worse than anticipated.

The patterns that we agreed with the group we should support them to disrupt were as follows:

- Improved connectivity and communication within and between the leadership group;
- More proactive and consistent external communication by the leadership group;
- Improving the conversion of information and data into knowledge in a nimble and agile way;
- Acting on the knowledge generated and following through on decision making in a timely fashion, and;
- Keeping responsibility and accountability for change at the leadership level

Framing Emergence

We also set out to frame expectations and create a reflective process prior to each event. Once we had completed the planning we sent an e-mail to the group laying out our thinking, identifying the themes for the workshop, acknowledging the emotional environment and asking them to think about certain things prior to the workshop.

Moving To Co-Creation

As the process progressed we were able to achieve more flow and coherence in the planning and commissioning of the events. Originally we were asked to run one event and then, when this was successful, we undertook a programme of development work, agreed with a planning group, in order to run a further event in March 2010. Following this the Senior Management Team commissioned two more events. In May we were able to process the material using the techniques and approaches we felt best suited the group and in June we co-created the event together having agreed the flow and outcomes of the programme. As we moved into more in depth appreciative inquiry the dialogue engaged more people and began to drive the process. The planning groups, having been formally chaired, flowed as action learning type discussions from which the focus for the event naturally emerged. The briefings with the Chief Executive reinforced and completed the flow of the programme and the Senior Management Team endorsed and supported the programme which the planning group came up with.

Figure 1 provides a summary of the work of the year, including the key themes addressed in each stage and a summary of the content of each session.

Initial Work: November 2009—February 2010

The initial workshop was difficult to deliver. The venue was not really big enough for us to work with the whole group, the over large Christmas tree didn't help either. However the group felt that progress had been made. For them, there had been a better degree of engagement than in previous sessions, and the drumming workshop had been a success. Would we facilitate a further set of development workshops in March, June and September?

We met with a planning group in the middle of January to review the December event and plan ahead for the March event. Following a meeting with the Chief Executive we agreed on an initial development programme to run on a monthly basis from March to June 2010 to focus on:

- Promoting further communication and connection within the group;
- Shifting the pattern of engagement and outcome so that the group could own responsibility and take decisive, cohesive action about the forthcoming budget reductions;
- Create a vision for the future of the Local Authority that they could use to navigate through the uncertainty of the next year;
- To keep up the momentum but hold people in a reflective process, and;

Overall Aims of Programme:

- Lack of connectivity as a leadership team
- Communicating with each other and externally differently as a group
- Devolving responsibility and then blaming
- Collecting too much information and data
- Not enough action and completion before we move onto the next thing

Disrupting established patterns

A Years Development with the Leadership Team

‘Being in the bubble’ - situational awareness

Trying to create a narrative to fit the financial envelope

Managing communication differently

Exploring our relationship with lead members

Implementing action plans—programme office

Resilience and capability

The Gathering Storm		The 50% Scenario	Exploring the minimum and the common sense councils							
04.12.09	**Jan & Feb**	**02.03.10**	**30.03.10**	**16.04.10**	**11.05.10**	**08.06.10**	**16.07.10**	**06.09.10**	**04.10.10**	**23.11.10**
Swot analysis		5 Key Ideas	Listening to the dialogue Change as dynamic	• Key Lines Of Equiry & cross cutting themes	Medium term financial planning timeline A leadership paradox			The 25 day plan	Agreeing a 20 day plan	
• Fish bowl • Drumming • Presenta-tions • Key inter-views • 50% scenar-ios		• The 50% scenario • Open space 9 questions • Feeling / thinking intent	• Commu-nity of practice • Social & intellectual capital • Impact on the day job	• Commu-nity of practice • Social & intellectual capital • 28 million • Impact on the day job	• The gallery • Equity dia-logue • World café • 22 million • Dance-floor mapping • Programme office • Knowing to doing	• News room • Situational awareness • Identity • Leadership • Redlines	• Anger • Intent • Member relations-hips & decision making • Communi-cation • Cycle of change	• Visioning • Deeper dia-logue	• 4 year plan • 20 day plan • Visioning • HR1 prep & comms	• New figures • Energy • Emotions • Supporting others • Intention • Terms & conditions

Figure 1 *A Year’s Development With The Leadership Team*

- To integrate the strategic thinking and planning with the feelings and emotional reactions this work raised.

We also agreed a preparatory programme prior to the first event in March which included:

- Interviews with five key leaders;
- Each Manager constructing a scenario, to an agreed template, about what they would do if they only had 50% of their current budget to spend. We agreed to process these templates and feed them back to the next leadership development day in March.

Creating A Community Of Practice: The March Event, A Taste Of Things To Come

Mindful Awareness

We prepared both the content and the process of the day carefully. We were aware that we still didn't know the group that well and that we needed to cater for a range of preferred learning styles and intelligences as well as ensure that we had a range of activities which enabled people to interact and connect with each other dynamically and purposefully.

Framing Emergence

We began with a set of ground rules which were designed to create a safe respectful container within which the group could confidently enjoy learning with and from each other and move outside individual and collective comfort zones whilst feeling safe enough to take some risks in order to explore ways in which normal routines and patterns of behavior could be disrupted within agreed boundaries of confidentiality. The ground rules sought to: establish a reflective space with a commitment to active listening; create a responsible environment where the perception of both feeling and thinking was encouraged; promote individual ownership of thoughts, experiences views and opinions and provide confidence in the group that we would call time out if the group dynamics or content of the conversations were causing too much distress, frustration or anger.

We also wanted to promptly introduce and present the conceptual frameworks we wanted to work within and to establish these as having relevance for the current and emerging circumstances they were facing. We chose four key conceptual frameworks to begin with:

- Stacey's approach to leadership in uncertainty, incoherent environments (see Figure 2);
- Mintzberg's emergent approach to creating strategy (Figure 3), and;
- Scharmer's Theory U (Figure 4).

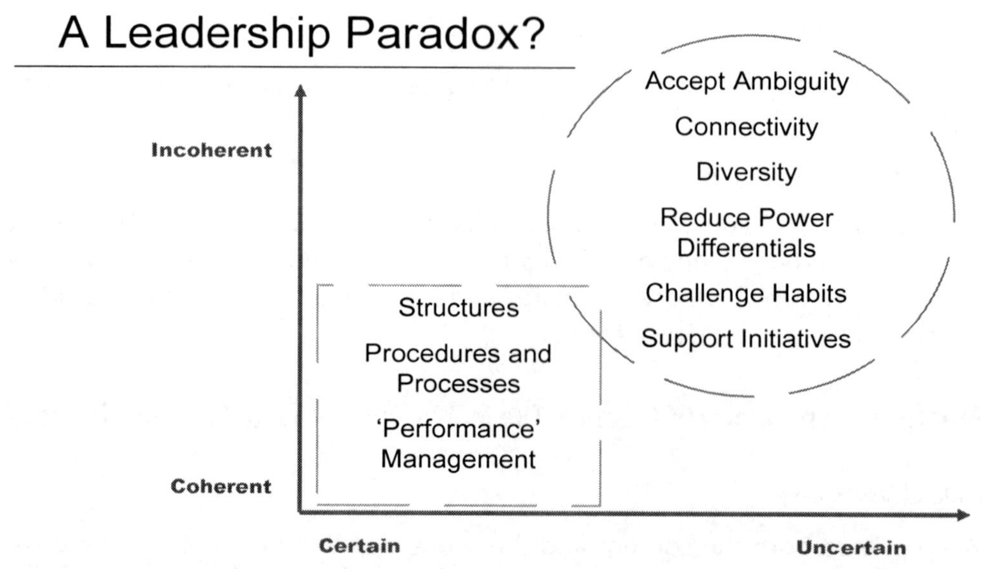

Figure 2 *A Leadership Paradox? (Adapted from the work of Ralph Stacy and Caryn Vanstone)*

We then went through a basic presentation on Wenger's concept of a community of practice looking at what a community is and why it might be a helpful construct to inform the development of our work on developing knowledge management and changing identity. Our final presentation was an overview of the individual work they had done on managing with 50% of their budgets which gave them sufficient information to work with as a community of practice, without being overwhelmed by the detail (see Figure 5).

Integral Hosting

In order to keep up the level of momentum and conversation we then asked the group to stand up and move about the room and to process their reactions and responses to the presentation in continually changing pairs to see if they could keep a dialogue progressing. From this conversational processing we then took them into a kinaesthetic space where we asked them to position themselves in the room in relation to how positively or negatively they were thinking and feeling about what they were about to lead and manage as a group. What we learned from this was that the group shared a high level of positive intent and commitment to see this process through despite acknowledging the potential impact on them.

After the first break we ran an open space session which helped the group identify and progress conversations with each other which were important for individuals to explore and to build a shared significance . We ended up with nine conversations. In the session after lunch we offered a 'framing the future' session which gave the group a choice of either creating a vision for the future in 2020

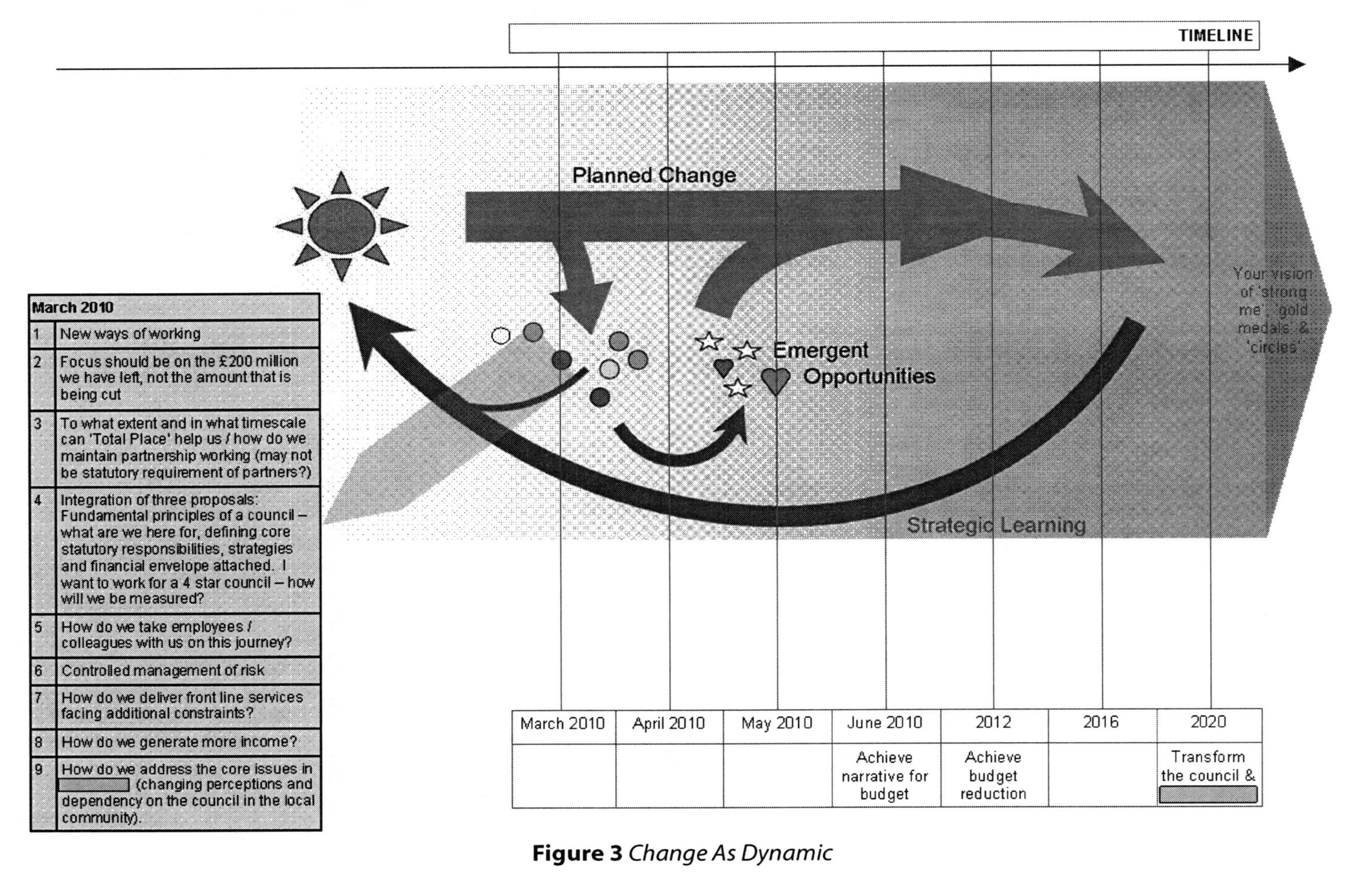

Figure 3 *Change As Dynamic*

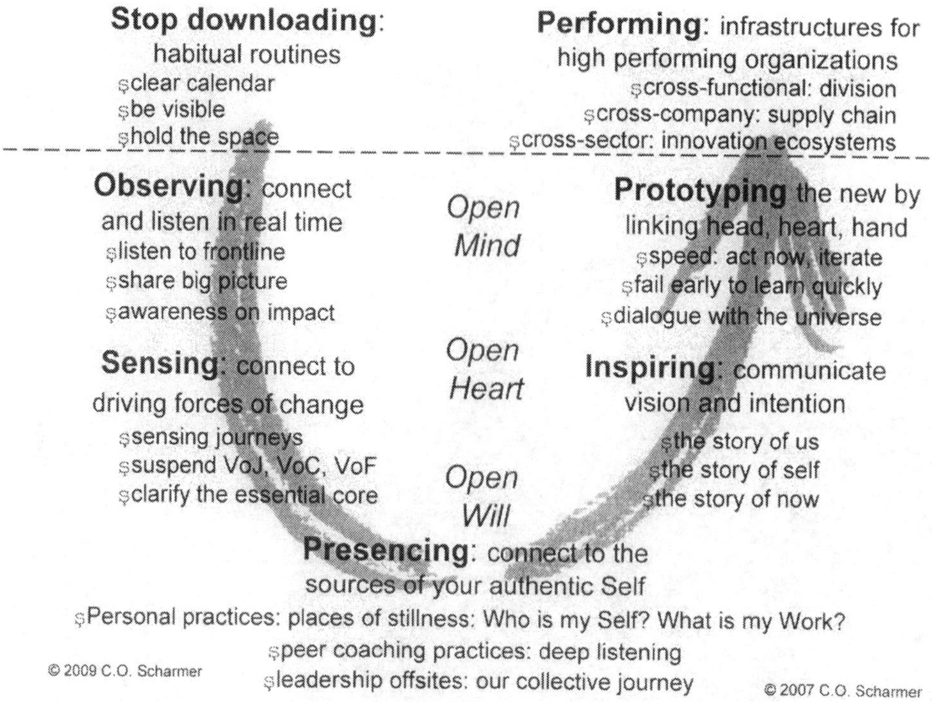

Stop downloading: habitual routines
- clear calendar
- be visible
- hold the space

Performing: infrastructures for high performing organizations
- cross-functional: division
- cross-company: supply chain
- cross-sector: innovation ecosystems

Observing: connect and listen in real time
- listen to frontline
- share big picture
- awareness on impact

Open Mind

Prototyping the new by linking head, heart, hand
- speed: act now, iterate
- fail early to learn quickly
- dialogue with the universe

Sensing: connect to driving forces of change
- sensing journeys
- suspend VoJ, VoC, VoF
- clarify the essential core

Open Heart

Inspiring: communicate vision and intention
- the story of us
- the story of self
- the story of now

Open Will

Presencing: connect to the sources of your authentic Self
- Personal practices: places of stillness: Who is my Self? What is my Work?
- peer coaching practices: deep listening
- leadership offsites: our collective journey

© 2009 C.O. Scharmer

© 2007 C.O. Scharmer

Figure 4 *Sharmer's Theory U.*

or running a scenario inquiry and reporting back on the outcomes from 2020. In the end four groups created visions and one group ran a scenario. In the final hour we ran a whole group dialogue bringing them close together in a circle, starting with the story from Dee Hock (2005) about the difficulty of disrupting patterns and behaviors and the difficulty of overcoming autopoietic states:

> *When everything changes around us and it becomes necessary to develop a new perception of things, a new internal model of reality, the problem is never to get new ideas in: the problem is to get the old ideas out. Every mind is filled with old furniture. It's familiar. It's comfortable. We hate to throw it out... there is nothing we fear more. We are our ideas concepts and perceptions* (p. 107).

Building And Creating Social And Intellectual Capital: March To June 2010

Mindful Awareness: Making Sense Of What Happened In The First Event

We learnt a lot from the first event:

- How to build on working together and working within the group so that there was more co-creation;
- That the level of engagement with, and commitment to, each other as well as their intellectual and emotional commitment to the task did not translate into effective or coherent working together, and;

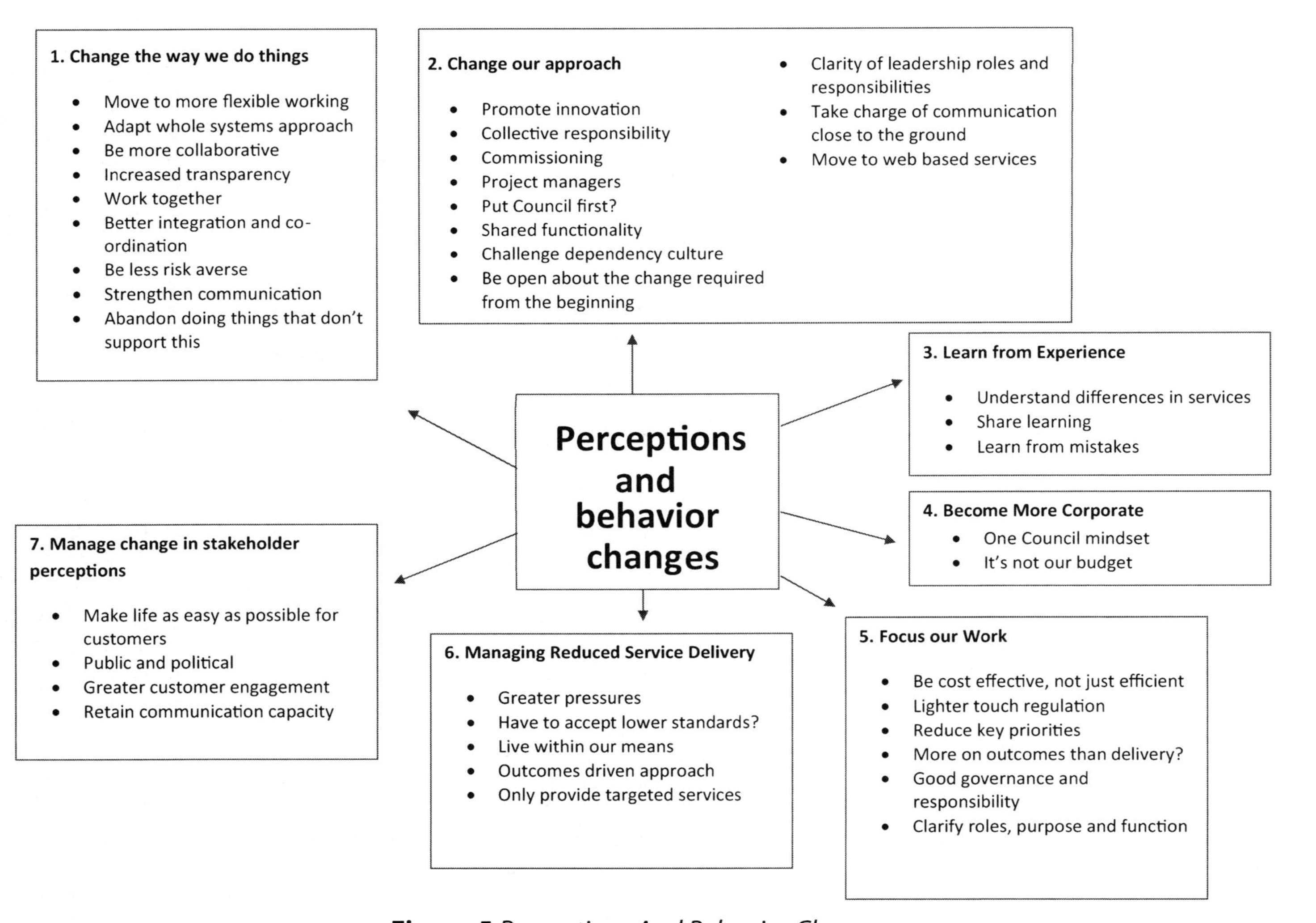

Figure 5 *Perceptions And Behavior Changes.*

- Something of the patterns in the group which enabled and prevented conversation and action.

Framing Emergence

The Senior Management Team took control of the agenda for the next sessions which they commissioned quickly, with a further event in March and another one in April. They directed the planning group to commission us to facilitate two further events which would involve Directorate presentations to each other to explore the concept of a 'Minimum Council' to support the development of key lines of enquiry into ways of saving money and rethinking the structure and provision of Council services within an articulated risk matrix. This would also build shared knowledge within the group about the functioning of the Council as a whole system.

We also explored the information and knowledge we created within the conceptual frameworks we used in the first session in March. We populated Mintzberg's emergent strategy with the work from the open space and vision sessions to create a frame for focusing on further information and knowledge management (see Figure 3)

We agreed a process which allowed us to facilitate the beginning and end of these days in relation to recapping and sharing some of the work which had been developed in the previous sessions. We were able to put some ground rules around the dialogue and challenge which followed each presentation and to facilitate the sense making which came out of each presentation. A pattern of small and large group dialogue which built through each of the sessions was established. The key outcomes from this part of the programme were:

- Building up a community of practice and inquiry around dialogue and standards for depth, style and presentation of material;
- A first view as to whether the proposals, aggregated together, would reach the budget reduction required, and;
- Keeping people engaged in dealing emotionally and intellectually with the complexity of the information which was emerging.

Integral Hosting

In May we felt we were ready to bring the knowledge we had created so far and present it as part of a strategic time-line. By this time three members of the group had begun to devote most of their time to supporting and challenging the work of the Directorates and work had also begun to shape cross cutting themes which were beginning to emerge. In this session we wanted to focus on a 'minimum' and a 'common sense' council within the budget reduction targets.

At the May event we began with a gallery-like display of the work of all the Directorates and then reflected on peoples' reactions to this. It was at this point that we began to disrupt the pattern of equity which characterized their conversations about change and reductions as key members began to realize that there would be differential impacts in each Directorate and that there was a degree of variation in how they had approached this. We explored with them whether this variance was within tolerable limits to sustain a community of practice or whether they needed to provide further internal challenge. They agreed that they needed to provide further internal challenge.

We assisted the flow, sharing and development of knowledge in the next session in the morning where we ran a world café style dialogue session based on the work of the cross cutting themes. It became clear from this process that two of the themes could be abandoned. The pattern of activity and priority between years one, two and three also became clear at this point.

The room we were working in doubled as a function room and at one of the events a dance floor was laid down for a function later in the evening. We asked the hotel to put the floor back down and used this as a visual frame to report on progress so far and agree the time-line and approaches within Directorates and in relation to cross cutting themes.

We gathered around the edge of the dance floor and looked at, and spoke about, the knowledge and strategy we had created as a whole group. We hadn't reached the target for the minimum. We had no 'wriggle room' to allow us to be flexible in the construction of a 'common sense' Council. We would have to challenge our sense of who we were becoming further, and realize that the 'Minimum Council' model was going to be part of the new reality.

Mindful Awareness

From the first workshop in March both the emotional and cognitive intent of the group remained high. There were consistent voices in the group, including the Chief Executive, reminding people of the emotional component of the task at hand. During the first workshop a majority of the group expressed a high level of positive affect and intent for the task in hand. People were realistic about the challenge of the task but remained positive that this climate gave them the opportunity to achieve a new vision—something that they felt they should have grasped a long time ago. We tested this further at the workshop in May, turning the dance floor into four sections at the conclusion of the presentations and asking them whether they still felt they could achieve the task and to place themselves in the quadrant which most aligned with their level of confidence. The majority of the group stayed in the strongly confident quadrant.

Containing uncertainty was also a key component of this transition. Prior to the election there was concern about the outcome both locally and nationally with a change probable in both. The result of the elections in May—a national Conservative and Liberal Democrat coalition and an enhanced majority locally for the current party did not resolve this uncertainty. By the time we met in June, we had some measure of the kind of additional budget cuts which were going to be asked for and had to find a way of containing this new anxiety. Not only had we not identified sufficient savings to reach the minimum council level, there were now going to have to be additional cutbacks and over four years rather than three as originally envisaged, with a weighting to more cuts in the first year.

The June Event

We agreed a clear process and set of outcomes for the workshop which was to take us from 'knowing to doing'. A briefing for Local Councillor's on this work and the additional challenges had been arranged for the 1st July. Our issue was that we did not really have a full understanding of how to support the group through this process. We planned, reflected and talked again on the morning of the event. All I knew as we started the event was that we wanted them to start by talking with each other over coffee and then come into a 'newsroom environment' to update each other.

It became clear that there had been further progress on bringing together a team to progress the challenge work. They had established a full time team of three working at senior level and all of the Directorates had progressed their work as well. We then asked them to think about Meg Wheatley's (1992) work on three critical questions around dialogue and identity: Can we talk? What just happened? And who are we now? Their work was displayed for viewing and we invited people informally to catch up and have 'water cooler' conversations about their responses and reactions to this newly emergent reality. Then we had a break.

After the break we decided to offer an input to the group on decision making under pressure and recognizing complex emergent environments. We then went back to the Stacey and Mintzberg diagrams we had been using and reviewed the importance of situational awareness in planning in highly emergent environments, particularly the importance of perception in this process.

We agreed to park that for a while and returned to the updates on the cross cutting themes. We discovered and explored some displacement anxiety as a conversation over saving money on car park passes took up time in the group over a conversation about the likely level of redundancies which would now be required. The presentations created their own dynamic for decision making, as well as landing the emotional reality that significant job losses could not be avoided.

After lunch we named some of the anger and distress that people were experiencing, realizing the likely impact on them, their staff and on the local area. This became the departure point and the place where they made choices about moving into responsible leadership and moving on. The group worked on checking the Directorate work and the cross cutting themes along with the implications for the changed environment. From this, we agreed a programme of key actions for the following ten days and confirmed these decisions. The moment for action had arrived.

From Knowing To Doing: July To September 2010

We agreed a further session for July to review these actions and from this a further session in September to allow us to keep up momentum over the summer holiday period.

Mindful Awareness

In the preparation for the July event we agreed that we should bring the emotional impact of the work to centre stage as we moved from preparation to implementation. We planned a whole session prior to lunch to address this. We realized that we would have to build on our work on managing uncertainty and situational awareness and so we also agreed that we should keep the newsroom format.

As we settled into the July event it became clear that significant members of the group were missing without a clear understanding of why. This became a crisis moment for the group and again their strong sense of intent and purpose meant that they resolved to continue with the session rather than cancel. A lengthy and very high quality dialogue as a whole group then emerged which continued for the whole of the morning and for the last session of the afternoon. Within this dialogue the group was able to:

- Develop their situational awareness of how the current situation was unfolding and how they were responding to these challenges emotionally;
- Review the initial presentations and dialogues with the politicians and reflect on the success of their preparation in disrupting previous patterns of preparation and engagement, and;
- Update each other in detail on what had happened since the last session and explore their emotional reactions to it.

In the afternoon we presented Prochaska and Di Clementi's model of the cycle of change and used this to explore how to avoid going into relapse and to maintain momentum and resilience over the summer. We also presented a model of managing risk and decision making where there are high levels of uncertainty (Funtowicz & Ravetz, 2008). This enabled the group to undertake some detailed exploration and planning of the work over the next nine months and to agree to expand the membership and work of the project office over the summer. At the end of this event we booked a follow up for early September to complete this phase of the work.

After the summer we felt the first impact of the budget cutbacks as the September event took place in the Town Hall—we returned 'home' after a tour of local Hotels and Community Centres. In this event we spent time looking at how to integrate the complex range of management tasks each person in the room would need to be able to fulfil over the coming months and to rehearse the skills and knowledge required to manage this well. We concentrated on three key areas of work: HR, Communications and Programme Management and created a 25 day programme to see the proposals that we had created ready to be launched for formal consultation at the beginning of October. We used this event to reflect more deeply on the comparative success of the work and to explore that we still did not have a narrative within our lines of enquiry which added up to the required saving over four years, nor were we able to reconceptualize our earlier visions to create a new view of what the Council of the future would look like. However we felt that we had created a very positive container, process and time-line which would be helpful in managing and minimizing the impact on individuals and that a shared approach to maximizing employment and retaining investment was emerging.

From Turbulence To Flow: October 2010 To January 2011

The next three events saw the group move from knowing to doing and from turbulence to flow. Over this time period we implemented a further three integrated action plans. The first centred on how to begin the statutory consultation process for redundancies and how to proactively manage the press coverage. We learned quickly from this process so that by the time we got to the December and January events we were managing and delivering on five separate consultation processes at various stages of completion as can be seen in figure one earlier in the paper. An example of one of the integrated action plans focused on situational awareness and managing a small time frame of projection can be found in Table 1.

As part of the preparation for the November event I sat in on the Senior Management Team meeting as they engaged in a dialogue about the future, informed by the work we had undertaken with them. As the conversation progressed I was able to see a picture of the future emerging (see Figure 6) which I presented to the confirmation meeting with the Chief Executive and secured agreement to share it with the wider group.

In November we set out to build the groups resilience and to emphasise that they were the strongest and most coherent part of the system able to deal with these unprecedented challenges. It emerged that there might be some hope of a better final settlement following extensive lobbying of the National Government. This hope was boosted by the postponement of the announcement of the final settlement figures until later in December.

Communication	Need to respond to staff suggestions at some point Try to identify where the PR story is in each KLOE	P to issue letter to all staff to advise of the process		Press FOI once we are confident staff consultation has gone	20th December—what is the actual cuts package?
KLOEs & Savings	A full picture of the scale of cuts may further engage members	5th Oct LT meeting to review additional 8.3M progress Use the 50% exercise to reassess	15 Oct: £8.3M has been looked at and decided with impacts noted—a clear mandate for the additional 8.3M so that we can consult with TUs	19th October: SMT to take stock of progress on the £8.3M	
Programme Office & Governance	Need consistency in attendance and feedback Continue weekly meetings—feedback & coordination role Regular reports back to SMT & LT where necessary	Need HR1 numbers with additional cuts for Friday Programme office meeting	12th Oct SMT to review progress on HR1 & financial plan Friday 15th October: All implementation plans to be completed and put on web for year 1 Discuss 8.3M gap for Y4	Friday 22nd October: Discuss comprehensive spending review	
HR1 Prep	Early retirement costs—need to look at changing Meeting every Wednesday with TUs; any feedback required—ask HR Face to face where possible Involve shop stewards ASAP If you can reach agreement early, do so	5th Oct LT meeting / clinic Friday 8th Oct—HR1 Issued	Detailed discussions and letters to staff Non staff savings explained Track all consultation notices issued—Crucial		November: HR catch ups with DMTs November 2010: Detailed discussions with service users and partners
Key Events		HR1 Issued	Consultation with Staff begins	Thursday 21st Oct: Comprehensive Spending Review	Half term

Table 1 *Session Four: Agreeing A 20 Day Plan*

We facilitated an exercise with them, designed to promote resilience which asked them to draw a situation which they had been in before and then on the back to draw the outcome. We used these drawings in the group to promote a number of discussions between people focusing on exploring what it meant to them and the impact, the positive experience and the learning they could bring to bear on this situation having seen it before. We displayed these images as a gallery before we had lunch and viewed the whole community's experience.

After lunch we focused on how they as a leadership group could identify and support the diverse range of needs of their service users, staff and other stakeholders who would all be differentially impacted. This enabled the group to articulate a coherent, integrated strategy for support.

In the afternoon we shared and explored the vision (see Figure 6) with the whole group which used three key metaphors for the journey of the group: climbing safely down the cliff of budget reductions; getting in to the four year financial envelope and then transforming services.

The December meeting began with a bitter and very angry Finance Director describing the sense of betrayal and outrage felt at the settlement, announced earlier in the week, which had resulted in poorer councils losing more and richer councils gaining more. He and the Chief Executive then gave an example of how they had seen this before and we would now have to get on and find the money. The group then had a dialogue and a debate about how best the Chief Executive should advise the Leader of the Council on the issue of potentially saving money through a variance in terms and conditions. The group was able to bring its' knowledge to bear on the issue and to provide sophisticated advice on the issues involved.

We assisted the group to make a complex plan as the new financial circumstances meant a new round of consultations about further job reductions. Our learning and rehearsal as a community of practice in October paid off as the group were able to organise to communicate with every member of staff before Christmas as well as run two more consultation processes and issue the first round of redundancy notices.

We were also able to help the group contextualize their immediate difficulty by presenting an overview of the journey of the group over the last year (see figure one again) and how they had responded to a highly uncertain and deteriorating financial situation over the year. This time frame then enabled us to evaluate the work of the year in relation to the extent which people felt that the patterns identified at the beginning of the year, as issues for the leadership group, had been disrupted (see Figures 7 and 8).

We met again in January, after the full Council had met to agree the redundancy notices and further consultation processes about job losses and other savings. After we had made a plan to take the work forward to March, the group took

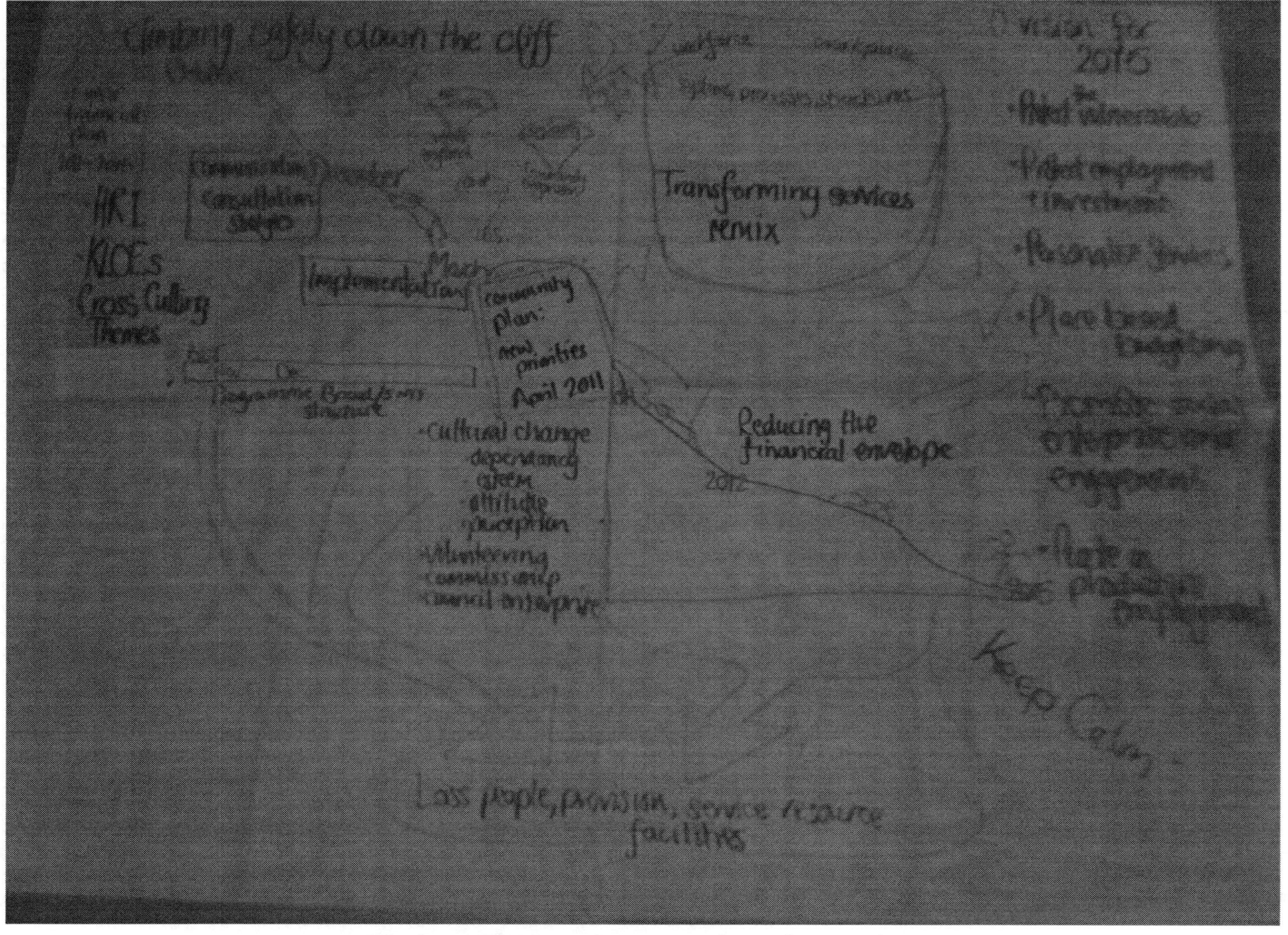

Figure 6 *The Group's Journey Through Metaphor*

time to consider some of the longer term issues that they would need to face and address in relation to culture and transformation. We formally rehearsed dialogic approaches to these issues with the group and the results were highly engaging with the real depth and challenge of the issues. There was agreement that further dialogue was required. When we went back to meet them for the preparation for the March event they had already had further dialogue sessions and had come up with both a vision and a process for transforming the Council and its local area. We will be working on this material in March.

Conclusion: Disturbing The Patterns?

We asked the participants to evaluate the programme over the year as to whether it had succeeded in disrupting the key patterns identified previously. The scores were from 1 (not at all) to 5 (yes the pattern has been shifted). The results in terms of actual and mean scores can be seen in Figures 7 and 8.

Both these results and the overall feedback were indicative of some success. We are going to follow this up by taking some external views of the programme as well as looking at the budget outturn and outcomes. They certainly seem to be doing better than their local comparators. Some quotes indicate the impact it has had on people:

> *The context of cuts, cuts, cuts was depressing but the process was truly engaging and absorbing.*
>
> *A job well done and a very good exercise in management team building. It will benefit this programme and other work.*
>
> *A high in the years I have worked for the Authority and they exceed 20 years.*

We will run a final session this year in March (2011) and are beginning to talk about creating a planned development programme for the next year: 'From Breakthrough to Transformation' with a focus on reframing and addressing dependency culture and extending the programme through the system and the organization with others in the group taking on a range of development roles, including with other partners.

Final Comment

We feel that the Consilient Approach as exemplified by this case study offers some qualified support to Higgs and Rowlands conclusion in relation to the efficacy of complexity approaches, based on complexity framing and containing emergence, having something to offer large scale and short timescale programmes of change. We have also seen that the maintenance of a high level of affective commitment in the leadership group, and a consistent focus on what it is important to do next about what is emerging, has also been important as has the refusal to be distracted into reactive unconsidered responses.

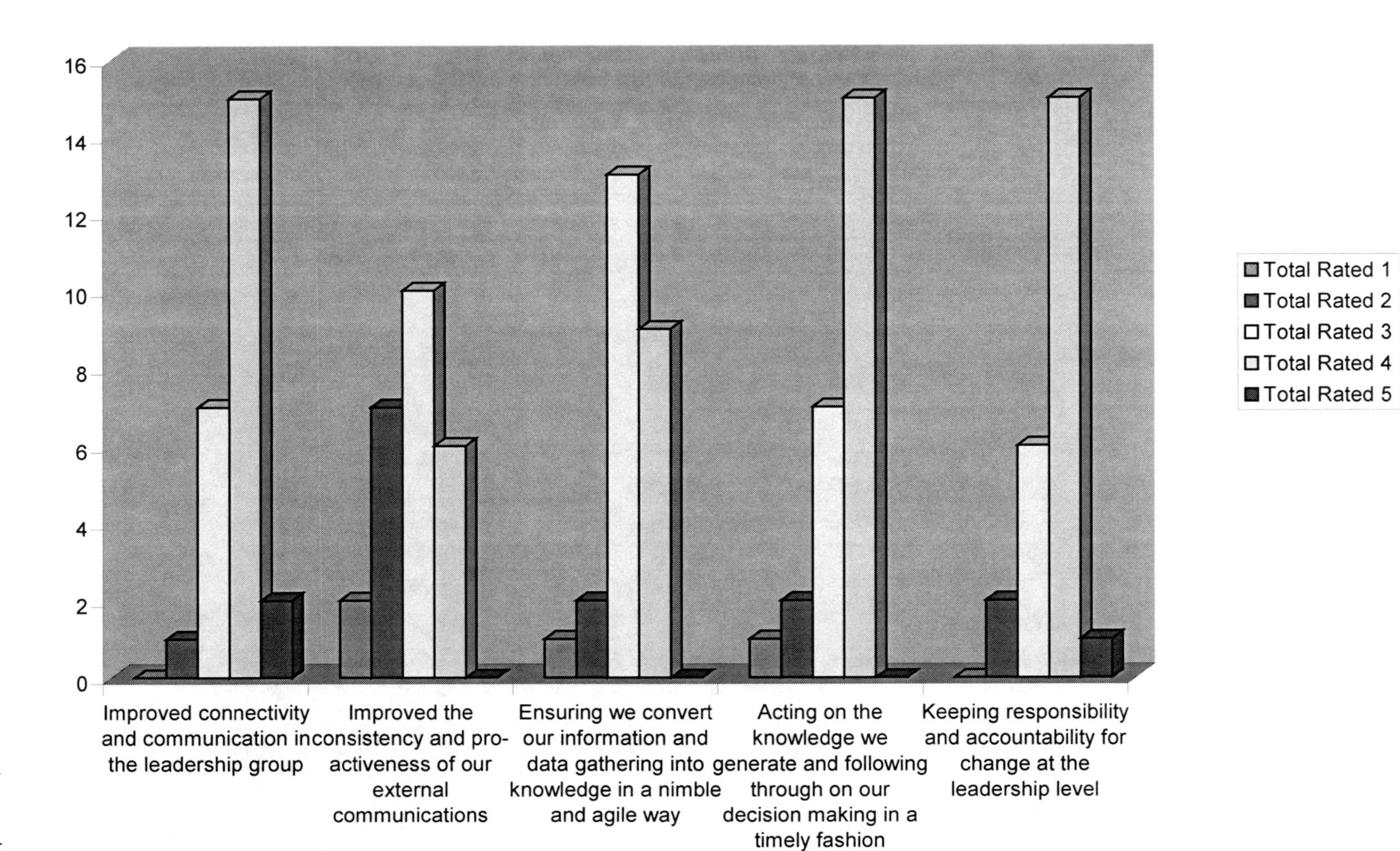

Figure 7 *Shared Significance: Achieving The Key Outputs From The Programme*

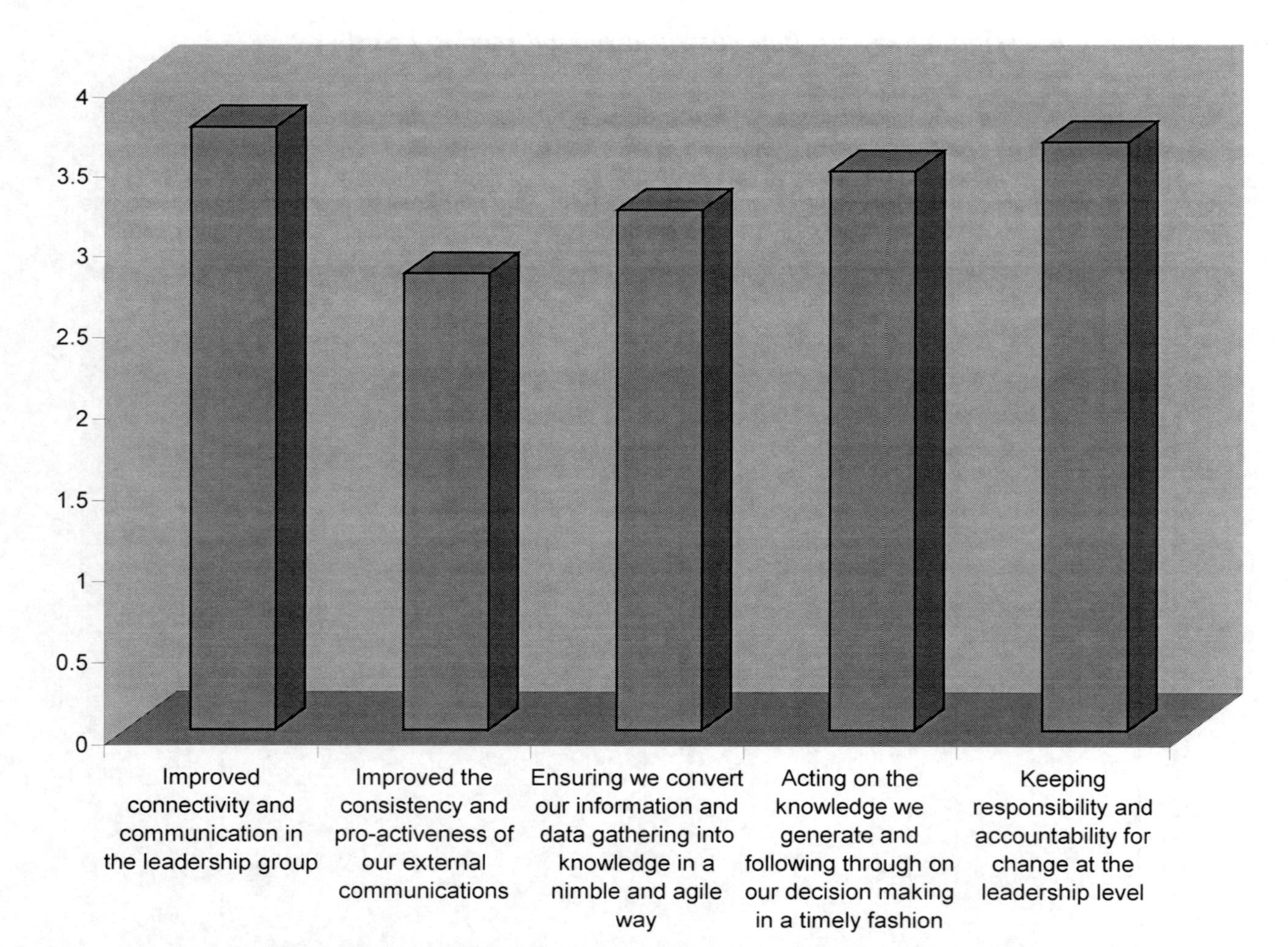

Figure 8 *Shared Significance: Achieving The Key Outputs From The Programme*

Building the intellectual and social capital of the leadership team through a highly engaging programme and grounding and promoting their emotional resilience all seem to have made a contribution to creating something which is more than the sum of its' parts.

As Steven Jay Gould (2004) concludes:

> *I too seek a consilience, a "jumping together" of science and the humanities into far greater and more fruitful contact and coherence-but a consilience of equal regard that respects the inherent differences, acknowledges the comparable but distinct worthiness, understands the absolute necessity of both domains to any life deemed intellectually and spiritually" full", and seeks to emphasise and nurture the numerous regions of actual overlap and common concern…Our richest form of unification emerges when we can agree a common set of principles and then derive our major strength for their realization from the different excellences of all cooperating components: e pluribus unum, or one from many. Let us compile a list of necessary components even longer than the effective and inherently different stratagems …with science and the humanities as the two great poles of support to raise the common tent of wisdom* (p. 259).

References

Battram, A. (1996). *Local Government Management Board: The Learning From Complexity Pack*, ISBN 9780748894802.

Bohm, D. (2004). *On Dialogue*, ISBN 9780415336413.

Brown, J. and Isaacs, D. (2005). *The World Café: Shaping Our Futures through Conversations That Matter*, ISBN 9781576752586.

Brown, S.L and Eisenhardt, K.M. (1997). *Competing on the Edge: Strategy as Structured Chaos*, ISBN 9780875847542.

Corrigan, C. Hosting in a hurry v 1.1; Putting the Art of Hosting into Practice, http://www.berkana.org/pdf/Hosting_in_a_Hurry.pdf.

Dawkins, R. (1976). *The Selfish Gene*, ISBN 9780192860927.

de Gues, A. (1997). *The Living Company*, ISBN 9780875847825.

Distin, K. (2011). *Cultural Evolution*, ISBN 9780521189712.

Dooley, K.J. (2008). "Research methods for studying the complexity dynamics of leadership," in M. Uhl-Bien and R. Marion (eds.), *Complexity Leadership: Conceptual Foundations*, ISBN 9781593117955.

Endsley, M.R. (2000). "Theoretical underpinnings of situation awareness: A critical review," in M.R. Endsley and D.J. Garland (eds.), *Situation Awareness Analysis and Measurement*, ISBN 9780805821345.

Fineman, S. (2003). *Understanding Emotion at Work*, ISBN 9780761947905.

Fitzgerald, L. and van Eijnatten, F. (2002). "Reflections: Chaos in organizational change," *Journal of Organizational Change Management*, ISSN 0953-4814, 15(4): 402-411.

Flin, R.H., Salas, E., Strub, M., Martin, I. (1997). *Decision Making under Stress: Emerging Themes and Applications*, ISBN 9780291398567.

Funtowicz, S. and Ravetz, J. (2008). "Post normal science," International Society for Ecological Economics, http://www.ecoeco.org/pdf/pstnormsc.pdf.

Gould, S.J. (2004). *The Hedgehog, the Fox, and the Magister's Pox: Ending the False War between Science and the Humanities*, ISBN 9781400051533.

Gunnlaugson, O. (2007). "Exploratory perspectives for an AQAL model of generative dialogue," *Integral Review*, ISSN 1553-3069, 4: 44-58.

Harrison, O. (2008). *Open Space Technology: A User's Guide*, ISBN 9781576754764.

Higgs, M.J. and Rowland, D. (2005). "All changes great and small: Exploring approaches to change and its leadership," *Journal of Change Management*, ISSN 1469-7017, 17(June): 121-151.

Hock, D. (1999). *One From Many: Visa and the Rise of Chaordic Organization*, ISBN 9781576753323.

IBM (2008). "Global CEO study: The enterprise of the future," http://www-935.ibm.com/services/us/gbs/bus/html/2008ghcs.html.

IBM (2010). "Global CEO study: Capitalizing on complexity," http://www-935.ibm.com/services/us/ceo/ceostudy2010/index.html.

Isern, J. and Pung, C. (2006). "Organizing for successful change management: A McKinsey Global Survey," *McKinsey Quarterly*, ISSN 0047-5394, (June): 1-8.

Kabat-Zinn, J. (1990). *Full Catastrophe Living: Using the Wisdom of Your Body and Mind to Face Stress Pain and Illness*, ISBN 9780385303125.

Keller, S. and Aiken, C. (2008). "The inconvenient truth about change management: Why it isn't working and what to do about it?" *McKinsey Quarterly*, ISSN 0047-5394, (November): 1-18.

Kolb, D. (1984). *Experiential Learning: Experience as the Source of Learning and Development*, ISBN 9780132952613.

Kotter, J. (1995). "Leading change: Why transformation efforts fail," *Harvard Business Review*, ISSN 0017-8012, (March-April): 59-6.

Lichtenstein, B.B., Uhl-Bien, M., Marion, R., Seers, A., Orton, J.D. and schreiber, C. (2006). "Complexity leadership theory: An interactive perspective on leading in complex adaptive systems," http://digitalcommons.unl.edu/managementfacpub/8.

Lissack, M. (1999). "Complexity: the science, its vocabulary, and its relation to organizations," *Emergence*, ISSN 1521-3250, 1(1): 110-126.

Luhmann, N. (1986). "The autopoeisis of social systems," *The Journal of Cybernetics*, ISSN 0022-0280, 6(2008): 84-95.

Marion, R. and Bacon, J. (2000). "Organizational extinction and complex systems," *Emergence*, ISSN 1521-3250, 1(4): 71-96.

Marion, R. and Uhl-Bien, M. (2001). "Leadership in complex organizations," *The Leadership Quarterly*, ISSN 1048-9843, 12: 398-418.

Maturana, H. and Varela, F. (1987). *The Tree of Knowledge: The Biological Roots of Human Understanding*, ISBN 9780877736424.

Menzies L.I. (1960). "Social systems as a defence against anxiety," *Human Relations*, ISSN 0018-7267, 13: 95-121.

Meyer, J.P. and Allen, N.J. (1990). "The measurement and antecedents of affective, continuance and normative commitment to the organization," *Journal of Occupational Psychology*, ISSN 0305-8107, 1-18.

Mintzberg, H. (1987). "Crafting strategy," *Harvard Business Review*, ISSN 0017-8012, (July-August): 66-75.

Mumford, A. (2003). "Leading change: A guide to whole systems working," *Industrial and Commercial Training*, ISSN 0019-7858, 35(6): 270-271.

Price, C. and Lawson, E. (2003). "The psychology of change management," *McKinsey Quarterly*, ISSN 0047-5394, (June):31-41.

Price, I. and Shaw, R. (1998). *Shifting the Patterns: Breaching the Memetic Codes of Corporate Performance*, ISBN 9781852522537.

Prochaska, J. and Di Clemente, C. (1982). "Transtheoretical therapy: Towards a more integrative model of change," *Psychotherapy: Theory, Research, Practice, and Training*, ISSN 0033-3204, 19(3): 276-288.

Scharmer, O. (2000). "Self transcending knowledge: Sensing and organizing around emerging opportunities," *Journal of Knowledge Management*, ISSN 1367-3270, 5(2): 137-151.

Scharmer, O. (2009). *Theory U: Leading From the Future as it Emerges*, ISBN 9781576757635.

Seddon, J. (2007). Watch Out for the Toolheads, www.superfactory.com.

Seddon, J. (2008). *Systems Thinking in the Public Sector: The Failure of the Reform Regime ... and the Manifesto for a Better Way*, ISBN 9780955008184.

Sibbet, D. (2006). Transforming Group Process with the Power of Visual Thinking, ISBN 9781879502574.

Smith, A. and Humphries, C. (2004). "Complexity theory as a practical management tool: A critical evaluation," Organization Management Journal, ISSN 1541-6518, 1(2): 91-106.

Stacey, R. (2007). *Strategic Management and Organizational Dynamics: The Challenge of Complexity to Ways of Thinking about Organizations*, ISBN 9780273708117.

Tetenbaum, T. (1998). "Shifting paradigms: From Newton to chaos," *Organizational Dynamics*, ISSN 0090-2616, 26(4): 21-32.

Wenger, E. (2002). *Cultivating Communities of Practice: A Guide to Managing Knowledge*, ISBN 9781578513307.

Wheatley, M. (1992). *Leadership and the New Science: Learning about Organization from an Orderly Universe*, ISBN 9781881052012.

Wheatley, M. (2000). *Turning to One Another: Simple Conversations to Restore Hope to the Future*, ISBN 9781576757642.

Whewell, W. (1840). *The Philosophy of the Inductive Sciences, Founded Upon Their History*, London, Parker.

Wilber, K. (2001). *A Brief History of Everything*, ISBN 9781570627408.

Brian Lawson is the Managing Director of Consilient Change, established in 2006, which focuses on the successful, sustainable, management of organizational transitions in the Public, Private and Not for Profit sectors. His recent work has included: support for the transformation of an International Charity Serving the Military to a more corporate, contract based, customer focused organization; work to support a leading interdisciplinary sustainability design practice to double its workforce and set a global vision; support for the Russell Group of University Research Libraries to launch and implement its strategy to develop a digital future and work with a Hospice to transform and integrate its systems. Brian has a BA (Hons) in Combined Studies and an MA (Hons) in Applied Social Science and is currently studying for a Doctorate in Business Administration at Hallam University in Sheffield in the United Kingdom. Brian has published articles in relation to Inter Agency Child Protection and coedited a book on Inter Agency approaches to what was then known as Munchausen Syndrome by Proxy. Brian started his career as a research chemist and has always kept up an interest in science, particularly in relation to chaos and complexity science and their relationship to social and organizational worlds. He has studied with Meg Wheatley and Fritjof Capra at Schumacher College in Devon. Brian has over twenty years experience of working in interdisciplinary settings in services for Children and Families.

Index

A

accountability 186, 281, 283, 362, 379-80
action research 137, 294, 329, 337, 346, 359
actors 24, 27, 170, 199, 201, 203-9, 294, 327
adaptation 8, 10, 37, 42, 52, 56, 59-60, 108, 156, 215-16, 226, 235, 294, 327, 332
adaptive stress 215-16, 218
agencies, public 139, 285-8, 290, 296, 300, 306
agents 2, 8-10, 12-15, 47, 51-7, 67, 72, 103, 150, 237, 256, 258, 261, 281, 312-13
  individual 13, 53, 55-6, 312
Agile software development 63, 66, 69-70, 74
agreement 108, 171, 176, 200, 204, 206, 263, 375, 378
application 19, 38, 63, 65, 74, 79, 97-100, 103, 105, 133-4, 136, 142-4, 201-6, 263, 353-4
approaches 27-8, 34, 43-4, 68, 95-6, 135-6, 169-70, 176-80, 182-3, 188-91, 239-40, 315-16, 328-30, 339-41, 351-3
archetypal images 116, 119, 123-4
archetypal models 113, 117-18
archetypes 113, 115-17, 119, 121, 123-4, 127-9, 197
assets, intangible 113, 120, 123, 127-8
assumptions, underlying 26, 30, 352
attractors 102, 108, 114, 228, 249, 257-8
authority 6, 31, 35-6, 89, 113, 199, 201, 204-7, 277, 283, 287, 291, 378
awareness 27, 66, 106, 126, 173, 251, 253, 358-9, 361-2
  situational 359-60, 372-4

B

behavior patterns 3, 116, 118, 126, 152, 257, 365
behaviors 2-3, 5-6, 13, 47-8, 51-3, 99-100, 106-8, 116-17, 150-1, 155-6, 159-64, 185, 255, 257, 311
  emergent 5-6, 13, 354
  human 30, 99, 115-16, 212, 239, 352, 357
beliefs 37, 71, 114-15, 121, 154, 173, 175, 180, 196, 199, 204-5, 208-9, 218, 230, 319
boundaries 31, 34, 56, 63, 102, 133, 136-44, 146-7, 173-4, 196, 210, 254, 257, 259, 355-6
boundary critique 136-8, 142-5, 272-3, 327, 331, 341, 347
  theory of 137-9, 147
boundary judgments 137, 211, 329, 341
budget 63-4, 221, 275, 279, 351, 366, 369, 372

C

cases 6, 9, 31, 78, 86, 88, 99, 113-15, 117-18, 195, 199-210, 259, 279, 317, 319-21
causality 65, 72, 98, 197
change management 129, 165, 351-2, 382
chaos 18, 58-9, 67, 99, 103, 105, 109-10, 141, 169, 211, 232, 238, 354, 381-2, 384
chaotic 71, 99, 114, 192
clusters 23, 26, 256, 286
co-creation 43-4, 153, 223, 228-9, 363, 368
coevolution 2-3, 24, 34, 37, 67, 98, 210-12, 292, 294, 302, 313
coevolve 2, 10, 23-4, 27-8, 67, 313
collective intelligence 221, 223, 226, 237-8, 360
commitment 121, 126, 158, 178-9, 215, 250, 258, 353, 365-6, 368
communication 30, 66, 123, 156, 203, 210, 234, 319, 334, 343, 356, 362-3, 375, 379-80
communities 40-1, 100, 104, 113-15, 119-21, 128, 139, 142-3, 228, 254, 285-93, 295-7, 306, 339-40, 365-6
community engagement 285-6, 288, 290, 292, 294-6, 298, 300-2, 304, 306
community sectors 296-8
competence, unconscious 105, 107
complex adaptive systems (CAS) 99-100, 103, 109, 169, 175, 178, 190, 232, 241, 265, 355, 382
complex engineered systems 77-9, 92
complex environments 4-5, 12, 138, 141, 145
complex issues 2, 11, 43, 188
complex problems 8, 21, 23, 32, 43-4, 66, 68-9, 141, 329
complex responsive processes 129, 169-70, 175, 177, 182-3, 189, 192
complex situations 5, 8, 10, 17, 119, 328-9, 334, 343-4
complex social systems 21-3, 35, 37, 43-5, 286, 306
complex systems 5-8, 10-11, 44, 56, 67, 72-3, 77-8, 92-3, 98-9, 133-4, 173-4, 212-13, 257, 273, 311-13
complexity 2-6, 12-13, 16-17, 57-60, 67-8, 95-6, 99-104, 135-6, 196-200, 203-5, 209-11, 215-16, 309-14, 320-4, 352-3
  level of 152, 204, 255, 334, 356
  levels of 152-3, 155
  perceived 208
  principles of 21-2, 30, 43, 197, 304
  real 208-10
complexity approaches 21, 345, 351, 378
complexity metaphors 249, 252, 258-9, 263
complexity practitioners 309, 317, 322
complexity science 1, 16, 19, 45, 60, 69, 98-9, 104-6, 108-9, 145-6, 265-6, 327-8, 332-4, 336, 354-5
  social 64, 69, 74
complexity theory 24-6, 29, 31, 38, 40-2, 44, 67-8, 98, 102, 104, 196-7, 211, 249, 252, 353-4

complexity thinking 34, 56, 78, 93, 95, 98, 133-4, 136, 138, 140-2, 144, 146-7, 332-3, 354-5, 358-9
computational models 195, 198-202, 205, 209
concepts 21, 66, 70, 106, 117-18, 122, 126, 128, 133, 136, 138-9, 141-3, 256-7, 263, 344-5
  boundary 136, 138, 142-3, 145
conceptual frameworks 143, 173, 359-60, 365, 370
conditions, boundary 108, 196, 199, 208
connections 5, 55, 69, 78, 139, 153, 177-8, 180-2, 185-6, 211, 225, 237, 256, 310-11, 319-20
connectivity 24, 26, 37, 56, 101, 104-5, 150
consciousness 99, 101-3, 105-6, 110-11, 226, 238, 253, 257
consensus 113, 119, 155-6, 241
consequences 5, 25, 39, 64, 71, 159, 185, 195, 198, 209, 215-16, 218-19, 225-6, 231-2, 235
consilience 358, 381
Consilient Approach 351-2, 354-6, 358-60, 362, 364, 366, 368, 370, 372, 374, 376, 378, 380, 382, 384
constructs 47-8, 52-5, 134, 336
  conceptual 327, 331, 334, 336
containers 258, 264, 278, 313, 352, 365
context 35, 39-40, 79, 108, 114-15, 123-4, 133-5, 138-9, 141, 143, 169, 184-5, 333-7, 339-43, 351-2
  organizational 116-17, 123, 150, 222, 265
contradictions 103, 115, 127, 150, 178, 180, 182, 185, 190
control 4-6, 53, 56, 91, 120, 136, 152, 165, 169, 173, 190, 226, 260-1, 271, 281
conversations 121, 134, 139, 142-3, 162-3, 174-5, 177-9, 181-4, 186-9, 238, 309-11, 313-14, 321-2, 365-6, 371-2
corporate governance (CG) 22, 24-8, 44-5, 101
costs 14-15, 44, 48, 53, 217-18, 254, 285, 369
creation 28, 37, 52, 54, 77, 79, 81, 89, 102, 155, 158, 164, 167, 227, 354-5
critical incidents 159-62
Cynefin model 69-70, 72, 118

D

decision makers 1, 22, 43, 195, 198, 205, 286
degree 15, 35, 42, 103, 111, 119-20, 167, 171, 181, 197, 225-6, 241, 246, 273, 293
design 1, 13-14, 19, 60, 77, 93, 155, 199, 210, 212, 216, 229, 246, 279, 281
dialogue 41, 99, 140, 145, 158, 162, 169, 176, 179, 182, 187-9, 191, 233, 370, 372-4
differences 29, 34-5, 43, 54, 81, 140, 157, 170, 200, 203, 285-7, 289-90, 297, 319-21, 323
dilemmas 25-6, 30, 74, 170, 175-6, 182, 185
dimensions 14, 21, 24, 26, 28, 30-1, 43, 177-8, 208, 231, 265, 292, 334
disruptive events 2-3, 5-7, 17, 67
distributed energy generation 51-2, 55
diversity 113, 117, 174, 182, 187, 207-9, 249, 256, 311, 329

domains 7, 63, 67, 70-1, 73, 104, 136, 153, 163, 177, 191, 227, 274, 332, 381
complex 7, 67-73
dynamics 3, 28, 35, 42, 47, 56, 102-3, 106, 108, 115, 117-19, 127-8, 211-12, 291-2, 354

E

ecosystems 28, 37, 104, 145, 343
emergence 44, 67-8, 98, 101-3, 105-6, 146-7, 174, 178, 211-12, 218-19, 222-3, 227-9, 291-3, 313-14, 354-5
enabling environment 21-4, 26, 28-32, 34, 36-8, 40, 42-4, 291, 294-5
engagement 39, 44, 105, 152, 169, 172-3, 175-7, 179, 226, 230, 235, 300-1, 337, 359, 363
entities 24, 103, 120, 150, 173, 222, 239, 276, 292, 330, 332-3, 343-4, 356
environment 2, 4, 6, 11-12, 17, 24, 28, 34, 41-2, 106-8, 120, 217-19, 227-8, 255-7, 358-9
equilibrium 2-3, 56, 67, 102, 309, 313
error 79, 91, 93, 199, 203, 205, 224
margins of 200, 203-6
ethics 95, 99-104, 110, 136, 328, 347
evidence 2-3, 6, 31, 92, 116, 135, 140, 204, 218, 286, 322, 351
experiences 54-5, 66-7, 105-8, 116-17, 120-1, 138-9, 182-4, 186-8, 190-1, 218-19, 233-4, 299-302, 321-2, 336-7, 357-9
extreme events 2-3, 6-7, 16, 67

F

feedback 73, 106, 223, 226, 231, 269, 345, 375, 378
active 34-5
negative 106-7, 173
fractals 78, 110, 221, 249, 259, 261, 263
framework 18, 27, 67, 73, 111, 151, 170, 188-9, 215, 220-2, 227, 234-6, 249-50, 271, 329-32

G

goals 5, 68, 70, 116, 141-2, 158, 201, 203-4, 206, 208-9, 213, 250, 254, 264, 318
political 203-4, 206-7
governance 25, 213, 269, 283, 290-1, 303, 346
governments 4, 19, 22, 37, 44-5, 49, 96, 110, 113, 140, 169, 201-2, 205-7, 213, 277-8
groups 120, 123-4, 126, 154-63, 177-83, 187, 189-91, 239-41, 258-64, 297, 358-63, 365-6, 368, 370-4, 376
stakeholder 24, 173, 341

H

hardware configurations 81, 85-6, 88
heuristics 6, 9, 86, 327, 331-2, 344

hierarchical, vertical 285, 289-91, 293, 295-6, 301-2
holism 97, 99, 102, 110, 327, 341, 344
Holistic Business Science 95-102, 104, 106, 108-11
holons 150, 155, 160
horizontal peer 285, 291-3, 295-7, 300-2
horizontal peer world 289, 294, 297, 300-2
horizontal systems 295-6, 299, 301
human systems 24, 193, 257, 261

I

identity 25, 30, 118, 123, 128, 177-81, 185-6, 223, 356-7, 359, 372
implementation 8, 64, 100, 164, 201-2, 206, 246-7, 269, 273, 276, 281, 362, 373
individuals 4, 31-2, 35, 42, 47, 52-3, 66, 100, 138-9, 152, 154-8, 162-3, 179-80, 286, 292-4
information 25, 39, 41, 51-2, 98, 104, 134, 177-82, 185-6, 196-7, 202, 204-7, 231, 370, 379-80
infrastructure 56, 60, 193, 271, 276
initiatives 39-42, 111, 113, 126, 153, 231-2, 279
  local 39-41, 43
inner mental context 100, 105, 107-8
innovation 22, 45, 109, 115, 130, 141, 143, 215, 219, 228-9, 235, 238, 354-5
inquiry 177, 179, 181-2, 184, 271, 324, 340, 361-2, 370
institutional fragmentation 269-74, 276, 278, 280-2
integration 22, 29, 31, 77, 90, 92, 105, 150, 172, 211, 259
intentions 104, 108-9, 123-4, 151, 155, 158, 160-4, 170, 177-81, 185-7, 250, 253, 343-4
interactions 2, 11, 67, 71, 152-3, 156, 174-5, 187, 251-2, 285-6, 293, 295-6, 310-12, 315-16, 354-5
interventions 47, 52-3, 57, 105, 137, 160, 163, 273, 312, 336, 361
interviews, semi-structured 22, 24-5, 29, 32
iterations 73, 156, 162-3, 198, 341-2

K

knowledge 7, 13-14, 38, 41, 53, 64, 120-1, 129-30, 137-8, 223-4, 251, 321, 362, 370-1, 379-80
knowledge management 65, 129, 147, 233, 265-6, 370, 383

L

landscape, fitness 292, 294-5, 301-2
leaders 29, 105-6, 113-14, 133-6, 138, 140-3, 145, 180-1, 190, 226, 231-4, 238-40, 244, 325, 352-3
  organizational 118, 215, 218, 251
leadership 108-9, 133-6, 141, 147, 172, 186-7, 189-92, 215-16, 218, 226-7, 232-5, 237, 243-6, 351-5, 381-4
  effective 135, 218, 224

leadership group 362, 376, 378-80
leadership teams 184-6, 215, 231, 351, 360, 362, 364, 381
learning 6, 34, 37, 39-40, 42, 54, 143-5, 177-8, 221-5, 235-6, 258, 345-6, 361, 376, 381-2
levels 13, 34, 38, 67, 69, 81, 84, 88-9, 151-2, 173, 182-3, 187, 221-2, 235, 256-7
  individual 157, 162-3, 329
  organizational 104, 215, 222, 234-5
local governments 130, 202, 205-7, 278, 283, 355

M

Managing Complexity 1-2, 4-8, 10, 12, 14, 16, 18
maturity 39, 103, 113-14, 116, 118-22, 124, 126, 128, 130
members 15, 26, 60, 66, 104, 130, 154, 158, 161-2, 184, 228, 234, 238, 277, 375-6
mental models 68, 152-3, 170, 185, 187, 190, 224-5, 236, 242-3, 329
  shared 224-5, 242
mergers 29, 31, 44, 117, 163, 165, 175
metaphors 116, 141, 145, 236, 244, 249-52, 254, 256-8, 260, 262-4, 266, 377
methodology 1-2, 7, 11, 17, 21-4, 29, 32, 38, 48, 72, 105, 146, 157, 176-8, 188-90
methods, paraplexity 249, 253, 258-9, 264
Mindful Awareness 361, 365, 368, 371, 373
modeling 11-12, 17, 52, 156, 195-6, 198-9, 203, 205, 210
models 10-11, 55-7, 63-5, 117-18, 121, 126, 169-70, 177, 179-80, 188, 195-6, 198-201, 203-10, 285-6, 301-2
  agent-based 30, 52, 55, 58, 309
  conceptual 157, 159, 163

N

narrative emergence 309-12, 314, 316, 318, 320, 322, 324
narratives 116, 243, 250-1, 263, 317-18, 322
networks 7-8, 14-15, 19, 40, 52, 56, 59, 115, 143, 145, 152, 210, 224, 293, 299-300
  relational 216, 224, 227

O

observations 98, 157-61, 209, 226, 238, 251, 322
operations 13, 53, 81, 83-6, 88-93, 121, 144, 156, 200, 269-71, 273, 276, 279, 281, 285
  existing 57, 81, 84-6
order 6, 42, 47-8, 50-1, 53-6, 99-100, 105, 114, 126-7, 155-6, 177-9, 181-2, 199, 278-9, 354-6
organizational behavior 103, 111, 117, 164, 167, 218
organizational change 130, 165, 265, 321, 329, 335, 382
organizational dynamics 166, 192, 288-9, 301, 324, 384
organizational learning 44-5, 164, 223-4, 242-4
organizational values 149-50, 155-9, 163-4, 166

organizational virtue 249, 251, 265
organizations 21-3, 31-8, 42-4, 99-103, 109-10, 116-22, 126-30, 141-5, 152-8, 169-74, 180, 191-3, 215-18, 222-35, 351-9
outcomes 34-5, 40, 52, 55, 65, 177-80, 182, 195-7, 200-6, 209, 230, 241-2, 361-3, 368-9, 371-2

P

parables 249-53, 258-9, 261, 263-5
paradox 115, 127, 129, 138, 170, 178, 181, 187-8, 353
participants 3, 27, 30, 70-1, 108, 119, 124, 126-7, 138-40, 145, 222, 257-9, 293, 313, 355
patterns 25, 67-8, 115-16, 118-21, 123-4, 126, 169, 221, 257, 259, 261, 357-9, 362-3, 370-1, 378
PCT (Personal Construct Theory) 36, 47-8, 50, 52, 54, 56-8, 60, 337
performance 13, 16, 156, 159, 177, 184, 219, 226, 230-1, 239, 277, 283, 340, 354
planning 1, 5, 65, 74, 84, 118, 144, 175, 185-6, 189, 245, 275, 328, 362-3, 372-3
pluralism 55, 272-3, 327, 341, 344, 347
policy 6, 22, 45, 144, 199, 203-4, 207, 210-11, 265, 270, 283, 285, 294-5, 297, 301-3
policy decisions 195, 199, 209
policy makers 42-3, 198, 201, 203-4, 209-10, 213, 286, 339
power 17, 97, 113, 115-16, 135, 140, 204, 206-8, 230, 233, 244-5, 254, 277, 288, 303
practice
  community of 140, 339, 358, 360-1, 366, 370-1, 376
  reflective 147, 236, 241, 329, 347
practitioners 64, 69, 116, 134-6, 138, 140, 143, 184, 286, 302, 316, 327, 330-3, 339, 341-3
predict-and-control 269, 271, 280, 282
predictions 1, 55, 92, 201-2, 205, 354
problem-space 21-3, 29, 31, 38-9, 42, 44
problems 21, 30-1, 39, 41-3, 63, 65-6, 96, 107-8, 119-20, 169-76, 178-81, 183-90, 209, 314-17, 356
  wicked 66, 68, 246, 333
process 25-7, 34-5, 42, 77-8, 105-7, 123-7, 133, 152, 174-5, 178-80, 195-7, 223-4, 357-61, 365-6, 370-2
  evolutionary 99, 253, 357
  value-crafting 157-9
Process Enneagram 169, 175-7, 179, 181, 183-5, 187-9
projects 12, 15, 17, 22, 24-5, 29, 31-2, 35, 63-6, 68, 72, 107, 213, 269-70, 279-82
properties 52, 55-6, 78, 98, 101, 115, 124, 227, 255, 261, 276, 312-13
  emergent 73, 117, 119, 124, 174, 250
public administration 53, 130, 195-6, 210-11, 213, 269, 271, 283, 304
public decision 195-6, 198-9, 208

Q

quadrants 103, 151-3, 155, 157, 160, 164, 360, 371
qualities 38, 74, 92-3, 102, 109, 117-21, 123-4, 126, 149, 182, 188, 206, 233, 277, 305
queues 239-40

R

Rapid Adaptability 215-16, 218, 220, 222, 224, 226, 228, 230, 232, 234, 236, 238, 240, 242, 244
real world applications 21, 104, 133-4, 227, 309, 324
realization 67, 126, 156-7, 171, 270-1, 277, 281, 381
relationships 23-4, 27-8, 30-1, 34, 117-19, 152-3, 172, 178, 180-3, 185-6, 209, 285, 287-8, 290-1, 355-6
  system of 291, 293, 296
requirements 10, 15, 65, 68, 84, 86, 89, 101, 108, 218, 230, 274
resilience 11, 50, 53, 57, 136, 141, 147, 152, 373, 376
resources 5, 9-10, 32, 36, 48-9, 56, 66, 68, 107, 120, 150-2, 155, 184-5, 253-4, 290
responsibility 8, 35-6, 41, 66, 77, 89, 96, 100, 103-4, 109, 120, 340-1, 362-3, 369, 379-80
roles 15, 19, 36-7, 44, 54, 66, 78, 91, 96, 100, 111, 153, 162-4, 299, 304-5
rules 2, 6, 8-9, 13, 47, 53-4, 66-7, 103, 120, 185, 197, 232, 311, 336, 341

S

scenarios 14, 86, 189, 202, 206, 229-30, 241-2, 245, 365, 368
self-awareness 121, 222-3, 237, 245, 250
self-organization 2, 5-7, 10, 24, 35-7, 41, 58, 67-8, 173-4, 178, 211-12, 215, 228, 232, 312-13
Self-Organization 127, 211-12, 244
services 15, 24, 34, 36-7, 53, 56, 60, 97, 133, 223, 228, 237, 275-7, 287-8, 369
Smart Grids 47-8, 50-4, 56-8, 60
social capital 224, 294, 381
social systems 24, 55, 180, 211-12, 251, 286, 297, 313, 357, 383
society 3, 23, 43, 58, 95, 97, 100-1, 104, 109, 116, 118, 120, 129, 139, 303
SODA 328-9, 334-5, 341
software development 63-5, 67-74
SSM (Soft Systems Methodology) 328-9, 335, 341
stakeholders 22-4, 26-8, 38, 113, 124, 172, 200, 225-8, 230-1, 235, 264, 273, 334-5, 340, 344
standards 177-8, 180, 182, 185-6, 370
stories 116, 123, 129, 139, 187, 229, 232-4, 244, 250-1, 255, 262-3, 292, 317, 320, 322
storytelling 232-4, 243
strange attractors 99, 116, 129-30, 227, 232, 257
strategies 36, 41, 58, 118, 128, 130, 136, 139, 141, 172, 180, 185-7, 216-19, 243-6, 359
structures 5, 24, 52, 86, 99, 117, 121, 126, 177-8, 180, 185-6, 197, 216-17, 252-3, 300

sub-systems 84-6, 91, 295-6, 301
subjective experience 99, 104, 109, 111
success 35, 37, 39-41, 72, 109, 123, 149, 174-5, 188, 191, 193, 229, 234, 292-4, 352-4
sustainability 23, 35, 37, 44, 97, 101, 111, 178-80, 184-6, 193, 222, 235, 337, 345
systemic change 102, 262, 327-8, 333
systems 2, 5-6, 55-7, 114-21, 139-41, 180-1, 187, 250-3, 256-63, 269-71, 288-95, 311-12, 329, 333-4, 340
  chaotic 102, 105, 116
  closed 197, 249, 254, 356
  complex infrastructure 269-70, 278
  economic 3, 96, 101, 263, 340
  horizontal peer 290, 297, 299-300
  non-linear 99, 117-18, 121
  vertical hierarchical 288, 290, 292, 294, 296-7, 299
systems approaches 286, 328, 330, 334-6, 339, 344-5, 369
systems practitioners 328, 333-4, 339-40, 343
systems theory, complex adaptive 170, 173-4, 177-8, 189
systems thinkers 147, 315, 333, 336, 346
systems thinking 135-6, 188, 229, 235, 244, 327-34, 336, 338, 340-6, 348, 383
  critical 175, 329, 345, 348

T

teams 1, 12, 26, 29-30, 34-5, 37, 64-6, 89, 91, 106, 156-63, 180, 184-6, 227, 246
technology 3, 9, 15, 43, 49, 57, 60, 67, 149, 164, 166-7, 211, 213, 246, 353
theory, integral 160, 162, 164-5
thinking, convergent 171, 173, 175, 221, 229
tools 11-12, 21, 28-30, 43-4, 77-9, 92-3, 119, 133, 149-50, 327-8, 330-3, 336-7, 339-40, 342-5, 356
  complexity-sensitive 271-2, 282
top-down 27, 34, 171, 287, 303-4
transformation 32, 95, 97-8, 101, 105-6, 108, 122, 129, 165, 351-2, 354, 359, 378, 384
transition 51-4, 57, 117, 138, 162, 262, 336, 358-9, 371

U

uncertainty 4-5, 37, 51, 101, 106, 170-1, 205, 229, 231, 241-2, 250-1, 255, 331, 361-3, 371-3
understanding 24-5, 37-8, 40, 42-4, 47, 55, 64, 68-70, 73, 139, 305-6, 339-40, 342-3, 351-2, 358-9

V

value crafting 149-50, 152, 154-60, 162-4, 166
value-crafting cycle 155-7, 160-2, 164
values 40, 97, 103, 134-5, 137-40, 152-60, 163-4, 191, 198-9, 204, 208-9, 217-18, 221-2, 244, 319
Virtual World 8, 11
vision 58, 100, 114, 134, 141, 177, 180-1, 219-22, 228-9, 236, 253, 257-8, 264, 306, 363
voluntary sector 295-6, 299-300, 302
VSM (Viable Systems Model) 328-9, 334-5, 337, 341, 345

W

work 73-4, 106-7, 133-8, 140-1, 143, 152-3, 169-70, 176-80, 185-8, 259-60, 317-19, 321-3, 358-63, 365-6, 369-74
  boundary 133, 136, 139, 304
work group 150, 155-6
world, real 7-9, 11, 57, 91, 98, 169-70, 172, 174-6, 178, 180, 182, 184, 186, 196-8, 273

Lightning Source UK Ltd.
Milton Keynes UK
UKOW010118150612

194425UK00003B/3/P